MICROBIAL ENERGETICS

Other Publications of the
*Society for General Microbiology**
THE JOURNAL OF GENERAL MICROBIOLOGY
THE JOURNAL OF GENERAL VIROLOGY

SYMPOSIA

* Published by the Cambridge University Press, except for the first Symposium, which was published by Blackwell's Scientific Publications Limited.

MICROBIAL ENERGETICS

EDITED BY

B. A. HADDOCK AND W. A. HAMILTON

TWENTY-SEVENTH SYMPOSIUM OF THE
SOCIETY FOR GENERAL MICROBIOLOGY
HELD AT
IMPERIAL COLLEGE LONDON
MARCH 1977

Published for the Society for General Microbiology

CAMBRIDGE UNIVERSITY PRESS

CAMBRIDGE

LONDON · NEW YORK · MELBOURNE

Published by the Syndics of the Cambridge University Press
The Pitt Building, Trumpington Street, Cambridge CB2 1RP
Bentley House, 200 Euston Road, London NW1 2DB
32 East 57th Street, New York, NY 10022, USA
296 Beaconsfield Parade, Middle Park, Melbourne 3206, Australia

Library of Congress Cataloguing in Publication Data
Society for General Microbiology.
Microbial energetics.
(Symposia – Society for General Microbiology; 27
Bibliography: p.
Includes index.
1. Micro-organisms – Physiology – Congresses. 2. Microbial metabolism – Congresses.
3. Bioenergetics – Congresses.
I. Haddock, Bruce Adrian, 1945– II. Hamilton, W. Allan.
III. Title. IV. Series:
Society for General Microbiology. Symposium; 27.
QR1.S6233 no. 27 [QR88]   576'.08s [576'.11'9121]   76–54367
ISBN 0 521 21494 7

First published 1977

Printed in Great Britain at the
University Press, Cambridge

CONTRIBUTORS

ALEEM, M. I. H., T. H. Morgan School of Biological Sciences, The University of Kentucky, Lexington, Ky 40506, USA

GARLAND, P. B., Department of Biochemistry, Medical Sciences Institute, University of Dundee, Dundee DD1 4HN, UK

GOTTSCHLICH, R., Institute for Biochemistry, University of Würzburg, BRD-87 Würzburg, Röntgenring 11, West Germany

HADDOCK, B. A., Department of Biochemistry, Medical Sciences Institute, University of Dundee, Dundee DD1 4HN, UK

HAMILTON, W. A., Department of Microbiology, University of Aberdeen, Aberdeen AB9 1AS, UK

HARTMANN, R., Institute for Biochemistry, University of Würzburg, BRD-87 Würzburg, Röntgenring 11, West Germany

JONES, C. W., Department of Biochemistry, School of Biological Sciences, University of Leicester, Leicester LE1 7RH, UK

JONES, O. T. G., Department of Biochemistry, University of Bristol, Bristol BS8 1TD, UK

JONES-MORTIMER, M. C., Department of Biochemistry, University of Cambridge, Tennis Court Road, Cambridge CB2 1QW, UK

KELLY, D. P., Department of Environmental Sciences, University of Warwick, Coventry CV4 7AL, UK

KNOWLES, C. J., Biological Laboratory, University of Kent at Canterbury, Canterbury CT2 7NZ, UK

KORNBERG, H. L., Department of Biochemistry, University of Cambridge, Tennis Court Road, Cambridge CB2 1QW, UK

KOSHLAND, D. E., JR, Department of Biochemistry, University of California, Berkeley, Ca 94720, USA

KRÖGER, A., Institut für Physiologische Chemie, Ludwig-Maximilians-Universität München, BRD-8000 München 2, Petten-Koferstrasse 14a, West Germany

MICHEL, H., Institute for Biochemistry, University of Würzburg, BRD-87 Würzburg, Röntgenring 11, West Germany

MITCHELL, P., Glynn Research Laboratories, Bodmin, Cornwall PL30 4AU, UK

OESTERHELT, D., Institute for Biochemistry, University of Würzburg, BRD-87 Würzburg, Röntgenring 11, West Germany

STOUTHAMER, A. H., Biological Laboratory, Free University, de Boelelaan 1087, The Netherlands

WAGNER, G., [Institute for Biochemistry, University of Würzburg, BRD-87 Würzburg, Röntgenring 11, West Germany

WHITTENBURY, R., Department of Biological Sciences, University of Warwick, Coventry CV4 7AL, UK

CONTENTS

EDITORS' PREFACE

It is our intention in this symposium to focus attention on the cross-fertilisation that is currently enriching the fields of bioenergetics and microbial physiology.

For many years the controversy over the mechanism of ATP synthesis in mammalian mitochondria has held the centre of the energetic stage. Increasingly, however, the great range of energy conservation mechanisms found among micro-organisms is capturing the imagination and holding the attention of experimenters. These studies both extend our knowledge of energy transduction in the living cell, and strengthen our understanding of the intricacies of microbial metabolism.

The elucidation of the mechanism of metabolic reactions and pathways in microbial physiology has led to the present efforts to understand the integration and control of these individual components of cellular function. Central to these studies is the charting of the ebb and flow of energy through the system.

After an introduction outlining the basic pattern of microbial energetics and highlighting some of the outstanding problems, this volume consists with seven chapters describing various energy conservation mechanisms developed by micro-organisms and their relationship to the habitat of the individual species and a further five chapters considering the energetics and control of transport, metabolism, growth and chemotaxis. The volume concludes with an epilogue that suggests the direction of future conceptual and experimental development.

We believe that the presentation and discussion of these papers within this format will be seen to be both timely and stimulating to a wide spectrum of teachers and researchers. Should this indeed prove to be the case, it will but reflect the commitment and enthusiasm with which the contributors have responded to our initial directives and any further editorial suggestions.

We should like to acknowledge also the help and direction provided by the staff of Cambridge University Press.

Department of Microbiology,   W. A. HAMILTON
University of Aberdeen, Aberdeen AB9 1AS

Department of Biochemistry,   B. A. HADDOCK
University of Dundee, Dundee DD1 4HN

ENERGY TRANSDUCTION AND TRANS-MISSION IN MICROBIAL SYSTEMS

P. B. GARLAND

Department of Biochemistry, Medical Sciences Institute,
University of Dundee, Dundee DD1 4HN, UK

INTRODUCTION

The main bioenergetic activities that are exhibited by a bacterium such as *Escherichia coli* and are dependent upon its cytoplasmic membrane are summarised below in scheme 1.

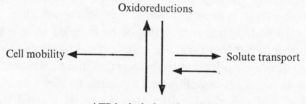

Scheme 1. Bioenergetic relationships in *E. coli*

I have in this scheme narrowed down the field of bioenergetics to exclude all those well-documented pathways of intermediary metabolism in which ATP and NADPH from catabolic processes can be used to drive biosynthetic pathways. What we are left with is a set of membrane-dependent activities whose interrelationships have not been, nor are likely to be, revealed by the classical methods used for studying inter-mediary metabolism in soluble or solubilised systems. The four activities shown in the scheme are respiratory chain oxidoreductions, ATP hydro-lysis and synthesis catalysed by the ATPase of the cytoplasmic mem-brane, the movement of solutes across the cytoplasmic membrane, and movement of the organism itself. With the exception of cell motility all of these activities have been shown to react reversibly with at least one of the others. Thus depending upon the thermodynamic poise of the pro-cesses, ATP hydrolysis may drive solute (H^+) transport across the cytoplasmic membrane, or H^+ movement may drive ATP synthesis (Maloney, Kashket & Wilson, 1974; Grinius, Slusnyte & Griniuviene, 1975; Maloney & Wilson, 1975; Wilson, Alderete, Maloney & Wilson, 1976). Similarly, the normally directed flow of electrons through the respiratory chain from redox complexes of lower to complexes of higher potential drives ATP synthesis, whereas under appropriate experimental conditions ATP hydrolysis will drive electrons in the reverse direction (Sweetman & Griffiths, 1971). The respiratory chain activity can also drive solute transport independently of ATP synthesis or hydrolysis, as shown for example by Schairer & Haddock (1972) for β-galactoside uptake in an ATPase-deficient mutant of *E. coli*. So far reversal of electron transport through the respiratory chain in response to solute movement has not been demonstrated.

THE PROBLEM

Although the processes of respiratory (or photosynthetic) chain oxido-reductions, of solute movement along specific pathways across membranes, and of ATP hydrolysis are all of great interest in their own right, the central question in bioenergetics is this – how are these processes interconnected so that one may drive, or be coupled to, another? One naturally presupposes that there are coupling mechanisms at work. It is the purpose of my introduction to this Symposium to summarise briefly some general ideas on coupling mechanisms, to clear away some of the semantic and other difficulties that so often bedevil discussions of bioenergetics, and to indicate how physiological and also molecular studies of bacterial systems can usefully, and perhaps critically, add to our knowledge of coupling mechanisms.

COUPLING MECHANISMS

In order that one process may drive another, there are three requirements. First, that the sum of the free energy changes for the two processes be negative. Second, that the two processes share a common intermediate. Third, that suitable reaction and diffusion pathways exist for the common intermediate to be shared by the two processes. It must be emphasised from the start that each process may independently have its own intermediates, conformational changes of proteins or anything else that may occur where enzymes are at work: but that is not a reason for identifying these intermediates, conformational changes and so on as being a part of the coupling mechanism.

Examples of states or intermediates that are common to and couple two processes can be readily brought to mind, and include not only chemical intermediates but also those that are mechanical or electrical. We are all familiar with the coupling of the pedal wheel to the rear wheel of a bicycle: the feature common to and shared by both is the chain. The flow of electrons through cables couples processes that generate electrical power to those that consume it. In a biochemical context, ATP is a chemical intermediate that permits catabolism to drive biosynthesis.

The search for the mechanism that couples respiratory chain or photosynthetic oxidoreductions to ATP synthesis and solute transport has extended for over a quarter of a century and produced numerous hypotheses. The most successful and useful of these hypotheses is that known as Mitchell's chemiosmotic hypothesis, of which an early account is to be found in Mitchell, 1961; full accounts in Mitchell, 1966 and 1968; and a recent appraisal in Mitchell, 1976. An article in an earlier Symposium of this series is particularly appropriate for those whose interest is primarily in transport rather than oxidative phosphorylation (Mitchell, 1970). Despite, or because of, these descriptions of the chemiosmotic hypothesis, the accounts of Mitchell's ideas to be found in standard and otherwise excellent texts of biochemistry are confusing for most except the initiated, who have less cause to read them (Mahler & Cordes, 1966; Lehninger, 1970; Stryer, 1975). A brief summary of the chemiosmotic coupling mechanism and its relationship to other hypotheses (Slater, 1953; Williams, 1961; Boyer, 1974) will therefore not be out of place. It is convenient to consider chemical coupling first (Slater, 1953).

Chemical coupling

This hypothesis (Slater, 1953) arose from analogy with the substrate-level phosphorylations of glycolysis, particularly those associated with the oxidation of glyceraldehyde-3-phosphate in which one metabolite of the glycolytic sequence, 1,3-diphosphoglycerate, is also a substrate and indeed the source of energy for an ATP synthesising reaction. Hence it was reasoned that, in the process of respiratory or photosynthetic oxidoreductions, an oxidoreduction carrier C would react with an unidentified compound I to yield an intermediate C∼I, where the 'squiggle' bond signifies a large and negative free energy of hydrolysis of the intermediate. It was then proposed that C∼I would drive the synthesis of a second intermediate, X∼I, which in turn would drive ATP synthesis or solute transport. It should be noted that of these intermediates only C∼I is common to and shared by both the oxidoreduction process and also the ATP synthesising process. C∼I is therefore a coupling intermediate whereas X∼I is not, because C, but neither X nor I, is an oxidoreduction carrier. Any number of reaction intermediates can be postulated to exist in the reaction mechanism of the ATPase without accepting that a respiratory carrier intermediate such as C∼I exists and reacts with them. Failure to recognise this point, allied with Mitchell's allowable but perhaps unfortunate use of X, I and X~I as shorthand notation for possible reaction intermediates in the ATPase, has led to confusion concerning the existence of 'high energy intermediates' in various coupling mechanisms. As indicated above, whenever we find enzyme-catalysed reactions we can expect to find intermediates if we look carefully enough. The hypothetical X~I is such an intermediate and is not to be confused with the common, shared chemical coupling intermediate, C~I. Inability to identify oxidoreduction carriers in the C~I form is a stumbling block to the acceptance of the chemical coupling hypothesis.

Conformational coupling

This hypothesis (Boyer, 1974) is really a special case of the chemical coupling hypothesis. The failure to discover C~I of the chemical coupling hypothesis is recognised by omitting I and replacing C~I with a conformationally energised form of the oxidoreduction carrier, C*. The relaxation of C* to C is then thought to drive a conformational change in the ATPase, relaxation of which then drives ATP synthesis.

Chemiosmotic coupling

This hypothesis (Mitchell, 1961) specifically recognises the role of the membrane and its supramolecular structure. Since the membrane in question may be mitochondrial, bacterial or thylakoid, I will refer to it as the coupling membrane. The hypothesis comes in four parts, and one cannot do better than quote the original (Mitchell, 1966, pp. 24–5), modified only to extend the sense to bacteria as well as mitochondria and chloroplasts.

1. The membrane-located ATPase systems. . .are hydro-dehydration systems with terminal specificities for water and ATP; and their normal function is to couple reversibly the translocation of protons across the membrane to the flow of anhydro-bond equivalents between water and the couple ATP/ (ADP+phosphate).
2. The membrane-located oxidoreduction chain systems. . .catalyse the flow of reducing equivalents, such as hydrogen groups and electron pairs, between substrates of different oxidoreduction potential; and their normal function is to couple reversibly the translocation of protons across the membrane to the flow of reducing equivalents during oxidoreduction.
3. There are present in membranes. . .substrate-specific exchange-diffusion carrier systems that permit the effective reversible trans-membrane exchange of anions against OH^- ions and of cations against H^+ ions; and the normal function of these systems is to regulate the pH and osmotic differential across the membrane, and to permit entry and exit of essential metabolites. . .without collapse of the membrane potential.

4. The systems of postulates 1, 2 and 3 are located in a specialised coupling membrane which has a low permeability to protons and to anions and cations generally.

In short, the postulates are for (1) a reversible proton-translocating ATPase, (2) a reversible proton-translocating respiratory or photosynthetic oxidoreduction chain, (3) a set of translocases that facilitate electroneutral exchange of anions for OH^- and cations for H^+ where it is necessary for these anions and cations to be transported, and (4) an ion-impermeable membrane incorporating the components of the previous three systems. Fig. 1 gives a diagrammatic representation of the various postulates.

Thermodynamics of the proton-translocating ATPase

The reaction catalysed by the proton-translocating ATPase can be written as

$$ATP^{4-}+H_2O+xH_R^+ \rightleftharpoons ADP^{3-}+\text{phosphate}^-+xH_L^+. \quad (1)$$

One consequence of moving H^+ from the bulk phase on one side of the membrane to the bulk phase on the other is to set up a membrane potential, $\Delta\psi$, measured as the electrical phase in the left (L) phase minus that in the right (R) phase (Fig. 1). Another consequence is the establishment of a pH difference, ΔpH, between the two phases, expressed as the pH in the left minus that in the right phase. It follows that the electrochemical activity of protons in the left phase will differ from that in the right phase, and that this difference in electrochemical activity, Δp, the protonmotive force, will describe the force tending to drive protons in the opposite direction to that of the ATPase-driven translocation. If $\Delta\psi$ is expressed in millivolts and ΔpH in pH units, then at 30 °C:

$$\Delta p = \Delta\psi - 60\,\Delta\text{pH}. \quad (2)$$

If the proton-translocating ATPase (reaction (1)) is allowed to proceed to equilibrium then a point will be reached where the protonmotive force is poised against the ATP/(ADP+phosphate) couple. The force with which this couple tends to drive protons through the proton-translocating ATPase can also be expressed in millivolts as the phosphate potential, ΔG_p. The relationships between the protonmotive force, the phosphate potential and the number of protons translocated per ATP hydrolysed to ADP+phosphate (i.e. x of reaction (1), or the $\rightarrow H^+/\sim P$ ratio) is given at equilibrium by

$$x\Delta p = \Delta G_p. \quad (3)$$

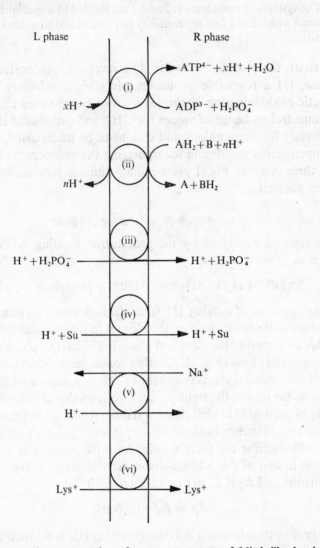

Fig. 1. Diagrammatic representation of some components of Mitchell's chemiosmotic hypothesis as applied to bacteria. The pair of parallel vertical lines represent the cytoplasmic membrane, generally impermeable to ions. The R phase represents the cytoplasmic space, the L phase represents the extracellular space including the periplasmic space. The large circles in the membrane represent (i) a proton translocating ATPase, (ii) a proton-translocating oxidoreduction, (iii) an electroneutral proton plus phosphate symporter, (iv) an electrogenic proton plus sugar (Su) symporter, such as that for β-galactosides in *E. coli* (West & Mitchell, 1973), (v) an antiporter for exchanging H^+ with Na^+ (West & Mitchell, 1974a), (vi) an electrogenic uniporter for lysine (Lys) as in *S. aureus* (Niven & Hamilton, 1974).

For typical intracellular conditions of 10 mM phosphate, equimolar ADP and ATP, and pH 7.0, the value of the phosphate potential is about 420 mV. Measurements have been made of the protonmotive force Δp in respiring *E. coli* (Collins & Hamilton, 1976; Padan, Zilberstein & Rottenberg, 1976), anaerobic *Streptococcus faecalis* (Harold & Papineau, 1972; Harold, Pavlasova & Baarda, 1970) and *Streptococcus lactis* (Kashket & Wilson, 1974), *Staphylococcus aureus* (Jeacocke, Niven & Hamilton, 1972; Collins & Hamilton, 1976) and in chromatophores of *Rhodopseudomonas capsulata* (Casadio, Melandri & Melandri, 1975) and *Rhodospirillum rubrum* (Schuldiner *et al.*, 1974). In addition, the $\Delta \psi$ component only of Δp has been measured in *E. coli* by Griniuviene, Chmieliauskaite & Grinius (1974) and in *S. faecalis* by Laris & Pershadsingh (1974). For *E. coli* respiring aerobically Δp was reported to be about 230 mV of which about 130 mV was contributed by $\Delta \psi$ and the remaining 100 mV by ΔpH (Collins & Hamilton, 1976). This suggests that the $\rightarrow$H$^+$/$\sim$P ratio would be two if a Δp of about 230 mV were to be poised against a phosphate potential of about 420 mV. Unfortunately direct measurements of the $\rightarrow$H$^+$/ATP ratio in bacterial systems are limited to one report only, where a value of 0.58 was obtained for fragments of *E. coli* membranes, and this value was considered to be an underestimate of the true value (West & Mitchell, 1974*b*).

Structural and genetic aspects of the proton-translocating ATPase

The structure and composition of the ATPase from coupling membranes appears to be similar in chloroplasts, bacteria and mitochondria. A conveniently brief review has recently been presented by Postma & van Dam (1976), to which the reader is referred for entry into the literature at large. I shall describe some features of the ATPase for the sake of completeness, and in order to give at least some structural framework to the concept of a proton-translocating ATPase. Isolation by mild procedures of a high molecular weight ATPase complex from coupling membranes is thought to yield a preparation representative of the transmembranous proton-translocating ATPase. The complex consists of two fractions. One is called the F_1 fraction or F_1-ATPase, and is readily removed from the membrane. The water-soluble F_1 fraction is composed of five types of polypeptide, has all of the ATPase activity of the complex, contains bound adenine nucleotides, and is the site at which azide or the antibiotic aurovertin inhibit ATPase activity. Several inhibitors of ATPase activity of the whole complex, such as oligomycin, Dio-9, rutamycin and *N,N'*-dicyclohexylcarbodiimide, do not inhibit the

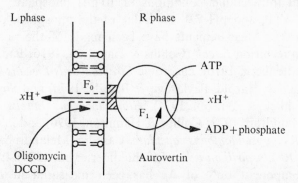

Fig. 2. Diagram of the membrane-bound ATPase complex. The membrane and the bulk phases on either side are as in Fig. 1. F_1-ATPase is shown as a spherical structure protruding into the R phase. The F_0 fraction is shown as a transmembranous structure surrounded by the phospholipid bilayer and providing a proton channel (interrupted lines) that is sealed at one end by the F_1-ATPase. The stalk (cross-hatched) holding the F_1-ATPase to the membrane is a specific polypeptide classified with the F_0-fraction. The sites of some inhibitors are shown. DCCD, N,N'-dicyclohexylcarbodiimide.

Factivity of the isolated $_1$-ATPase. The other fraction of the ATPase complex is the F_0 fraction: it is less well characterised than F_1, requires detergent for solubilisation from the membrane, contains four or more types of polypeptide, lacks ATPase activity, and if reassociated with F_1 it restores to the F_1-ATPase sensitivity towards the inhibitors oligomycin, Dio-9, rutamycin and N,N'-dicyclohexylcarbodiimide. Detachment of F_1 from the membrane causes increased proton permeability that is reversed by treatment of the membrane with oligomycin or N,N'-dicyclohexylcarbodiimide. Mutations giving lesions in the ATPase complex also increase the permeability of the membrane to protons, and this leakiness is abolished by N,N'-dicyclohexylcarbodiimide. The phenotype of these mutations in *E. coli* is that of inability to grow unless energy can be derived from substrate-level phosphorylation, but the cells are not respiratory deficient. Two main classes of mutant can be identified, the *unc* A group lacking ATPase activity and carrying a lesion in the F_1-ATPase, and the *unc* B group possessing an ATPase activity but possessing a lesion presumably in F_0 (Cox & Gibson, 1974). The usual interpretation of these properties in terms of the proposed proton-translocating role of the ATPase is that the F_1 fraction possesses the enzymatic machinery for ATP hydrolysis and synthesis whilst the F_0 fraction provides a proton pathway through the membrane to the F_1-ATPase. Fig. 2 gives a diagrammatic representation of this interpretation. A detailed review is given by Simoni & Postma (1975).

Proton-translocating oxidoreduction chain

The ability of both respiratory and photosynthetic oxidoreduction chains to translocate protons at high rates is not in doubt, whereas the underlying mechanism is. In general, a proton-translocating oxidoreduction between two redox couples A/AH_2 and B/BH_2 can be written as

$$AH_2 + B + nH_R^+ \rightleftharpoons A + BH_2 + nH_L^+. \tag{4}$$

The consequences of this process are as for the proton-translocating ATPase, namely, setting up across the membrane of a protonmotive force Δp composed of $\Delta \psi$ and ΔpH terms. If reaction (4) is allowed to proceed to equilibrium, the protonmotive force will become poised against the electronmotive force, which is given by the difference ΔE between the redox potentials of the couples A/AH_2 and B/BH_2. Thus, at equilibrium,

$$n\Delta p = 2\Delta E, \tag{5}$$

where n is the number of protons translocated per electron pair proceeding between the two couples, or the $\rightarrow H^+/2e^-$ ratio. Values for $\rightarrow H^+/2e^-$ vary with the electron transport pathway under study, which means that for a given respiratory chain the value may vary with the substrate and that for a given substrate the value may vary with the chain, there being considerable variation in bacteria of respiratory chains both within a given organism and between different organisms (Haddock & Jones, 1977). In *E. coli* with either oxygen or nitrate as the terminal acceptor the $\rightarrow H^+/2e^-$ ratio is approximately 4.0 for the oxidation of NADH or formate and 2.0 for the oxidation of succinate (Lawford & Haddock, 1973; Garland, Downie & Haddock, 1975). More recently $\rightarrow H^+/2e^-$ ratios of 3.0 have been observed for the oxidation by *E. coli* of ubiquinol₁ (A. Lamont & P. B. Garland, unpublished observations).

Coupling of oxidoreductions to ATP synthesis

If reaction (1) for the proton-translocating ATPase and reaction (4) for the proton-translocating oxidoreduction are valid, then if the two reactions occur in one membrane they can be coupled by virtue of their common substrate, the translocated proton. The two reactions can then be added to yield an overall reaction (6) in which proton translocation does not appear.

$$AH_2 + B + (n/x)ADP^{3-} + (n/x)\text{phosphate} \rightleftharpoons A + BH_2 + (n/x)ATP^{4-} + (n/x)H_2O. \tag{6}$$

The ratio n/x is the ratio of $\rightarrow H^+/2e^-$ to $\rightarrow H^+/\sim P$, and is therefore the $\sim P/2e^-$ ratio.

Exchange diffusion carriers

The third of Mitchell's (1966) postulates seems to exclude electrophoretic solute movement down the membrane potential, on the grounds that the membrane potential would thereby be collapsed. This restraint no longer operates, as illustrated by the mechanism of β-galactoside uptake in *E. coli* where co-transport of a proton with the sugar by the *lac* permease causes the protonmotive force to drive uptake of the sugar against its concentration gradient (West & Mitchell, 1973). The protonmotive force also drives the concentrative uptake of glycine and leucine by co-transport with H^+ in *S. aureus* (Niven & Hamilton, 1974), whereas the $\Delta\psi$ component alone drives uptake of lysine (Niven, Jeacocke & Hamilton, 1973).

The ion-impermeable coupling membrane

Although there is no reason to doubt that the coupling (cytoplasmic) membrane is generally impermeable to ions except those with delocalised charge, and to most solutes except those that are of low molecular weight or are lipophilic, the permeability of the membrane to H^+ and OH^- is of particular interest when the protonmotive force is of the magnitude, about 230 mV, that might be needed to equilibrate with a phosphate potential of about 420 mV if the $\rightarrow H^+/\sim P$ ratio is 2.0. At this level of protonmotive force the electrical resistance of the membrane is probably not far short of becoming non-ohmic, at which point a small increase in Δp would result in a disproportionately large increase in leakage of H^+ across the membrane (Nicholls, 1974a). Thus the voltage and proton conductance characteristic of the membrane resembles the voltage and electron conductance of a zener diode. In at least one coupling membrane (brown adipose tissue mitochondria), the position of the 'knee' in the voltage and proton-conductance characteristic is under physiological control (Nicholls, 1974b). The possibility that bacteria might also physiologically shift the position of the knee of the proton-conductance characteristic in order to balance the supply and demand of reducing equivalents, carbon skeletons and ATP is unexplored.

The coupling hypothesis of Williams

Williams (1961) proposed that the respiratory and photosynthetic oxidoreduction chains were devices for separating protons from electrons, and that the resulting localised increase in the electrochemical proton activity within the membrane could be channelled to the ATPase. The free energy change of the reaction

$$ADP^{3-} + phosphate^{2-} \rightleftharpoons ATP^{4-} + OH^- \tag{7}$$

approaches zero at about pH 1.0 (Johnson, 1960), and a localised proton accumulation driven by an oxidoreduction could be envisaged as driving reaction (7) in the direction of ATP synthesis. Although Williams's mechanism shares some common ground with Mitchell's in that there is an emphasis on the charge-separating and proton-generating capacity of oxidoreductions, it differs in that the scale of these processes need not be other than microscopic and confined to the domains of the proteins concerned. Perhaps the crucial distinction between Mitchell's mechanism and that of Williams lies not in energy transduction, where both invoke the formation and discharge of proton electrochemical activity differentials, but in energy transmission. In Mitchell's mechanism, transmission of energy between the oxidoreduction enzymes and ATPase is by a flow of protons in the bulk phase on either side of the membrane, using conventional electrophysiology. Williams argues that although proton movements may indeed be demonstrated in the bulk phases under appropriate experimental conditions designed to permit such observations, the normal flow would be within the membrane, conceivably at very fast rates in the ordered water associated with polypeptides (Williams, 1975). The question that arises then is this – if a proton is produced as a consequence of an oxidoreduction, which route offers the least resistance to its transmission to the ATPase? Mitchell's route is from the membrane to the double layer to the bulk phase and back again and at least one of these steps has been shown experimentally to present a significant diffusion barrier (Auslander & Junge, 1974). According to Williams, protons take the pathway of least resistance which is proposed to be intra- rather than extramembranous. Clearly a Williams type of mechanism could co-exist with Mitchellian types of bulk phase proton phenomena if membrane-located protons can under appropriate experimental conditions escape into or equilibrate with the bulk phase.

Which coupling mechanism?

The considerable predictive success of the chemiosmotic hypothesis and its ability to rationalize the relationships between transport, ATP synthesis and hydrolysis, and oxidoreductions, must surely recommend it to the microbial physiologist. The hypothesis as initially formulated (see above) is relatively independent of mechanism, and does not stand or fall by the validity or otherwise of Mitchell's provocative explanations of how the ATPase (Mitchell, 1974) or respiratory chain (Mitchell, 1975) might translocate protons.

PHYSIOLOGICAL STUDIES OF BACTERIAL BIOENERGETICS

It seems reasonable to assume that the underlying mechanisms for the coupling of transport, photosynthetic and respiratory chain oxidoreductions, and ATP synthesis and hydrolysis, are essentially the same over a wide range of organisms. The student of bacterial bioenergetics is fortunate in having at his disposal a great variety of different micro-organisms adapted in some instances to quite extreme ecological niches. From a bioenergetic viewpoint four classes of environment are of particular interest. They are those that support (a) barotolerant bacteria, such as *Pseudomonas bathycetes*, whose native habitat in the Mariana Trench provides a pressure in excess of 1000 atmospheres (Smith, Landau & Pope, 1976); (b) acidophilic bacteria that grow in environments at pH 3 or less (Brock, 1969); (c) alkalophilic bacteria that grow at pH 9 or higher (Brock, 1969); and (d) chemolithotrophic bacteria where unconventional electron transfer pathways and mechanisms may be needed to achieve respiration-driven proton translocation (Cobley, 1976). In the present context the acidophiles and alkalophiles merit further discussion if it is assumed that their intracellular pH is about 7.

Acidophiles

If the internal pH is 7 then at an external pH of 3 the ΔpH term of the protonmotive force will yield 240 mV, sufficient to drive ATP synthesis provided that $\rightarrow H^+/\sim P$ is approximately 2 or higher. In the absence of any respiratory mechanism to pump out the protons that have entered via the ATPase, the membrane potential arising from inward diffusion of protons would be expected to fall to -240 mV (extracellular *minus* intracellular), and the protonmotive force would fall to zero. The capacitance of biological membranes is so low that only an insignificant change of ΔpH need occur before it is poised against an equal and opposite $\Delta\psi$. These discussions define the limits of protonmotive force against which the respiratory chain would have to translocate protons outwardly in order to achieve a chemiosmotic mechanism. In an even more extreme environment of, say, pH 1 the ΔpH term of the protonmotive force would rise to 360 mV, and it is conceivable that a negative membrane potential could be used in the steady (respiratory) state as an offset to bring Δp down to a value against which the respiratory chain could drive protons. Alternatively, respiratory chain proton translocation could proceed by utilising a lower value for the $\to H^+/2e^-$ ratio or a higher gap of oxidoreduction potential ΔE between redox carriers (see reaction (5)).

Alkalophiles

In this case the problem from a chemiosmotic viewpoint shifts from the proton-translocating respiratory chain to the proton-translocating ATPase. In the case of a bacterium growing aerobically in an environment at pH 9 with an internal pH of 7, the ATPase would not be able to synthesize ATP unless the $\Delta\psi$ term of the protonmotive force rose to about 350 mV, which would bring Δp to 230 mV, sufficient to drive ATP synthesis provided that $\to H^+/\sim P$ is about 2 or greater. It is debatable whether a membrane can withstand a potential of 350 mV across it without becoming permeable to protons. An alternative way whereby a cell could live chemiosmotically with ΔpH of 2 or more units (inside acid) would be to operate at a lower protonmotive force and higher $\to H^+/\sim P$ ratio. Failing this, the cell could neatly sidestep the problem, and with it chemiosmosis, by utilising an intramembranous energy conservation mechanism of the Williams type. It should be clear from these few remarks that an investigation of extreme alkalophiles could be most rewarding, especially if directed towards the relationships between Δp, $\Delta\psi$, ΔpH and ΔGp.

The transfer of bioenergetic parameters between different systems

Values for Δp, ΔE and ΔG_p in bacteria can be obtained, it seems, without meeting insuperable difficulties. The same cannot be said for the $\to H^+/\sim P$ ratio, where only one report is available (West & Mitchell, 1974b), nor for the $\sim P/2e^-$ ratio, where there are many reports but few in agreement (e.g. Hempfling, 1970; van der Beek & Stouthamer, 1973). The significance of measurements of $\to H^+/2e^-$ ratios is not beyond criticism either, for there may be not only difficulty in identifying exactly the substrate (e.g. intracellular NADH or ubiquinol or menaquinol) for oxidation during brief bursts of respiration, but also underestimates may be made of the number of protons translocated if there is a rapid back diffusion of H^+ with anions such as phosphate (Brand, Reynafarje & Lehninger, 1976) or formate (Garland et al., 1975). One way out of these experimental difficulties is to assume that the values for $\to H^+/\sim P$ obtained with a convenient system such as submitochondrial particles are applicable to other, less convenient, systems such as inside–out vesicles of the cytoplasmic membrane of *E. coli*. Similarly, it is tempting to apply $\sim P/2e^-$ and $\rightarrow H^+/2e^-$ ratios obtained from well-documented oxidoreductions in mitochondria to less well-documented bacterial systems. Such extrapolations from mitochondrial to bacterial systems may well be erroneous. This is because of a fundamental difference in the physiology of a mitochondria as compared to a free-living bacterium such as *E. coli*. A mitochondrion is obliged to import ADP and phosphate from its environment and to export ATP (Klingenberg, 1970). *E. coli* is under no such obligation, and the interconversions of ATP with ADP and phosphate remain on only one side, the intracellular side, of the coupling membrane. Fig. 3 shows diagrammatically these differences between mitochondria and bacteria. If we use the ionic species and stoichiometries shown in Fig. 3 then the overall reaction in a eukaryotic cell for the production by mitochondrial oxidative phosphorylation of a molecule of extramitochondrial ATP from one of extramitochondrial ADP and one of extramitochondrial phosphate is:

$$H_2PO_4^- + ADP^{3-} + 3H_L^+ \rightarrow ATP^{4-} + 3H_R^+ + H_2O. \tag{8}$$

Thus we see that the effect of the electrogenic exchange of ADP^{3-} for ATP^{4-}, with electroneutral phosphate movement, is to increase the number of protons required to move across the membrane from two (the $\rightarrow H^+/\sim P$ ratio for the mitochondrial proton translocating ATPase, reaction (1), Moyle & Mitchell, 1973), to three (the $\rightarrow H^+/\sim P$ ratio for the overall reaction (8)). To what extent the exchange of ADP for ATP is fully electrogenic is not known for all conditions, and if the exchanging species were ADP^{3-} and ATP^{3-} then the exchange would be electroneutral (Klingenberg, 1970) and the overall $\rightarrow H^+/\sim P$ ratio of reaction (8) would fall to 2, that of the ATPase. However, reaction (8) is in keeping with the finding by Brand *et al.* (1976) that the $\rightarrow H^+/2e^-$ ratio per energy conservation site of the mitochondrial respiratory chain is 3.0, and not 2.0 as is usually supposed (Mitchell & Moyle, 1965; Lawford & Garland, 1972). It also follows from reactions (8) and (3) that the value of the ratio $\Delta G_p/\Delta p$ (i.e. the ratio of the extramitochondrial phosphorylation potential to the protonmotive force at equilibrium) should also be 3.0. This is very close to the values of 3.04 and 3.18 measured by Wiechmann, Beem & van Dam (1975).

These considerations arising from the physiological necessity for the transport of ATP, ADP and phosphate across the mitochondrial coupling membrane do not apply to bacteria, although one must leave open the question of whether, in bacteria, phosphate movements have interfered with measurements of $\rightarrow H^+/2e^-$ ratios (Haddock & Jones, 1977).

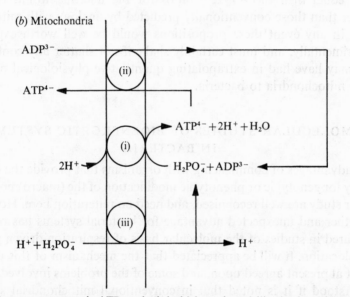

Fig. 3. Proton movements in ATP synthesis in (a) bacteria and (b) mitochondria. For bacteria the R phase corresponds to the cytoplasm and the L phase to the extracellular space. For mitochondria the R phase corresponds to the matrix space and the L phase to the extramitochondrial space. The membrane, shown by two parallel vertical lines, is the cytoplasmic membrane for bacteria and the inner membrane for mitochondria. The large circles in the membrane are (i) proton-translocating ATPase with an $\to$H$^+$/$\sim$P ratio of 2.0 (ii) adenine nucleotide translocase, (iii) electroneutral phosphate carrier. Conventional measurements of ATP synthesis and $\sim$P/2e$^-$ ratios of mitochondria measure the conversion of extramitochondrial ADP and phosphate to extramitochondrial ATP. By contrast, measurements of $\sim$P/2e$^-$ ratios in bacteria (Hempfling, 1970) are not complicated by adenine nucleotide transport.

Interestingly, if one were to regard isolated mitochondria as a suspension of bacteria and assay the $\sim$P/2e$^-$ ratio by the method of Hempfling (1970), then the values so obtained for the conversion of intramito-chondrial ADP and phosphate to ATP should, if the preceding paragraph is correct, lead to values that are 50% higher than those normally obtained for extramitochondrial ADP and phosphate. Fig. 3 shows how the overall $\rightarrow$H$^+$/$\sim$P ratios for bacterial and intramitochondrial ATP synthesis differ from that for the complete mitochondrial system.

The physiology of Paracoccus denitrificans

The close resemblances between the respiratory chains of *P. denitrificans* and of mitochondria have been emphasised by John & Whatley (1975). Nevertheless this free-living microbe does not have the mitochondrial obligation to export ATP made from imported ADP and phosphate. This difference could be reflected in any of several ways, but if all else were equal then the $\sim$P/2e$^-$ ratios for the microbe would be 50% higher than those conventionally predicted by analogy with mitochondria. In any event these propositions would be well worth exploring experimentally, and must certainly shake if not shatter any confidence one may have had in extrapolating quantitative physiological findings from mitochondria to bacteria.

MOLECULAR STUDIES OF BIOENERGETIC SYSTEMS IN BACTERIA

The advantages of commencing with organisms that provide the opportunity for genotypic or phenotypic modification of the (macro)molecules under study are well recognised, and need no reiteration here. However, a further and unexpected advantage for bacterial systems has recently appeared in studies of the molecular basis of respiration-driven proton translocation. It will be appreciated that the mechanism of this process is not at present agreed upon, and some of the problems involved can be understood if it is noted that in conventional mitochondrial systems there are about 30 individual polypeptides possibly involved in the oxidation of NADH by oxygen. Sixteen belong to the complex that reduces ubiquinone with NADH (Ragan, 1976), eight to the complex that reduces cytochrome *c* with ubiquinol (Gellerfors & Nelson, 1975), and six to the complex that oxidises cytochrome *c* with oxygen (Eytan, Carrol, Schatz & Racker, 1975). Until recently bacterial respiratory systems had not been characterised in similar detail, but in at least one area, that of the anaerobically induced cytochrome *b*-containing formate dehydrogenase and nitrate reductase complexes of the cytoplasmic membrane of *E. coli*, rapid progress is being made. These two anaerobic complexes are proton-translocating, with an $\rightarrow$H$^+$/2e$^-$ ratio of 2.0 for the sequence from formate to ubiquinone, and 2.0 for the sequence from ubiquinol to nitrate (Garland *et al.*, 1974; A. Lamont & P. B. Garland, unpublished observations). Each of the two complexes contains only three polypeptides, and at least two of the three polypeptides in each case can be identified with an oxidoreduction function (Enoch & Lester, 1974, 1975; MacGregor, 1975; Clegg, 1976). Furthermore, the cytochrome b-containing nitrate reductase complex has been shown to span the cytoplasmic membrane by both structural (Boxer & Clegg, 1975) and functional (Jones & Garland, 1977) criteria. Both the formate dehydrogenase and nitrate reductase enzyme complexes of $E.$ $coli$ are being studied from various viewpoints in several laboratories, and it seems likely that these bacterial systems will provide considerable insight into the molecular mechanism of respiration-driven proton translocation.

I am indebted to many colleagues but particularly Dr B. A. Haddock for stimulating discussions and for drawing my attention to the interesting physiology of acidophiles and alkalophiles.

REFERENCES

AUSLANDER, W. & JUNGE, W. (1974). The electric generator in photosynthesis of green plants. II. Kinetic correlation between protolytic reactions and redox reactions. *Biochimica et Biophysica Acta*, **357**, 285–98.

BOXER, D. H. & CLEGG, R. A. (1975). A transmembrane-location for the proton-translocating reduced ubiquinone-nitrate reductase segment of the respiratory chain of *Escherichia coli*. *FEBS Letters*, **60**, 54–7.

BOYER, P. D. (1974). Conformational coupling in biological energy transductions. *Biochimica et Biophysica Acta Library*, **13**, 289–301.

BRAND, M. D., REYNAFARJE, B. & LEHNINGER, A. L. (1976). Stoichiometric relationship between energy-dependent proton ejection and electron transport in mitochondria. *Proceedings of the National Academy of Sciences, U.S.A.*, **73**, 437–41.

BROCK, T. D. (1969). Microbial growth under extreme conditions. *Symposia of the Society for General Microbiology*, **19**, 15–41.

CASADIO, R., MELANDRI, A. B. & MELANDRI, B. A. (1975). The degree of coupling of ATP synthase in bacterial photophosphorylation. In *Electron Transfer Chains and Oxidative Phosphorylation*, ed. E. Quagliariello, S. Papa, F. Palmieri, E. C. Slater & N. Siliprandi, pp. 407–10. Amsterdam: North-Holland.

CLEGG, R. A. (1976). Purification and some properties of nitrate reductase (E.C. 1.7.99.4) from *Escherichia coli* K12. *Biochemical Journal*, **153**, 533–41.

COBLEY, J. G. (1976). Reduction of cytochromes by nitrite in electron transport particles from *Nitrobacter winogradskyi*. Proposal of a mechanism for H^+ translocation. *Biochemical Journal*, **156**, 493–8.

COLLINS, S. H. & HAMILTON, W. A. (1976). Magnitude of the proton-motive force in respiring *Staphylococcus aureus* and *Escherichia coli*. *Journal of Bacteriology*, **126**, 1224–31.

18   P. B. GARLAND

Cox, G. B. & Gibson, F. (1974). Studies on electron transport and energy-linked reactions using mutants of *Escherichia coli*. *Biochimica et Biophysica Acta*, **346**, 1–25.

Enoch, H. G. & Lester, R. L. (1974). The role of a novel cytochrome *b*-containing nitrate reductase in the *in vitro* reconstruction of formate–nitrate reductase activity of *E. coli*. *Biochemical and Biophysical Research Communications*, **61**, 1234–41.

Enoch, H. G. & Lester, R. L. (1975). The purification and properties of formate dehydrogenase and nitrate reductase from *Escherichia coli*. *Journal of Biological Chemistry*, **250**, 6693–705.

Eytan, G. D., Carroll, R. C., Schatz, G. & Racker, E. (1975). Arrangement of the subunits in solubilized and membrane-bound cytochrome c oxidase from bovine heart. *Journal of Biological Chemistry*, **250**, 8598–603.

Garland, P. B., Downie, J. A. & Haddock, B. A. (1975). Proton translocation and the respiratory nitrate reductase of *Escherichia coli*. *Biochemical Journal*, **152**, 547–59.

Gellerfors, P. & Nelson, B. D. (1975). Analysis of the peptide composition of purified beef-heart complex III by dodecyl-sulfate electrophoresis. *European Journal of Biochemistry*, **52**, 433–43.

Grinius, L., Slusnyte, R. & Griniuviene, B. (1975). ATP synthesis driven by proton-motive force imposed across *Escherichia coli* cell membranes. *FEBS Letters*, **57**, 290–3.

Griniuviene, B., Chmieliauskaite, V. & Grinius, L. (1974). Energy-linked transport of permeant ions in *Escherichia coli* cells. Evidence for membrane potential generation by proton pump. *Biochemical and Biophysical Research Communications*, **56**, 206–13.

Haddock, B. A. & Jones, C. W. (1977). Bacterial respiration. *Bacteriological Reviews* (in press).

Harold, F. M. & Papineau, J. M. (1972). Cation transport and electrogenesis by *Streptococcus faecalis*. I. The membrane potential. *Journal of Membrane Biology*, **8**, 27–44.

Harold, F. M., Pavlasova, E. & Baarda, J. R. (1970). The transmembrane pH gradient in *Streptococcus faecalis*. Origin and dissipation by proton conductors and *N,N'*-dicyclohexylcarbodiimide. *Biochimica et Biophysica Acta*, **196**, 235–44.

Hempfling, W. P. (1970). Studies of the efficiency of oxidative phosphorylation in intact *Escherichia coli* B. *Biochimica et Biophysica Acta*, **205**, 169–82.

Jeacocke, R. E., Niven, D. F. & Hamilton, W. A. (1972). The proton-motive force in *Staphylococcus aureus*. *Biochemical Journal*, **127**, 57p.

John, P. & Whatley, F. R. (1975). *Paracoccus denitrificans* and the evolutionary origin of the mitochondrion. *Nature, London*, **254**, 495–8.

Johnson, M. (1960). Enzymic equilibria and thermodynamics. In *The Enzymes*, ed. P. D. Boyer, H. Lardy & K. Myrback, vol. **3B**, pp. 407–41. New York & London: Academic Press.

Jones, R. W. & Garland, P. B. (1977). Topography and specificity of the interaction of bipyridylium compounds with the anaerobic respiratory chain of *Escherichia coli*. *Biochemical Journal* (in press).

KASHKET, E. R. & WILSON, T. H. (1974). Proton-motive force in fermenting *Streptococcus lactis* 7962 in relation to sugar accumulation. *Biochemical and Biophysical Research Communications*, **59**, 879–86.

KLINGENBERG, M. (1970). Metabolite transport in mitochondria: an example for intracellular membrane function. In *Essays in Biochemistry*, ed. P. N. Campbell & F. Dickens, vol. **6**, pp. 118–59. London & New York: Academic Press.

PARIS, P. C. & PERSHADSINGH, H. A. (1974). Estimation of membrane potentials in *Streptococcus faecalis* by means of a fluorescent probe. *Biochemical and Biophysical Research Communications*, **57**, 620–6.

LAWFORD, H. G. & GARLAND, P. B. (1972). Proton translocation coupled to quinone reduction by reduced nicotinamide adenine-dinucleotide in rat liver and ox heart mitochondria. *Biochemical Journal*, **130**, 1029–44.

LAWFORD, H. G. & HADDOCK, B. A. (1973). Respiration-driven proton translocation in *Escherichia coli*. *Biochemical Journal*, **136**, 217–20.

LEHNINGER, A. L. (1970). *Biochemistry*, 1st edn, p. 404. New York: Worth.

MACGREGOR, C. H. (1975). Anaerobic cytochrome b_1 in *Escherichia coli*: association with and regulation of nitrate reductase. *Journal of Bacteriology*, **121**, 1111–16.

MAHLER, H. R. & CORDES, E. H. (1966). *Biological Chemistry*, 1st edn, p. 609. New York & London: Harper & Row.

MALONEY, P. C., KASHKET, E. R. & WILSON, T. H. (1974). A proton motive force drives ATP synthesis in bacteria. *Proceedings of the National Academy of Sciences USA*, **71**, 3896–900.

MALONEY, P. C. & WILSON, T. H. (1975). ATP synthesis driven by a proton-motive force in *Streptococcus faecalis*. *Journal of Membrane Biology*, **25**, 285–310.

MITCHELL, P. (1961). Coupling of phosphorylation to electron and hydrogen transfer by a chemiosmotic type of mechanism. *Nature, London*, **191**, 144–8.

MITCHELL, P. (1966). *Chemiosmotic Coupling in Oxidative and Photo-synthetic Phosphorylation*. Bodmin, England: Glynn Research Ltd.

MITCHELL, P. (1968). *Chemiosmotic Coupling and Energy Transduction*. Bodmin, England: Glynn Research Ltd.

MITCHELL, P. (1970). Membranes of cells and organelles: morphology, transport and metabolism. *Symposia of the Society for General Microbiology*, **20**, 121–66.

MITCHELL, P. (1974). A chemiosmotic molecular mechanism for proton-translocating adenosine triphosphatases. *FEBS Letters*, **43**, 189–94.

MITCHELL, P. (1975). The proton-motive Q cycle: a general formulation *FEBS Letters*, **59**, 137–9.

MITCHELL, P. (1976). Vectorial chemistry and the molecular mechanics of chemiosmotic coupling: power transmission by proticity. *Biochemical Society Transactions*, **4**, 399–430.

MITCHELL, P. & MOYLE, J. (1965). Stoichiometry of proton translocation through the respiratory chain and adenosine triphosphatase systems of rat liver mitochondria. *Nature, London*, **208**, 147–51.

20 P. B. GARLAND

MOYLE, J. & MITCHELL, P. (1973). Proton translocation quotient for the adenosine triphosphatase of rat liver mitochondria. *FEBS Letters*, **30**, 317–20.

NICHOLLS, D. G. (1974a). The influence of respiration and ATP hydrolysis on the proton-electrochemical gradient across the inner membrane of rat liver mitochondria as determined by ion distribution. *European Journal of Biochemistry*, **50**, 305–15.

NICHOLLS, D. G. (1974b). Hamster brown-adipose-tissue mitochondria. The control of respiration and the proton electrochemical gradient by possible physiological effectors of the proton conductance of the inner membrane. *European Journal of Biochemistry*, **49**, 573–83.

NIVEN, D. F. & HAMILTON, W. A. (1974). Mechanisms of energy coupling to the transport of amino acids by *Staphylococcus aureus*. *European Journal of Biochemistry*, **44**, 517–22.

NIVEN, D. F., JEACOCKE, R. E. & HAMILTON, W. A. (1973). The membrane potential as the driving force for the accumulation of lysine by *Staphylococcus aureus*. *FEBS Letters*, **29**, 248–52.

PADAN, E., ZILBERSTEIN, D. & ROTTENBERG, H. (1976). The proton electrochemical gradient in *Escherichia coli* cells. *European Journal of Biochemistry*, **63**, 533–41.

POSTMA, P. W. & VAN DAM, K. (1976). The ATPase complex from energy transducing membranes. *Trends in Biochemical Sciences*, **1**, 16–17.

RAGAN, C. I. (1976). The structure and subunit composition of the particulate NADH-ubiquinone reductase of bovine heart mitochondria. *Biochemical Journal*, **154**, 295–305.

SCHAIRER, H. U. & HADDOCK, B. A. (1972). β-Galactoside accumulation in Mg^{2+}-, Ca^{2+}-activated ATPase deficient mutant of *E. coli*. *Biochemical and Biophysical Research Communications*, **48**, 544–51.

SCHULDINER, S., PADAN, E., ROTTENBERG, H., GROMET-ELHANAN, Z. & AVRON, M. (1974). pH and membrane potential in bacterial chromatophores. *FEBS Letters*, **49**, 174–7.

SIMONI, R. D. & POSTMA, P. W. (1975). The energetics of bacterial active transport. *Annual Reviews of Biochemistry*, **44**, 523–54.

SLATER, E. C. (1953). Mechanism of phosphorylation in the respiratory chain. *Nature, London*, **172**, 975–8.

SMITH, W. P., LANDAU, J. V. & POPE, D. H. (1976). Specific ion concentration as a factor in barotolerant protein synthesis in bacteria. *Journal of Bacteriology*, **126**, 654–60.

STRYER, L. (1975). *Biochemistry*, 1st edn, pp. 347–8. San Francisco: Freeman.

SWEETMAN, A. J. & GRIFFITHS, D. E. (1971). Studies on energy-linked reactions. Energy-linked reduction of oxidized nicotinamide-adenine dinucleotide by succinate in *Escherichia coli*. *Biochemical Journal*, **121**, 117–24.

VAN DER BEEK, E. G. & STOUTHAMER, A. H. (1973). Oxidative phosphorylation in intact bacteria. *Archiv für Mikrobiologie*, **89**, 327–39.

WEST, I. C. & MITCHELL, P. (1973). Stoichiometry of lactose–proton symport across the plasma membrane of *Escherichia coli*. *Biochemical Journal*, **132**, 587–92.

WEST, I. C. & MITCHELL, P. (1974a). Proton/sodium antiport in *Escherichia coli*. *Biochemical Journal*, **144**, 87–90.

WEST, I. C. & MITCHELL, P. (1974b). The proton-translocating ATPase of *Escherichia coli*. *FEBS Letters*, **40**, 1–4.

WIECHMANN, A. H. C. A., BEEM, E. R. & VAN DAM, K. (1975). The relationship between H^+ translocation and ATP synthesis in mitochondria. In *Electron Transfer Chain and Oxidative Phosphorylation*, ed. E. Quagliariello, S. Papa, F. Palmieri, E. C. Slater & N. Siliprandi, pp. 335–42. Amsterdam: North-Holland.

WILLIAMS, R. J. P. (1961). Possible functions of chains of catalysts. *Journal of Theoretical Biology*, **1**, 1–17.

WILLIAMS, R. J. P. (1975). Protein connections between protons, electrons and ATP. In *Electron Transfer Chains and Oxidative Phosphorylation*, ed. E. Quagliariello, S. Papa, F. Palmieri, E. C. Slater & N. Siliprandi, pp. 417–22. Amsterdam: North-Holland.

WILSON, D. M., ALDERETE, J. F., MALONEY, P. C. & WILSON, T. H. (1976). Proton-motive force as the source of energy for adenosine-5′-triphosphate synthesis in *Escherichia coli*. *Journal of Bacteriology*, **126**, 327–37.

AEROBIC RESPIRATORY SYSTEMS IN BACTERIA

COLIN W. JONES

*Department of Biochemistry, School of Biological Sciences,
University of Leicester, Leicester LE1 7RH, UK*

INTRODUCTION

Bacterial respiration and energy transduction have been the subject of several recent reviews. Most of these have concentrated upon the mechanism of oxidative phosphorylation and of associated energy-dependent processes such as active transport, or on the organisation of the respiratory chain (Harold, 1972, 1976; Boos, 1974; Simoni & Postma, 1975; Haddock & Jones, 1977), laying particular emphasis on the vectorial aspects of energy transduction as defined by chemiosmotic theory (Mitchell, 1966, 1968; West & Mitchell, 1972). Almost all of these reviews have centred around the facultative anaerobe *Escherichia coli*, the pre-eminence of this organism for studying bacterial energetics probably reflecting both its capacity for growth under a variety of conditions and the relative ease with which mutants can be prepared which are defective in electron transfer or energy coupling (Gibson & Cox, 1973; Haddock, this volume).

Although the literature also contains a massive amount of data on respiration and energy conservation in other bacteria, much of it is rather fragmentary and recent comparative reviews of this work have largely been confined to detailed surveys of the distribution and properties of the membrane-bound ATPase complex (Abrams & Smith, 1974) and of various redox carriers, particularly cytochromes (Bartsch, 1968; Horio & Kamen, 1970; Lemberg & Barrett, 1973; Meyer & Jones, 1973a; Jurtshuk, Mueller & Acord, 1975; Jones & Meyer, 1976; but see also Smith, 1968; White & Sinclair, 1971; Gel'man, Lukoyanova & Ostrovskii, 1975); energetic aspects have been largely ignored. This paper will therefore present a more general review of the comparative energetics of bacterial respiratory chains, with particular emphasis on those factors which appear to be responsible for the variation in electron transfer pathways and energy conservation efficiencies which are characteristic of bacterial systems. It is purposely limited to the aerobic respiratory chains of heterotrophic bacteria, since those of autotrophic and photosynthetic organisms, together with anaerobic respiration using alternative electron acceptors, are described elsewhere in this volume (Aleem, O. T. G. Jones, and Kröger, this volume).

ELECTRON TRANSFER

Respiratory chain patterns

The membrane-bound respiratory chains of aerobically grown bacteria contain the same basic types of redox carriers as those present in the mitochondria of higher organisms, viz. iron–sulphur proteins, flavoproteins, quinones, cytochromes and cytochrome oxidases. However, a closer inspection of the distribution and properties of the individual redox components in these systems reveals that this apparent unity is largely superficial, and that the respiratory chains of very few species of heterotrophic bacteria are truly similar to those of mitochondria (Table 1). Representative of these organisms are the aerobes *Paracoccus* (*Micrococcus*) *denitrificans* and *Alcaligenes eutrophus* (*Hydrogenomonas eutropha*), both of which exhibit pseudomitochondrial respiratory chains in terms of their component redox carriers and of their sensitivities to classical inhibitors of mitochondrial electron transfer such as rotenone, antimycin A and cyanide (Asano, Imai & Sato, 1967*a*; Scholes & Smith, 1968; Ishaque & Aleem, 1970; Shipp, 1972*b*; Drozd & Jones, 1974; Lawford *et al.*, 1976). Indeed, partly as a result of these and other similarities, it has been suggested that the inner membrane of the present-day mitochondrion probably evolved from the plasma membrane of an ancestral relative of *P. denitrificans* via endosymbiosis with a primitive host cell (John & Whatley, 1975).

In contrast, the majority of bacterial respiratory systems are significantly, and sometimes grossly, different from those of mitochondria; indeed, perhaps the major characteristics of bacterial respiratory systems as a whole are the immense variety of their redox carrier patterns and their relative insensitivity to inhibitors (see, for example, Meyer & Jones, 1973*a*; Gel'man *et al.*, 1975; Jones & Meyer, 1976: see also Table 1). The variations in redox pattern between different species usually reflect the replacement of one carrier by another (which may or may not have similar properties) or the addition/deletion of specific carriers. The effect of such changes on electron transfer and associated energy conservation are quite varied. Thus, the replacement of ubiquinone by menaquinone in most Gram-positive bacteria, or the presence of both quinones in many enteric bacteria (Bishop, Pandya & King, 1962), have virtually no effect on aerobic respiration since both

Table 1. The respiratory chain composition of selected bacteria following heterotrophic growth under highly aerobic conditions

Organism	TH	Ndh	Q	MK	b	b	c_1	c	c	aa_3	o	a_1	d	c_550CO	References
Mitochondria	TH	Ndh	Q	—	b_T	b_K	c_1	c	—	aa_3	—	—	—	—	—
Paracoccus (Micrococcus) denitrificans	TH	Ndh	Q	—	b_{556}	b_{562}	c_{546}	c_{549}	—	aa_3	(o)	—	—	—	Scholes & Smith (1968); Shipp (1972b); Lawford et al. (1976)
Alcaligenes eutrophus (Hydrogenomonas eutropha)	TH	Ndh	Q	—	b	—	—	c	—	aa_3	(o)	—	—	—	Drozd & Jones (1974); Jones et al. (1975)
Micrococcus lysodeikticus	(TH)	Ndh	—	MK	b_{560}	b_{556}	—	—	c_{552}	aa_3	(o)	—	—	—	Fujita, Ishikawa & Shimazono (1966); Tikhonova (1974)
Pseudomonas ovalis Chester	TH	Ndh	Q	—	b	—	—	c	—	—	o	—	—	—	Francis et al. (1963); Jones et al. (1975)
Arthrobacter globiformis	—	Ndh	—	MK	b	—	—	c	—	aa_3	(o)	—	—	—	J. M. Brice & C. W. Jones (unpublished data)
Mycobacterium phlei	TH	Ndh	—	MK·2H	b_{561}	—	—	c_{548}	c_{554}	aa_3	o	—	—	—	Asano & Brodie (1964); Murthy & Brodie (1964); Revsin & Brodie (1969)
Micrococcus luteus (Sarcina lutea)	—	Ndh	—	MK	b_{562}	b_{557}	—	c_{549}	c_{553}	aa_3	o	a_1	d	—	Erickson & Parker (1969)
Haemophilus parainfluenzae	—	Ndh	—	DMK	b_{562}	b_{557}	—	(c_{550})	c_{552}	—	o	a_1	d	—	White (1962); White & Smith (1962); Smith et al. (1970)
Azotobacter vinelandii	—	Ndh	Q	—	b_{560}	—	$c_{4(551)}$	$c_{5(555)}$	—	—	o	a_1	d	—	Jones & Redfearn (1967); Campbell, Ormer-Johnson & Burris (1973)
Beneckea natriegens	—	Ndh	Q	—	b_{562}	b_{557}	c_{547}	c_{550}	c_{554}	—	o	(a_1)	(d)	$c_{550}CO$	Weston & Knowles (1973, 1974); Linton, Harrison & Bull (1975)
Chromobacterium violaceum	(TH)?	Ndh	Q?	—	—	b_{558}	—	—	—	—	o	—	—	—	Niven, Collins & Knowles (1975)
Escherichia coli	(TH)	Ndh	Q	(MK)	(b_{552})	b_{556}	—	(c_{548})	—	—	o	—	—	cCO	Shipp (1972a); Haddock & Schairer (1973); Ashcroft & Haddock (1975); Haddock, Downie & Garland (1976)
Klebsiella pneumoniae (Aerobacter aerogenes)	TH	Ndh	Q	—	b_{563}	b_{559}	—	—	—	—	o	—	—	—	Harrison (1973); Jones et al. (1975)
Acinetobacter lwoffi	TH	Ndh	Q	—	b	b	—	—	—	aa_3	(o)	—	—	—	Jones et al. (1975)
Bacillus megaterium	—	Ndh	—	MK	b	b	—	—	—	aa_3	(o)	—	—	—	Kröger & Dadak (1969); Downs & Jones (1975a)
Bacillus licheniformis	—	Ndh	—	MK	b	b	—	—	—	aa_3	(o)	—	—	—	Jones et al. (1975)

TH, nicotinamide nucleotide transhydrogenase; Ndh, NADH dehydrogenase (containing iron–sulphur proteins and flavoproteins); Q, ubiquinone; MK menaquinone; DMK demethyl menaquinone; cytochromes are referred to by letter plus, where possible with the b- and c-type cytochromes, their wavelength maxima in low temperature, reduced minus oxidised difference spectra. Brackets indicate redox carriers of low concentration or activity.

species of quinone are lipophilic hydrogen carriers, albeit with different standard redox potentials (Kröger & Klingenberg, 1970). Similarly, the replacement of cytochrome oxidase, aa_3 by the carbon-monoxide-binding cytochromes (and potential oxidases) o, a_1, d and/or $c_{(CO)}$ causes certain changes in respiratory properties (e.g. alterations in electron transfer capacity, affinity for molecular oxygen or sensitivity to cyanide), which are probably of little significance during growth under highly aerobic conditions (see, for example, Meyer & Jones, 1973a; Pudek & Bragg, 1974). Addition or deletion of redox carriers applies principally to nicotinamide nucleotide transhydrogenase and the c-type cytochromes (Jones et al., 1975; Jones & Meyer, 1976); deletion of the former probably affects the rate and extent of NADP(H)/NAD(H) inter-conversion within the cell, but the absence of cytochrome c appears to have little effect on the rate of respiration although it may significantly alter the vectorial organisation of the respiratory chain within the coupling membrane and hence lower the efficiency of oxidative phos-phorylation (see later). These changes in redox pattern are accompanied by a general loss of sensitivity to inhibitors, such that most of the respiratory systems in the lower half of Table 1 are markedly resistant to both rotenone and antimycin A; their resistance to cyanide largely reflects the concentration and activity of cytochrome oxidase d (Jones, 1973; Pudek & Bragg, 1974).

The effect of growth conditions

Variations in redox carrier patterns can also be induced within single species by altering the growth conditions, particularly with respect to the availability of molecular oxygen; these variations occur principally amongst the terminal oxidases and quinones.

At this point it is necessary to interpose a word of caution concerning the identification of various cytochromes as potential oxidases. Classic-ally, this involves measuring their abilities to bind carbon monoxide in a light-sensitive manner (carbon monoxide difference spectra and photo-dissociation spectra), to carry a significant proportion of the total electron flux (action spectra) and to react rapidly with molecular oxygen (stopped flow redox spectrophotometry). Unfortunately, most studies have been limited to measuring difference spectra, or less frequently action spectra, both of which are insufficient (see Harrison, 1976). The few rapid-reaction, kinetic studies which have been carried out have generally confirmed oxidase roles for cytochromes aa_3, o (except in P. denitrificans) and d, but not for cytochrome a_1 (Smith et al., 1970; Haddock, Downie & Garland, 1976; Lawford et al., 1976), although a_1

presumably has some oxidase activity in those few species of bacteria where it is the sole carbon-monoxide-binding cytochrome (Castor & Chance, 1959; Jones & Meyer, 1976); no kinetic studies on cytochrome $c_{(CO)}$ have yet been reported.

As can be seen from the selected examples shown in Table 2, oxygen deprivation tends to cause the complete or partial replacement of cytochrome oxidase aa_3 by o (e.g. *P. denitrificans* and *A. eutrophus*) or the enhanced synthesis of cytochrome oxidase d (plus a_1) relative to o and/or $c_{(CO)}$ (e.g. in *E. coli, Klebsiella pneumoniae, Haemophilus parainfluenzae* and *Beneckea natriegens*; see Jurtshuk *et al.*, 1975). These changes undoubtedly reflect attempts by these organisms to combat oxygen insufficiency via the synthesis of elevated concentrations of alternate oxidases which may have higher turnover numbers and/or exhibit increased affinities for molecular oxygen (White, 1963; Sinclair & White, 1970; Meyer & Jones, 1973b; Weston, Collins & Knowles, 1974: see also Harrison, 1972, 1976). However, it should be noted that enhanced synthesis of cytochrome oxidase d has also been observed during the aerobic growth of *K. pneumoniae* at very low rates (Harrison, 1973) and following the exposure of *Azotobacter* sp. to very high oxygen tensions (Drozd & Postgate, 1970; Jones, Brice, Wright & Ackrell, 1973).

The aerobic growth of organisms such as *Achromobacter, E. coli* and *Chromobacterium violaceum* in the presence of low concentrations of cyanide, or under conditions which promote cyanide evolution, tends to stimulate the synthesis of cytochrome oxidase d (plus a_1) at the expense of cytochrome o and $c_{(CO)}$ (Arima & Oka, 1965; Oka & Arima, 1965; Niven *et al.*, 1975; Ashcroft & Haddock, 1975). These changes once again reflect moves to combat the potential diminution of respiratory activity, the appearance of cytochrome d under these conditions providing an alternative route to oxygen since this oxidase is relatively resistant to cyanide (Oka & Arima, 1965; Jones, 1973; Kauffman & Van Gelder, 1973; Pudek & Bragg, 1974). In contrast, the growth of *Bacillus cereus* in the presence of cyanide leads only to the enhanced synthesis of cytochrome oxidase aa_3, since this organism is apparently incapable of synthesising cytochrome d (McFeters, Wilson & Strobel, 1970).

Significant variations in the relative concentrations of ubiquinone and menaquinone can also be induced in those organisms whose respiratory systems contain both types of quinone (i.e. species within the genera *Escherichia* and *Proteus*). Thus, ubiquinone appears to be the major quinone component during aerobic growth, whereas anaerobic or

Table 2. Variation in respiratory chain composition of selected species of bacteria following growth under different conditions with respect to the availability of molecular oxygen

Organism	Growth conditions	Respiratory chain composition											References
Paracoccus denitrificans	Aerobic	TH	Ndh	Q	—	b_{558}	b_{563}	c_{546}	c_{549}	aa_3		(o)	—	Lam & Nicholas (1969); Shipp (1972b); Sapshead & Wimpenny (1972); Lawford et al. (1976)
	Oxygen-limited	TH	Ndh	Q	—	b	b_{563}	—	c	(aa_3)		o	—	
Escherichia coli	Aerobic	(TH)	Ndh	Q	(MK)	b_{556}	b_{558}	—	(c_{548})	—		o	—	Shipp (1972a); Haddock & Schairer (1973); Pudek & Bragg (1974, 1976); Hendler, Towne & Shrager (1975); Ashcroft & Haddock (1975); Haddock & Garland (1976)
	Aerobic + cyanide (150 μM)	(TH)	½Ndh	Q	MK	b_{556}	b_{558}	—	—	(a_1)		—	d	
	Anaerobic (or oxygen-limited)	(TH)	Ndh	Q	MK	b_{556}	b_{558}	—	—	(a_1)		o	d	
Haemophilus parainfluenzae	Aerobic	—	Ndh	—	DMK	b_{557}	b_{562}	c_{550}	c_{552}	(a_1)		o	d	White & Smith (1962); White (1962); Sinclair & White (1970); Smith et al. (1970); White & Sinclair (1971)
	Oxygen-limited	—	Ndh	—	DMK	b	b	c	c	—		—	d	

Soluble c-type cytochromes have been detected in *E. coli* following anaerobic growth (Gray, Wimpenny, Hughes & Ranlett, 1963; Fujita, 1966). Brackets indicate redox carriers of low concentration or activity. Abbreviations as for Table 1.

oxygen-limited conditions tend to enhance the level of menaquinone (Whistance & Threlfall, 1968; Kröger, Dadák, Klingenberg & Diemer, 1971; Haddock & Schairer, 1973), as also does aerobic growth in the presence of low concentrations of cyanide (Ashcroft & Haddock, 1975).

Respiratory chain branching

All known respiratory systems exhibit some evidence of branching, but the extent and physiological importance of this phenomenon is extremely variable (see White & Sinclair, 1971). Extensive branching obviously occurs at the level of the primary dehydrogenases, since all respiratory chains are able to oxidise directly a variety of substrates in addition to NADH, and formal branching may also occur in association with any protonmotive quinone cycle which may be present (Mitchell, 1975a, b; see later). In addition, branching is invariably present at the terminal end of the respiratory system during anaerobic growth on fumarate or nitrate, since the appropriate reductases are accompanied by significant amounts of cytochrome oxidase(s) (Haddock et al., 1976).

Under strictly aerobic growth conditions, the high-potential end of the bacterial respiratory chain may be either linear or branched. Linear terminal pathways are relatively rare and are limited to the few respiratory systems which contain only one functional species of cytochrome oxidase (e.g. most strains of E. coli and K. pneumoniae cultured under highly aerobic conditions, and Pseudomonas ovalis Chester; Fig. 1). Paracoccus denitrificans also falls into this category since although it contains cytochromes aa_3 and o, the latter does not exhibit significant oxidase activity (Lawford et al., 1976).

Many bacterial respiratory chains probably contain more than one functional species of cytochrome oxidase (although insufficient kinetic studies have been carried out to make this an absolute certainty) and terminal branching is therefore likely to be very common. In most cases the branching is probably limited to simple electron transfer from a common, penultimate b- or c-type cytochrome to molecular oxygen via the available oxidases (e.g. Micrococcus lysodeikticus, Alcaligenes eutrophus, Haemophilus parainfluenzae and Bacillus megaterium; Fig. 2). However, in a limited number of systems, branching is much more complex and appears to involve b- and/or c-type cytochromes as well as multiple cytochrome oxidases; e.g. E. coli grown under oxygen-limited conditions, Azotobacter vinelandii, Beneckea natriegens, Chromobacterium violaceum (grown on minimal medium such that there is little evolution of cyanide), and possibly also Pseudomonas aeruginosa (Fig. 3). Each of these respiratory systems contains at least two terminal

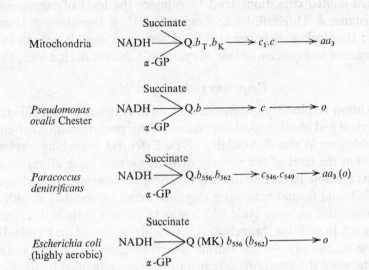

Fig. 1. Bacterial respiratory chains with linear terminal pathways. Redox components which are present in low concentrations or are kinetically inactive are enclosed in brackets. Abbreviations: α-GP, α-glycerophosphate; other abbreviations as for Table 1.

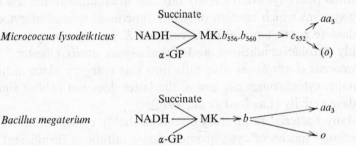

Fig. 2. Bacterial respiratory chains with terminal pathways which are branched at the cytochrome oxidase level. Abbreviations and symbols as for Fig. 1.

branches, one of which is considerably less sensitive to cyanide than the other. In *A. vinelandii*, *B. natriegens*, *C. violaceum*, and *P. aeruginosa*, the cyanide-sensitive pathway catalyses the oxidation of artificial substrates such as ascorbate-TMPD (*N,N,N′,N′*-tetramethyl-*p*-phenylenediamine) or ascorbate-DCPIP (2,6-dichlorophenol indophenol) (I_{50} for cyanide $\leqslant 10$ μM), whilst the cyanide-resistant pathway ($I_{50} \leqslant 0.5$ mM) is mainly associated with the oxidation of physiological substrates (Jones & Redfearn, 1967; Paterson, 1970; Yates & Jones, 1974; Kauffman & Van Gelder, 1974; Weston *et al.*, 1974; Niven *et al.*,

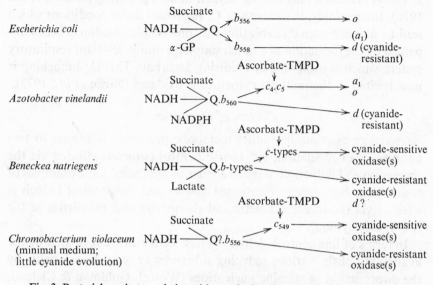

Fig. 3. Bacterial respiratory chains with terminal pathways which are extensively branched. Abbreviations and symbols as for Fig. 1.

1975). There is reasonably good evidence that in *A. vinelandii*, the cyanide-sensitive pathway is mediated via cytochromes b c_4 $c_5 \rightarrow o$ (plus a_1?) whereas cytochromes $b \rightarrow d$ appear to constitute the cyanide-resistant branch. Cytochrome oxidases o and d also appear to terminate cyanide-sensitive and cyanide-'resistant' pathways respectively in *E. coli*, where cytochrome a_1 is present but inactive as an oxidase (Pudek & Bragg, 1974; Haddock *et al.*, 1976). Since *B. natriegens* and *C. violaceum* exhibit an approximately similar redox carrier pattern (plus c-type cytochrome oxidases) to *A. vinelandii*, it is possible that their respiratory systems are also similarly organised, but this clearly requires more detailed investigation. Furthermore, the existence of branched respiratory pathways in isolated membranes need not necessarily reflect the presence of such pathways *in vivo* (Linton, Harrison & Bull, 1975, 1976), although they probably do so in *A. vinelandii* (Downs & Jones, 1975*b*).

Although the availability of molecular oxygen (as determined by the stage of growth in batch culture) clearly affects the precise redox carrier compositions of these organisms, the terminal respiratory systems of *A. vinelandii*, *B. natriegens* and *C. violaceum* remain branched, albeit with slightly modified properties (Ackrell & Jones, 1971*b*; Weston *et al.*, 1974; Niven *et al.*, 1975). In contrast, the terminal respiratory system of

E. coli is branched only during oxygen-limited growth (Haddock *et al.*, 1976). Interestingly, the growth of *C. violaceum* under conditions which lead to massive cyanide evolution (i.e. on complex medium) is accompanied by the formation of a much simpler, cyanide-resistant respiratory system which is incapable of oxidising ascorbate-TMPD; branching is now limited to the multiple cytochrome oxidases (Niven *et al.*, 1975).

Control of respiration

Bacterial respiration, like other metabolic processes, is subject to two basic types of regulation, viz. coarse control (which is effected via the repression and induction of redox carrier biosynthesis, some examples of which have been briefly mentioned above) and fine control (which is effected via the inherent kinetic and thermodynamic properties of the respiratory system).

In terms of fine control, dehydrogenase activities are regulated by the availability of the various reducing substrates or, very occasionally, by the direct action of adenine nucleotides (Worcel, Goldman & Cleland, 1965; Ackrell, Erickson & Jones, 1972), whilst cytochrome oxidase activities are influenced by the ambient concentration of molecular oxygen. Several bacterial respiratory systems also exhibit classical respiratory control as evidenced by the ability of ADP (in the presence of inorganic phosphate), uncoupling agents and ionophorous antibiotics to stimulate respiration by collapsing the transmembrane proton-motive force (Eilermann, Pandit-Hovenkamp & Kolk, 1970; Hempfling, 1970*a*, *b*; John & Whatley, 1970; John & Hamilton, 1970, 1971; Jones, Ackrell & Erickson, 1971; Beatrice & Chappell, 1974). The latter is probably the major control process in linear respiratory systems under highly aerobic conditions, although kinetic regulation at the cytochrome oxidase level may be more important during oxygen deprivation (for a detailed analysis of respiratory regulation *in vivo* see Harrison, 1976).

The major function of a branched respiratory system is undoubtedly to allow some flexibility in the exact route of electron transfer, thus enabling the cell to minimise the potentially deleterious effects of certain growth environments and to take maximum advantage of others. Hence in the presence of low concentrations of molecular oxygen or cyanide, electron transfer would tend to be routed via the terminal branch which is most capable of maintaining a high potential rate of respiration; under these conditions regulation would tend to remain the prerogative of the energy transduction system. The presence of multiple cytochrome oxidases obviously allows this re-routing to take place immediately, each oxidase carrying a fraction of the total electron flux which is determined by its concentration and kinetic properties relative to those of the other oxidases; *H. parainfluenzae* and *E. coli* are well documented examples of this type of system (see for example White & Sinclair, 1971; Pudek & Bragg, 1974).

In the more complex branched systems, where cytochrome *c* appears to be a constituent of one terminal pathway but not the other (e.g. *A. vinelandii*, *B. natriegens* and *C. violaceum*), the two pathways may exhibit unequal phosphorylation efficiencies (Ackrell & Jones, 1971*a*; Downs & Jones, 1975*b*) as well as different affinities for molecular oxygen (Weston *et al.*, 1974). Differential electron flow through these pathways would therefore be determined not only by the kinetic properties of the respective oxidases and of the redox carriers at the point of branching, but also by the energy state of the cell. A good example of such a system is afforded by nitrogen-fixing cultures of *A. vinelandii* in which the terminally branched respiratory system helps to protect the nitrogenase complex from the deleterious effects of molecular oxygen; under highly aerobic conditions, electron transfer is probably routed via the non-phosphorylating, cytochrome $b \rightarrow d$ pathway, thus causing an increase in cellular respiratory activity and a concomitant decrease in the ambient oxygen concentration (see Yates & Jones, 1974).

ENERGY CONSERVATION

In comparison with the relatively large amount of information which is available in the literature concerning the composition and organisation of bacterial respiratory chains, much less is known about the efficiencies with which they conserve energy. The main reasons for this inbalance are undoubtedly the experimental difficulties of assaying oxidative phosphorylation in bacterial cells and in respiratory membranes derived from them.

The bacterial plasma membrane, unlike the inner membrane of the eukaryote mitochondrion (see Klingenberg, 1970), lacks an adenine nucleotide translocase and is therefore normally impermeable to exogenous ATP and ADP. Thus, since the reversible, proton-translocating ATPase (ATP synthetase) is located on the cytoplasmic side of the coupling membrane (see Salton, 1974; Haddock & Jones, 1977; Harold, 1976) whole cells will neither hydrolyse exogenous ATP nor utilise ADP as a phosphoryl acceptor for oxidative phosphorylation. Oxidative phosphorylation in whole cells is therefore assayed by measuring either respiration-induced changes in the composition of endogenous adenine nucleotide pools (see Hempfling, 1970*a*; Baak & Postma, 1971; van der

Beek & Stouthamer, 1973), respiration-linked proton ejection (see Scholes & Mitchell, 1970) or molar growth yields (see Stouthamer & Bettenhaussen, 1973; Stouthamer, this volume).

In contrast, the esterification of ADP and inorganic phosphate can be used to assay oxidative phosphorylation in bacterial respiratory membranes, provided that the latter are in the form of closed vesicles with an inside-out configuration (i.e. where the orientation of the membrane is opposite to that present in the intact cell, such that the ATPase and the primary dehydrogenases are on the outer surface and are thus accessible to exogenous adenine nucleotides and oxidisable substrates respectively). Unfortunately, such preparations are usually contaminated with open membrane fragments (and right-side-out vesicles) which are capable of respiration but not energy conservation; P/O ratios of bacterial respiratory membranes are therefore invariably lower than those observed with intact mitochondria or well prepared submitochondrial particles. Nevertheless, they are often high enough to allow the unambiguous detection and location of individual energy coupling sites (see Imai, Asano & Sato, 1967; John & Whatley, 1970; Ackrell & Jones, 1971a), particularly when supplemented with measurements of other reactions which reflect membrane energisation at the expense of either ATP hydrolysis or respiration, e.g. reversed electron transfer (including transhydrogenation), respiratory control or the quenching of various fluorescent probes (see, for example, Eilermann, 1970; Sweetman & Griffiths, 1971a, b; Bragg & Hou, 1974; Houghton, Fisher & Sanadi, 1975).

Isolated respiratory membranes

P/O ratios

P/O ratios of respiratory membranes derived from heterotrophic bacteria are rarely greater than 1.0–1.5 for the oxidation of NADH, although higher efficiencies have occasionally been observed using membranes prepared from the structurally more complex chemolithotrophs (see Gel'man et al., 1975). Absolute values are therefore generally of limited use in determining the number of energy coupling sites, and it is necessary instead to compare P/2e ratios measured during electron transfer to and from substrates of different redox potential; e.g. using NAD(P)H, succinate or reduced cytochrome c as electron donors, and fumarate, quinones, oxidised cytochrome c or molecular oxygen as electron acceptors.

This approach has been used with some success to detect energy coupling at sites 1 (NADH dehydrogenase) and 2 (the quinone–cytochrome b region) in respiratory membrane vesicles from a wide range of bacteria, including *Mycobacterium phlei* (Asano & Brodie, 1965). *Paracoccus denitrificans* (Imai *et al.*, 1967; John & Whatley, 1970), *Alcaligenes eutrophus* (Ishaque & Aleem, 1970), *Azotobacter vinelandii* (Eilermann *et al.*, 1970; Ackrell & Jones, 1971*a*), *Acinetobacter lwoffi* and *Pseudomonas ovalis* Chester (Jones *et al.*, 1975). The unambiguous detection of ATP synthesis at site 3 is considerably more difficult since physiological substrates do not donate reducing equivalents to the respiratory chain at this level. Furthermore, exogenous mammalian cytochrome c is by no means capable of interacting universally with endogenous bacterial cytochrome c, and artificial electron donors such as ascorbate-TMPD or ascorbate-DCPIP can tap into the respiratory chain at different points according to the experimental conditions employed. Nevertheless, oxidative phosphorylation at site 3 has been detected with some certainty in several heterotrophically grown bacteria, including *M. phlei* (Asano & Brodie, 1965) and *A. vinelandii* (Ackrell & Jones, 1971*a*). Interestingly, site 3 has also been detected in respiratory membranes of *A. eutrophus* and *P. denitrificans* grown autotrophically but not heterotrophically (Ishaque & Aleem, 1970; Knoblock, Ishaque & Aleem, 1971), whereas the converse is true for *Pseudomonas saccharophila* (Ishaque, Donawa & Aleem, 1971); however, these differences could merely reflect different orientations of the membrane vesicles when prepared from cells cultured under different conditions (Burnell, John & Whatley, 1975).

As far as energy conservation at site 0 (nicotinamide nucleotide transhydrogenase) is concerned, the difference in standard redox potential between the $NADPH+H^+/NADP^+$ and $NADH+H^+/NAD^+$ couples is so small (approximately 4 mV) that ATP synthesis concomitant with the oxidation of NADPH by NAD^+ is only possible in the presence of a very high $[NADPH][NAD^+]/[NADP^+][NADH]$ ratio. Indeed, significant ATP synthesis has recently been detected under such conditions in membrane vesicles from *A. lwoffi* and *P. ovalis* Chester (Jones *et al.*, 1975), as well as in beef heart submitochondrial particles (Van de Stadt, Nieuwenhuis & Van Dam, 1971).

ADP/O ratios

One of the classical methods of assessing the efficiency of oxidative phosphorylation in eukaryotes is via the measurement of ADP/O ratios from respiratory control cycles initiated by the addition of small amounts of ADP to state 4 mitochondria (Chance & Williams, 1956).

Unfortunately, this method can only be applied to well coupled respiratory systems and is thus of little general use with membrane vesicles from heterotrophic bacteria, even though preparations from several organisms have shown some evidence of respiratory control (Eilermann *et al.*, 1970; John & Whatley, 1970; Jones *et al.*, 1971). However, this approach has been applied successfully to respiratory membranes from *P. denitrificans*, where the resultant ADP/O ratios of approximately 1.3 indicate the presence of a minimum of two sites of ATP synthesis between NADH and molecular oxygen (John & Hamilton, 1970, 1971).

Reversed electron transfer

Since oxidative phosphorylation is a reversible process, free energy released by forward electron transfer (e.g. from succinate to oxygen) can be conserved via the synthesis of ATP; conversely, energy released by ATP hydrolysis can be used to drive reversed electron transfer (e.g. from succinate to NAD^+, or from NADH to $NADP^+$). In both cases energy transduction is mediated via the formation and utilisation of the transmembrane protonmotive force. Energy-dependent, reversed electron transfer may therefore occur either anaerobically (at the expense of ATP hydrolysis) or aerobically (at the expense of forward electron transfer to oxygen via a second energy coupling site), and is clearly a powerful experimental tool with which to investigate the presence and location of potential coupling sites in inside-out membrane vesicles.

ATP-dependent reduction of $NADP^+$ by NADH (site 0) has been observed with respiratory membranes from several bacteria including *E. coli* (Murthy & Brodie, 1964; Bragg & Hou, 1968; Sweetman & Griffiths, 1971*b*), *Paracoccus denitrificans* (Asano *et al.*, 1967*a*), *Klebsiella pneumoniae* and *Alcaligenes eutrophus*, but not *Arthrobacter globiformis*, *Arthrobacter ramosus* or *Bacillus megaterium* (J. M. Brice, C. Edwards & C. W. Jones, unpublished data). In addition, aerobic transhydrogenation driven by the oxidation of succinate and/or NADH has been observed in membranes from *E. coli* (Fisher & Sanadi, 1971; Bragg, Davies & Hou, 1972; Bragg & Hou, 1974) and *P. denitrificans* (Asano *et al.*, 1967*a*), thus indicating the presence also of energy coupling sites 1 and (probably) 2. Site 1 has also been detected in respiratory membranes from *E. coli*, *P. denitrificans* and *A. eutrophus*, through their ability to catalyse reversed electron transfer from succinate (or other flavin-linked substrates) to NAD^+ at the expense of ATP hydrolysis (Asano, Imai & Sato, 1967*b*; Ishaque & Aleem, 1970; Sweetman & Griffiths, 1971*a*; Poole & Haddock, 1974). Furthermore, the ability of respiratory membranes from *A. eutrophus* to catalyse the

ATP-dependent reduction of NAD^+ by ascorbate or ferrocytochrome c indicates the additional presence of site 2 in this organism (Ishaque & Aleem, 1970). However, the aerobic reversal of electron transfer across sites 0 and 1 using energy generated by forward electron transfer through site 3 has not yet been demonstrated with bacterial systems (although it is readily exhibited by eukaryote mitochondria), and once again reflects the experimental difficulty of measuring energy coupling at this terminal site using bacterial membrane vesicles.

Other methods

Respiration- or ATP-dependent energisation of inside-out membrane vesicles (or whole cells) can also be detected from the resultant fluorescence changes of dyes such as atebrin, 9-amino-6-chloro-methoxy-acridine (ACMA) or 1-anilino-8-naphthalene sulphonate (ANS). The results point to the presence of energy coupling sites 1 and/or 2 in E. coli, Mycobacterium phlei, Micrococcus lysodeikticus and Azotobacter vinelandii (Eilermann, 1970; Reeves, Lombardi & Kaback, 1972; Nieuwenhuis et al., 1973; Haddock & Downie, 1974; Aithal, Kalra & Brodie, 1974; Tikhonova, 1974), but as yet there has been little or no attempt to detect other potential energy conservation sites using this method.

Sites 1 and (probably) 2 have also been detected in M. lysodeikticus by following the respiration-dependent uptake of phenyldicarbaundecarborane anions into inside-out membrane vesicles (Tikhonova, 1974).

Whole cells

Changes in intracellular nucleotide concentrations

In 1970 Hempfling introduced a novel method for the estimation of P/2e ratios in whole bacteria which is based upon monitoring the intracellular concentrations of NAD^+, AMP, ADP and ATP following the addition of small aliquots of oxygen-saturated buffer to anaerobic cell suspensions. Using this technique, P/NADH ratios close to 3 have been obtained with E. coli B (Hempfling, 1970a) and A. vinelandii (Baak & Postma, 1971). Unfortunately, this potentially important innovation has been severely criticised on the grounds that re-reduction of NAD^+ by residual endogenous substrates could lead to an underestimate of the amount of NADH oxidised and hence to erroneously high P/NADH ratios; it also fails to take into account oxidation of the endogenous quinol pool which in certain cases could be significantly large. Indeed, when expressed in terms of oxygen consumption rather than NADH oxidation, whole cell P/2e ratios are often as low as $\leqslant 1$ (van der Beek & Stouthamer, 1973), although values of 2 and above have been reported for *A. vinelandii* (Knowles & Smith, 1970; Baak & Postma, 1971).

It would appear, therefore, that the accuracy of this technique is in some doubt and that its reproducibility between different laboratories is far from perfect. Nevertheless, Hempfling and his colleagues have used this method extensively in their studies of *E. coli* B (Hempfling, 1970*a, b*; Hempfling & Beeman, 1971) for which they have reported a P/NADH ratio of < 0.4 following batch growth in the presence of excess glucose, compared with > 3.0 after the disappearance of glucose from the growth medium or following growth on oxidisable substrates such as lactate or acetate. The ability of glucose to effect catabolite repression of the energy conservation apparatus is reversed by the addition of cyclic AMP to the growth medium, as evidenced by the resultant increases in the P/NADH ratio (Hempfling & Beeman, 1971), ATPase activity (Bragg *et al.*, 1972) and molar growth yield (Hempfling & Mainzer, 1975; Mainzer & Hempfling, 1976).

Respiration-linked proton translocation

Following the pioneering work of Scholes & Mitchell (1970) on *Paracoccus denitrificans*, increasing use has been made of respiration-linked proton translocation as an index of the efficiency of respiratory chain energy conservation in whole bacteria. Assuming an $\rightarrow H^+/P$ ratio of 2 g ion H^+ per mole ATP (Moyle & Mitchell, 1973*a*; but see also West & Mitchell, 1974; Brand, Reynafarge & Lehninger, 1976) the measured $\rightarrow H^+/O$ (g ion H^+ per g atom 0) is approximately equal to twice the number of potential energy coupling sites (Mitchell, 1966, 1968). Furthermore, the location of the proton-translocating segments within the respiratory chain can be determined with some precision by measuring the $\rightarrow H^+/O$ ratios of cells which have been starved of endogenous substrates and then loaded with defined substrates which are capable of donating reducing equivalents to different respiratory carriers, e.g. isocitrate ($NADP^+$), malate (NAD^+), succinate (flavin) and ascorbate-TMPD (cytochrome *c*). This approach was first used by Lawford & Haddock (1973) and has subsequently been applied to a wide range of bacterial systems, examples of which are shown in Table 3.

Through the use of this type of experiment it is becoming increasingly clear that bacterial respiratory chains are organised into two, three or four proton-translocating segments according to species. Segments 1 and 2 appear to be present in all of the organisms which have so far been investigated, with the possible exceptions of *Bacillus megaterium* M

Table 3. →H⁺/O ratios (as g atom H⁺ per g atom O) of whole bacteria following starvation and exposure to selected substrates

State of cells	Substrate	Immediate electron donor	→H⁺/O ratio			
			Escherichia coli EMG-2[a]	Acinetobacter lwoffi 4B[b]	Pseudomonas ovalis Chester[b]	Alcaligenes eutrophus H16[c]
Unstarved	Endogenous					7.82±0.16(8)
Starved	Endogenous		3.36±0.32(6)	5.69±0.15(4)	6.56±0.18(9)	6.45±0.13(12)
	+Isocitrate	NADPH		4.40±0.23(5)	5.90±0.31(4)	7.62±0.18(3)
	+Glutamate	{NADPH+NADH		5.44±0.08(3)		
	+Malate	NADH	3.86±0.12(10)	4.49±0.26(4)	5.82±0.15(5)	6.06±0.17(3)
	+Pyruvate	NADH				
	+Lactate	NADH				
	+Glycerol	FpH₂				
	+Succinate	FpH₂	2.26±0.08(4)		3.91±0.28(3)	4.49±0.16(6)
	+Ascorbate-TMPD	Cytochrome c (Fe²⁺)			2.55±0.15(3)	2.56±0.21(3)
	Proton-translocating segments		1, 2	0, 1, 2	0, 1, 2, 3	0, 1, 2, 3

[a] Lawford & Haddock (1973).
[b] Jones et al. (1975).
[c] C. Edwards & C. W. Jones (unpublished data). See also Drozd & Jones (1974).

(Downs & Jones, 1975a), E. coli grown under anaerobic conditions (Downie, 1974), and A. vinelandii during highly aerobic growth (Downs & Jones, 1975b); in contrast, proton-translocating segments 0 and/or 3 are predicated upon the presence of a membrane-bound nicotinamide nucleotide transhydrogenase and a high-potential cytochrome c respectively (Jones et al., 1975). However, it should be noted that the redox activities of these latter two components must be commensurate with that of the remainder of the respiratory chain in order for proton translocation at segments 0 and 3 to be readily observed (Table 4). This can be appreciated by comparing the efficiency of respiration-linked proton translocation in Acinetobacter lwoffi, Paracoccus denitrificans, Alcaligenes eutrophus and Pseudomonas ovalis Chester (which exhibit very high transhydrogenase activities) with that in E. coli and Klebsiella pneumoniae (where the activities are very much lower); loop O is easily detected in the former group but not in the latter, although there is no doubt that both E. coli and K. pneumoniae are capable of energy transduction at the level of site 0 since membrane vesicles prepared from these organisms readily catalyse ATP- and/or respiration-dependent transhydrogenation. Furthermore, these latter reactions are considerably diminished in mutants of E. coli which have lesions in the F_0 and/or F_1 components of the ATPase complex such that their coupling membranes are leaky to protons (Rosen, 1973; Altendorf, Harold & Simoni, 1974). Thus it is likely that the membrane-bound transhydrogenases of bacterial respiratory chains, like those of eukaryote mitochondria (Moyle & Mitchell, 1973b), are capable of reversible proton translocation.

Only A. vinelandii of the organisms with putatively branched, terminal respiratory systems has so far been investigated with respect to proton translocation (Downs & Jones, 1975b); the results confirm that site 3 is located on the cyanide-sensitive, cytochrome c mediated branch and that electron transfer down the cytochrome $b \rightarrow d$ pathway is not associated with proton ejection.

Molar growth yields

The concept of assessing the efficiency of bacterial energy conservation by measuring molar growth yields has been appreciated for several decades (for a review see Stouthamer, 1969; Stouthamer, this volume). Initially, measurements were confined to anaerobic cultures but were subsequently extended to include organisms growing under aerobic conditions where oxidative phosphorylation replaces substrate-level phosphorylation as the major energy conserving process.

Table 4. *The relationship between respiratory chain composition and the efficiency of respiration-linked proton translocation in selected bacteria*

Organism	Simplified respiratory system									→H⁺/O ratio (endogenous; g ion H⁺ per g atom O)	Proton-translocating segments		References
		Ndh	Q	MK	b		aa₃	(o)				2	Downs & Jones (1975a)
Bacillus megaterium D440	—	Ndh	—	MK	b	—	aa₃	(o)	—	3.95±0.15(5)	—	2	Downs & Jones (1975a)
Bacillus subtilis D473	—	Ndh	—	MK	b	—	aa₃	(o)	—	3.97±0.16(3)	—	2	Jones et al. (1975)
Bacillus licheniformis A5	—	Ndh	—	(MK)	b	—	aa₃	o	—	3.89±0.09(11)	—	2	Jones et al. (1975)
Escherichia coli EMG-2	(TH)	Ndh	Q	(MK)	b	—	—	o	—	3.53±0.22(6)	(0)?	2	Lawford & Haddock (1973); see also West & Mitchell (1972)
Escherichia coli W	(TH)	Ndh	Q	(MK)	b	—	—	o]	—	4.00±0.07(7)	(0)?	2	Brice et al. (1974); see also Jones et al. (1975)
Escherichia coli E24	(TH)	Ndh	Q	(MK)	b	—	—	]o	—	3.63±0.32(3)	(0)?	2	J. M. Brice & C. W. Jones (unpublished data)
Klebsiella pneumoniae NCTC 5055	(TH)	Ndh	Q	—	b	—	—	o	—	3.92±0.22(7)	(0)?	2	Jones et al. (1975)
Klebsiella pneumoniae NCTC 9633	(TH)	Ndh	Q	—	b	—	—	o	—	4.14±0.18(6)	(0)?	2	J. M. Brice & C. W. Jones (unpublished data)
Micrococcus lysodeikticus	—	Ndh	—	MK	b	c	aa₃	(o)	—	5.02±0.22(4)	—	2	Jones et al. (1975)
Arthrobacter globiformis NCIB 8602	—	Ndh	—	MK	b	c	aa₃	(o)	—	4.73±0.22(3)	—	2	J. M. Brice & C. W. Jones (unpublished data)
Azotobacter vinelandii NCIB 8660	TH	Ndh	Q	—	[b]	c	—	o	a₁	3.94±0.16(7)	0	[3(br.)]	Downs & Jones (1975b)
Acinetobacter lwoffi 4B	TH	Ndh	Q	—	b	c	—	o	d	5.69±0.15(4)	0	2	Jones et al. (1975)
Paracoccus denitrificans NCIB 8944	TH	Ndh	Q	—	b	c	aa₃	(o)	—	7.57(2)	0	2	C. Edwards & C. W. Jones (unpublished data)
Paracoccus denitrificans ATCC 13543	TH	Ndh	Q	—	b	c	aa₃	(o)	—	8.0±0.1	0	2	Scholes & Mitchell (1970); see also Lawford et al. (1976)
Alcaligenes eutrophus H16	TH	Ndh	Q	—	b	c	aa₃	(o)	—	7.82±0.16(8)	0	2	Jones et al. (1975); see also Drozd & Jones (1974); Beatrice & Chappell (1974)
Pseudomonas ovalis Chester	TH	Ndh	Q	—	[b	c	—	o]	—	6.56±0.18(9)	0	3	Jones et al. (1975)

Respiratory chain components present in low concentrations or exhibiting low activities are shown in brackets. Abbreviations: br, branched pathway; other abbreviations as for Table 1. Cells were grown in batch culture, prepared and assayed for →H⁺/O ratios using essentially standard procedures.

During anaerobic growth, cell yields are most usefully expressed in terms of the amount of carbon substrate which is fermented to provide a source of energy, e.g. $Y_{glucose}$ (g cells per mol glucose). This value obviously reflects both the efficiency of energy conservation via substrate-level phosphorylation and the efficiency with which this energy is subsequently used to drive biosynthesis and other energy-dependent cellular reactions (Y_{ATP}; g cells per mol ATP equivalents). Since the substrate-level phosphorylation reactions of bacterial catabolism are well documented, and can be readily assessed experimentally from fermentation balances, it is relatively easy to calculate Y_{ATP} from $Y_{glucose}$. In contrast, the relationship of oxidative phosphorylation to the efficiency of aerobic growth is rather more complicated. The aerobic growth yields of simple heterotrophic organisms are most usefully expressed in terms of the amount of oxygen consumed during in-vivo respiration (Y_{O_2}; g cells per mol O_2), which is equal to the product of N (the overall efficiency of aerobic energy conservation, i.e. oxidative phosphorylation plus a small contribution from substrate-level phosphorylation; mol ATP equivalents per mol O_2) and Y_{ATP}. Thus, in order to calculate the efficiency of oxidative phosphorylation from Y_{O_2} it is necessary to know the value of Y_{ATP} and also to make a small correction for substrate-level phosphorylation.

Unfortunately, much of the early work in this field yielded suspect values for $Y_{glucose}$, Y_{O_2} and Y_{ATP} since it took no account of energy requirements for cell maintenance rather than growth and was carried out in the main using batch cultures of changing nutritional status (see Stouthamer, 1969). More recently, however, the use of nutrient-limited continuous cultures has enabled observed growth yields to be corrected for the influence of maintenance energy (Pirt, 1965). Thus, for example, the growth efficiency of a glucose-limited anaerobic culture is described by the equation

$$1/Y_{glucose} = 1/Y_{glucose}^{max} + \frac{M_{glucose}}{\mu},$$

where $Y_{glucose}^{max}$ is the *true* molar growth yield (g cells per mol glucose), $M_{glucose}$ is the maintenance requirement (mol glucose h^{-1} g cells^{-1}) and μ is specific growth rate (h^{-1}; equivalent to the dilution rate, D, of a continuous culture). The assumption implicit in this equation is that the maintenance requirement is independent of the growth rate; this is probably true for energy-limited cultures, but not for growth under conditions where the carbon source is in excess and where energy is undoubtedly wasted by a variety of metabolic 'slip' reactions (Neijssel &

Tempest, 1975, 1976). $Y_{glucose}^{max}$ is thus readily calculated from the intercept of the straight-line plot of $1/Y_{glucose}$ versus $1/\mu$ ($\equiv 1/D$) and can be used to calculate Y_{ATP}^{max} (the *true* molar growth yield with respect to ATP equivalents; g cells per mol ATP equivalents) provided that the fermentation pattern is known. Using this approach, Y_{ATP}^{max} values of 12.4 and 14.0 g cells per mol ATP equivalents have recently been determined for anaerobic, glucose-limited cultures of *E. coli* and *K. pneumoniae* respectively (Stouthamer & Bettenhaussen, 1975; Mainzer & Hempfling, 1976). These values are thus significantly higher than the average Y_{ATP} of 10.5 g cells per mol ATP equivalents reported earlier for a selection of bacteria growing anaerobically in batch culture (Bauchop & Elsden, 1960; Stouthamer, 1969).

Similarly, the efficiency of aerobic growth under energy-limited, continuous culture conditions can be described by the equation

$$1/Y_{O_2} = 1/Y_{O_2}^{max} + \frac{M_{O_2}}{\mu}$$

where $Y_{O_2}^{max}$ is the *true* molar growth yield (g cells per mol O_2) and M_{O_2} is the maintenance respiration rate (mol O_2 h^{-1} g cells^{-1}). This equation is perhaps more commonly used in its transformed state (Harrison & Loveless, 1971), viz.

$$Q_{O_2} = \frac{\mu}{Y_{O_2}^{max}} + M_{O_2}$$

where Q_{O_2} is the *in situ* respiratory activity of the growing culture (mol O_2 h^{-1} g cells^{-1}). $Y_{O_2}^{max}$ is thus readily calculated from the slope of the straight-line plot of Q_{O_2} versus μ ($\equiv D$).

In order to calculate the value of N from $Y_{O_2}^{max}$ it is, of course, necessary to assume a value for Y_{ATP}^{max} (or vice versa). This problem is easily resolved for glucose-limited, aerobic cultures of *E. coli* and *K. pneumoniae* since it is possible to use values of Y_{ATP}^{max} already determined for anaerobic growth. The average $Y_{O_2}^{max}$ values for these organisms of 59.3 and 49.1 g cells per mol O_2 respectively (Table 5) thus reflect N values of 4.8 and 3.5 mole ATP equivalents per mol O_2. Since the glucose is essentially completely oxidised to carbon dioxide and water under these growth conditions, the contribution of substrate-level phosphorylation to total energy conservation is relatively slight; the N values thus indicate the presence of only two sites of ATP synthesis in the respiratory chains of these two organisms.

The interpretation of $Y_{O_2}^{max}$ values for growth on non-fermentable carbon sources (e.g. glycerol, lactate, acetate) is rather more difficult

Table 5. *The efficiencies of growth and energy conservation of the facultative anaerobes* Escherichia coli *and* Klebsiella pneumoniae *growing aerobically in continuous culture*

Growth-limiting carbon source	Organism	$Y^{max}_{O_2}$ (g cells per mol O_2)	Average $Y^{max}_{O_2}$ (g cells per mol O_2)	Y^{max}_{ATP} (g cells per mol ATP equivalents)	N (mol ATP equivalents per mol O_2)	References
Glucose	Escherichia coli W	59.7				Farmer & Jones (1976)
	Escherichia coli B	55.8				Mainzer & Hempfling (1976)
	Escherichia coli B (+cAMP)	62.4[d]	59.3	12.4]	4.8	Hempfling & Mainzer (1975)
	Klebsiella pneumoniae NCTC 418	52.1[c]				Neijssel & Tempest (1976)
	Klebsiella pneumoniae NCIB 8017	49.3[a]				Harrison & Loveless (1971)
	Klebsiella pneumoniae NCTB 8017	45.9	49.1	14.0	3.5	Stouthamer & Bettenhaussen (1975)
Glycerol	Escherichia coli W	50.9				Farmer & Jones (1976)
	Escherichia coli B	29.8[d]				Hempfling & Mainzer (1975)
	Escherichia coli E24	42.5	41.1	13.3	3.1	J. M. Brice & C. W. Jones (unpublished data)
	Klebsiella pneumoniae NCTC 9633	57.2				J. M. Brice & C. W. Jones (unpublished data)
	Klebsiella pneumoniae NCTC 418	41.5[a,c]	49.4	13.3	3.7	Neijssel & Tempest (1976)
Lactate	Escherichia coli W	35.0				Farmer & Jones (1976)
	Escherichia coli E24	36.5	35.8	8.4	4.3	J. M. Brice & C. W. Jones (unpublished data)
	Klebsiella pneumoniae NCTC 9633	31.0				J. M. Brice & C. W. Jones (unpublished data)
	Klebsiella pneumoniae NCTC 418	27.8[b,c]	29.4	8.4	3.5	Neijssel & Tempest (1975)

Growth temperature 30 °C unless otherwise stated.

[a] $Y^{max}_{O_2}$ calculated from the data presented.
[b] $Y^{max}_{O_2}$ calculated from the data presented using a value for M_{O_2} of 0.00073 mol O_2 h^{-1} g cells^{-1} (Farmer & Jones, 1976).
[c] Growth temperature 35 °C.
[d] Growth temperature 37 °C.

since it is of course impossible to determine corresponding anaerobic Y_{ATP}^{max} values. The problem is further complicated by several reports that $Y_{O_2}^{max}$ values for *E. coli* and *K. pneumoniae* vary dramatically with the nature of the growth-limiting carbon source (Hempfling & Mainzer, 1975; Neijssel & Tempest, 1975, 1976; Farmer & Jones, 1976), ranging from an average of approximately 50 g cells per mol O_2 for growth on glycerol or glucose down to as low as 20 g cells per mol O_2 for growth on acetate. Several explanations have been put forward to explain these variations. One possibility is that the nature of the carbon source affects the number of energy conservation sites (Hempfling & Beeman, 1971; Hempfling & Mainzer, 1975) but this is not supported by the $\rightarrow H^+/O$ ratios, which indicate that two coupling sites are always present irrespective of the carbon source (Farmer & Jones, 1976). An alternative possibility is that these changes in $Y_{O_2}^{max}$ result predominantly from alterations in the value of Y_{ATP}^{max} rather than of N. This view is strongly supported by theoretical considerations of the ATP requirements for cell biosynthesis (Stouthamer, 1973) which clearly indicate that the energy requirements for growth on minimal medium plus, for example, acetate are almost three-fold higher than for growth on glycerol or glucose.

Since the calculated energy requirement for the biosynthesis of cell materials from glucose (excluding active transport) is equivalent to a Y_{ATP}^{max} of 33.9 g cells per mol ATP equivalents (Stouthamer, 1973; Farmer & Jones, 1976), the experimentally determined anaerobic Y_{ATP}^{max} values of 12.4 and 14.0 g cells per mol ATP equivalents for *E. coli* and *K. pneumoniae* respectively indicate that approximately 60% of the available free energy is wasted and/or is used to drive a variety of energy-dependent processes such as active transport, cellular organisation, etc. Assuming that this percentage is relatively constant for growth on a variety of non-fermentable carbon sources, experimental Y_{ATP}^{max} values of 13.3 and 8.4 g cells per mol ATP equivalents can be predicted for aerobic growth on glycerol and lactate respectively. When these values are substituted into the $Y_{O_2}^{max}$ values obtained for glycerol- and lactate-limited continuous cultures of *E. coli* and *K. pneumoniae* (Table 5), the resultant N values are again consistent with the presence of only two sites of oxidative phosphorylation.

The apparent success of this latter approach with these two facultative anaerobes opens up the possibility of investigating quantitatively the efficiency of respiratory chain energy conservation in obligately aerobic organisms (Table 6). Thus, *B. megaterium* and *A. lwoffi* exhibit $Y_{O_2}^{max}$ values which are commensurate with the presence of only two potential

Table 6. *The efficiencies of growth and energy conservation of selected aerobic bacteria growing in continuous culture under glycerol- or lactate-limited conditions*

Growth-limiting carbon source	Organism	$Y_{O_2}^{MAX}$ (g cells per mol O_2)	Average $Y_{O_2}^{MAX}$ (g cells per mol O_2)	Y_{ATP}^{MAX} (g cells per mol ATP equivalents)	N (mol ATP equivalents per mol O_2)
Glycerol	Bacillus megaterium D440	53.8			
	Bacillus megaterium M	47.6	50.7	13.3	3.8
	Arthrobacter ramosus NCIB 9066	69.5			
	Arthrobacter globiformis NCIB 8602	97.8	80.1	13.3	6.0
	Paracoccus denitrificans NCIB 8944	73.0			
Lactate	Acinetobacter lwoffi 4B	27.8	27.8	8.4	3.3
	Arthrobacter globiformis NCIB 8602	40.8			
	Paracoccus denitrificans NCIB 8944	41.3	49.1	8.4	5.9
	Alcaligenes eutrophus H16	65.2			

From Downs & Jones (1975a) and J. M. Brice, C. Edwards & C. W. Jones (unpublished data).

sites of oxidative phosphorylation, whereas *Arthrobacter globiformis*, *Arthrobacter ramosus*, *P. denitrificans* (but see also Van Verseveld & Stouthamer, 1976; Stouthamer, this volume) and *A. eutrophus* grow with efficiencies which are compatible with the presence of an additional energy conservation site (site 3) commensurate with their ability to synthesise a high-potential cytochrome *c*. Since these differences are also reflected in the whole cell → H^+/O ratios (ignoring the contribution, if any, of the proton-translocating transhydrogenase), the results indirectly confirm that the reversible, membrane-bound ATPases of these organisms retranslocate protons with → H^+/ATP ratios of approximately 2 g ion H^+ per mol ATP.

CONCLUDING REMARKS

It is clear from the data presented in this review that the respiratory systems of heterotrophic bacteria exhibit great diversity not only in their redox carrier composition, but also in the number of energy coupling sites which they contain. These two phenomena appear to be intimately related since the pseudomitochondrial respiratory chains of *P. denitrificans* and *A. eutrophus* appear to be organised into four proton-translocating segments (energy-transduction sites), whereas the more truncated respiratory chains of *B. megaterium*, *E. coli*, *K. pneumoniae*, *M. lysodeikticus*, etc. contain only two or three segments. These differences appear to be caused by the inability of the latter group to synthesise a membrane-bound transhydrogenase (segment 0) and/or a high-potential cytochrome *c* (segment 3); most organisms appear to contain proton-translocating segments 1 and 2 during normal aerobic growth.

It is unlikely that the transhydrogenase plays a significant role in respiratory chain phosphorylation since there is no evidence that cells contain the requisite $[NADPH][NAD^+]/[NADP^+][NADH]$ ratio *in vivo* during balanced growth. Indeed, it is much more probable that the transmembrane protonmotive force built up by forward electron transfer through segments 1, 2 and (where present) 3 serves to drive the transhydrogenase reaction in an energy-consuming direction, i.e. towards the reduction of $NADP^+$ by NADH. The NADPH thus produced probably serves as a source of reducing power for biosynthesis, particularly of certain amino acids (Bragg *et al.*, 1972). It must be assumed that those organisms which lack an energy-dependent transhydrogenase have alternative means of generating sufficient NADPH, e.g. by increasing the complement or activities of their $NADP^+$-linked dehydrogenases, or by using a soluble, energy-independent transhydrogenase (Ragland,

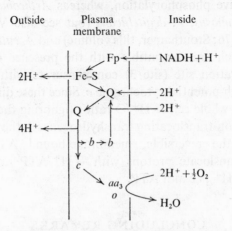

Fig. 4. The possible vectorial organisation of bacterial respiratory chains which contain cytochrome *c* (e.g. *P. denitrificans*, *M. lysodeikticus*). In some organisms ubiquinone is replaced by menaquinone. The transhydrogenase segment is omitted. Abbreviations: Fp, flavoprotein; Fe–S, iron–sulphur protein; other abbreviations and symbols as for Table 1 and Fig. 1. (After Garland *et al.*, 1975.)

Kawasaki & Lowenstein, 1966). The limited amounts of accurate growth yield data which are currently available do not clearly indicate whether the energy-dependent transhydrogenase reaction has any significant effect on the efficiency of bacterial growth. Further investigations into the functions of site 0 in bacterial systems are obviously required, possibly using transhydrogenase-negative (*nut*⁻) mutants of suitable organisms (see Haddock, this volume).

By analogy with mitochondria, the quinone moieties of the more complex bacterial respiratory systems are probably involved in a protonmotive quinone cycle (Mitchell, 1975*a, b*; Garland *et al.*, 1975). This ingenious reaction sequence circumvents repeated failures to detect a hydrogen carrier for proton translocation at segment 3 by combining segments 2 and 3 into a complex redox cycle which serves to translocate $4H^+$ concomitant with the oxidation of one molecule of QH_2 (or MKH_2) by molecular oxygen; cytochrome *c* is thus required in order to complete the vectorial organisation of the terminal electron-carrying limb (a simplified version of this scheme is shown in Fig. 4). Such a scheme is compatible with the general concept that quinones act as mobile hydrogen acceptors for the primary dehydrogenases (Kröger & Klingenberg, 1970), with the proposed location of the redox carriers in *M. lysodeikticus* (Tikhonova, 1974) and with the multiplicity of *b*-type cytochromes in many bacterial respiratory chains, including those of *P. denitrificans* and *M. lysodeikticus* (Shipp, 1972*b*; Lawford *et al.*, 1976).

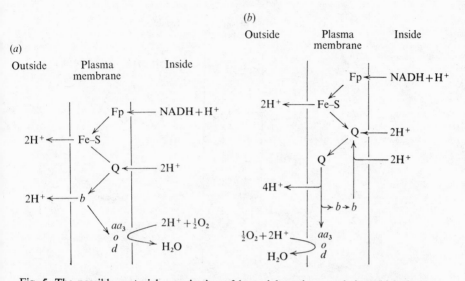

Fig. 5. The possible vectorial organisation of bacterial respiratory chains which do not contain cytochrome *c* (e.g. *B. megaterium, E. coli*). In some organisms ubiquinone is replaced by menaquinone. The transhydrogenase segment is omitted. Without (*a*) and with (*b*) a protonmotive quinone cycle. Abbreviations and symbols as for Fig. 4. (After Garland *et al.*, 1975; Haddock & Jones, 1977.)

In those bacterial respiratory chains which lack cytochrome *c*, proton translocation at segment 2 can easily be explained without reference to the protonmotive quinone cycle (Fig. 5*a*). As Garland *et al.* (1975) have pointed out, a protonmotive quinone cycle would become necessary in such organisms only if the oxidase reaction occurred on the outer surface (Fig. 5*b*). However, it should be noted that oxygen, unlike many inorganic electron acceptors (e.g. NO_3^-), is not excluded from the interior of the cell by the prevailing membrane potential and can therefore easily reach the cytoplasmic surface by simple diffusion; consequently there is no *a priori* need for reduction to occur on the outer surface. On the other hand, a protonmotive quinone cycle would be compatible with the presence of multiple *b*-type cytochromes in many of the simpler bacterial systems including aerobically grown *E. coli* (Shipp, 1972*a*; Haddock & Schairer, 1973; Hendler *et al.*, 1975; Pudek & Bragg, 1976).

Respiratory systems which are branched such that the terminal pathways are of unequal energy conservation efficiency (e.g. *A. vinelandii*) are possibly organised as shown in Fig. 6. However, it must be appreciated that relatively little is known about either the detailed

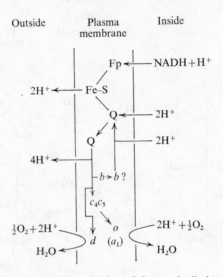

Fig. 6. The possible vectorial organisation of the terminally branched respiratory chain of *A. vinelandii*. Abbreviations and symbols as for Fig. 4.

vectorial organisation of bacterial respiratory chains or the exact location of the individual redox carriers within the coupling membrane (with the exception of the ATPase complex and some dehydrogenases), even in such relatively simple and intensively studied systems as those of *E. coli* and *M. lysodeikticus* (see Futai, 1974; Salton, 1974; Tikhonova, 1974). But in view of the successful topological studies which are currently being carried out on both the mitochondrial inner membrane and the nitrate reductase complex in the plasma membrane of anaerobically grown *E. coli* (see Racker, 1970; Harold, 1972; Garland *et al.*, 1975), it is clearly only a matter of time before similarly detailed studies are extended to the coupling membranes of aerobic bacteria. The results of such investigations should enable us to define more accurately the protonmotive aspects of energy transduction by the various types of bacterial respiratory chains, and to rationalise the complex redox carrier and branching patterns of some of these systems. Armed with this information, it may then be possible to augment recent outline ideas on the origin of biological energy conservation processes (Raven & Smith, 1976) with a more detailed description of the evolutionary history of proton-translocating, bacterial respiratory chains.

The author is indebted to Dr C. Edwards for many helpful discussions during the preparation of this review, and the Science Research Council for financial support (B/SR/59102).

REFERENCES

ABRAMS, A. & SMITH, J. B. (1974). Bacterial membrane ATPase. In *The Enzymes*, ed. P. D. Boyer, vol. 10, pp. 395–429. New York & London: Academic Press.

ACKRELL, B. A. C., ERICKSON, S. K. & JONES, C. W. (1972). The respiratory chain NADPH dehydrogenase of *Azotobacter vinelandii*. *European Journal of Biochemistry*, **26**, 387–92.

ACKRELL, B. A. C. & JONES, C. W. (1971*a*). The respiratory system of *Azotobacter vinelandii*. I. Properties of phosphorylating respiratory membranes. *European Journal of Biochemistry*, **20**, 22–8.

ACKRELL, B. A. C. & JONES, C. W. (1971*b*). The respiratory system of *Azotobacter vinelandii*. II. Oxygen effects. *European Journal of Biochemistry*, **20**, 29–35.

AITHAL, H. N., KALRA, V. K. & BRODIE, A. F. (1974). Temperature-induced alterations in 8-anilino-1-naphthalene sulphonate fluorescences with membranes from *Mycobacterium phlei*. *Biochemistry*, **13**, 171–8.

ALTENDORF, K., HAROLD, F. M. & SIMONI, R. D. (1974). Impairment and restoration of the energised state in membrane vesicles of a mutant of *Escherichia coli* lacking adenosine triphosphatase. *Journal of Biological Chemistry*, **249**, 4587–93.

ARIMA, K. & OKA, T. (1965). Cyanide resistance in *Achromobacter*. I. Induced formation of cytochrome a_2 and its role in cyanide-resistant respiration. *Journal of Bacteriology*, **90**, 734–43.

ASANO, A. & BRODIE, A. F. (1964). Oxidative phosphorylation in fractionated bacterial systems. XIV. Respiratory chains of *Mycobacterium phlei*. *Journal of Biological Chemistry*, **239**, 4280–91.

ASANO, A. & BRODIE, A. F. (1965). Oxidative phosphorylation in fractionated bacterial systems. XVIII. Phosphorylation coupled to different segments of the respiratory chain of *Mycobacterium phlei*. *Journal of Biological Chemistry*, **240**, 4002–10.

ASANO, A., IMAI, K. & SATO, R. (1967*a*). Oxidative phosphorylation in *Micrococcus denitrificans*. II. The properties of pyridine nucleotide transhydrogenase. *Biochimica et Biophysica Acta*, **143**, 477–86.

ASANO, A., IMAI, K. & SATO, R. (1967*b*). Oxidative phosphorylation in *Micrococcus denitrificans*. III. ATP-supported reduction of NAD$^+$ by succinate. *Journal of Biochemistry, Tokyo*, **62**, 210–14.

ASHCROFT, J. R. & HADDOCK, B. A. (1975). Synthesis of alternative membrane-bound redox carriers during aerobic growth of *Escherichia coli* in the presence of potassium cyanide. *Biochemical Journal*, **148**, 349–52.

BAAK, J. M. & POSTMA, P. W. (1971). Oxidative phosphorylation in intact *Azotobacter vinelandii*. *FEBS Letters*, **19**, 189–92.

BARTSCH, R. G. (1968). Bacterial cytochromes. *Annual Reviews of Microbiology*, **22**, 181–200.

BAUCHOP, T. & ELSDEN, S. R. (1960). The growth of microorganisms in relation to their energy supply. *Journal of General Microbiology*, **23**, 457–69.

BEATRICE, M. & CHAPPELL, J. B. (1974). Respiration-driven proton translocation in *Hydrogenomonas eutropha* H16. *Biochemical Society Transactions*, **2**, 151–3.

BISHOP, D. H. L., PANDYA, K. P. & KING, H. K. (1962). Ubiquinone and vitamin K in bacteria. *Biochemical Journal*, **83**, 606–14.

BOOS, W. (1974). Bacterial transport. *Annual Reviews of Biochemistry*, **43**, 123–46.

BRAGG, P. D., DAVIES, P. L. & HOU, C. (1972). Function of energy-dependent transhydrogenase in *Escherichia coli*. *Biochemical and Biophysical Research Communications*, **47**, 1248–55.

BRAGG, P. D. & HOU, C. (1968). Oxidative phosphorylation in *Escherichia coli*. *Canadian Journal of Biochemistry*, **46**, 631–41.

BRAGG, P. D. & HOU, C. (1974). Energisation of energy-dependent transhydrogenase of *Escherichia coli* at a second site of energy conservation. *Archives of Biochemistry and Biophysics*, **163**, 614–16.

BRAND, M. D., REYNAFARJE, B. & LEHNINGER, A. L. (1976). Stoichiometric relationship between energy-dependent proton ejection and electron transport in mitochondria. *Proceedings of the National Academy of Sciences, USA*, **73**, 437–41.

BRICE, J. M., LAW, J. F., MEYER, D. J. & JONES, C. W. (1974). Energy conservation in *Escherichia coli* and *Klebsiella pneumoniae*. *Biochemical Society Transactions*, **2**, 523–6.

BURNELL, J. N., JOHN, P. & WHATLEY, F. R. (1975). The reversibility of active sulphate transport in membrane vesicles of *Paracoccus denitrificans*. *Biochemical Journal*, **150**, 527–36.

CAMPBELL, W. H., ORME-JOHNSON, W. H. & BURRIS, R. H. (1973). A comparison of the physical and chemical properties of the four cytochromes c from *Azotobacter vinelandii*. *Biochemical Journal*, **135**, 617–30.

CASTOR, L. N. & CHANCE, B. (1959). Photochemical determinations of the oxidases of bacteria. *Journal of Biological Chemistry*, **234**, 1587–92.

CHANCE, B. & WILLIAMS, G. R. (1956). The respiratory chain and oxidative phosphorylation. *Advances in Enzymology*, **17**, 65–134.

DOWNIE, J. A. (1974). Energy conservation associated with differing oxidases in *Candida utilis* and *Escherichia coli*. PhD thesis, University of Dundee.

DOWNS, A. J. & JONES, C. W. (1975a). Energy conservation in *Bacillus megaterium*. *Archives of Microbiology*, **105**, 159–67.

DOWNS, A. J. & JONES, C. W. (1975b). Respiration-linked proton translocation in *Azotobacter vinelandii*. *FEBS Letters*, **60**, 42–6.

DROZD, J. W. & JONES, C. W. (1974). Oxidative phosphorylation in *Hydrogenomonas eutropha* H16 grown with and without iron. *Biochemical Society Transactions*, **2**, 529–31.

DROZD, J. W. & POSTGATE, J. R. (1970). Effects of oxygen on acetylene reduction, cytochrome content and respiratory activity of *Azotobacter chroococcum*. *Journal of General Microbiology*, **63**, 63–73.

EILERMANN, L. J. M. (1970). Oxidative phosphorylation in *Azotobacter vinelandii*: atebrine as a fluorescent probe for the energised state. *Biochimica et Biophysica Acta*, **216**, 231–3.

EILERMANN, L. J. M., PANDIT-HOVENKAMP, H. G. & KOLK, A. H. J. (1970).

Oxidative phosphorylation in *Azotobacter vinelandii* particles: phosphorylation sites and respiratory control. *Biochimica et Biophysica Acta*, **197**, 25–30.

ERICKSON, S. K. & PARKER, G. L. (1969). The electron transport system of *Micrococcus lutea* (*Sarcina lutea*). *Biochimica et Biophysica Acta*, **180**, 56–62.

FARMER, I. S. & JONES, C. W. (1976). The energetics of *Escherichia coli* during aerobic growth in continuous culture. *European Journal of Biochemistry*, **67**, 115–22.

FISHER, R. J. & SANADI, D. R. (1971). Energy-linked nicotinamide adenine dinucleotide transhydrogenase in membrane particles from *Escherichia coli*. *Biochimica et Biophysica Acta*, **245**, 34–41.

FRANCIS, M. J. O., HUGHES, D. E., KORNBERG, H. L. & PHIZACKERLEY, P. J. R. (1963). The oxidation of L-malate by *Pseudomonas* sp. *Biochemical Journal*, **89**, 430–7.

FUJITA, T. (1966). Studies on soluble cytochromes in Enterobacteriaceae; cytochromes b_{562} and c_{550}. *Journal of Biochemistry, Tokyo*, **60**, 329–34.

FUJITA, M., ISHIKAWA, S. & SHIMAZONO, N. (1966). Respiratory chain and phosphorylation site of the sonicated membrane fragments of *Micrococcus lysodeikticus*. *Journal of Biochemistry, Tokyo*, **59**, 104–14.

FUTAI, M. (1974). The orientation of membrane vesicles from *Escherichia coli* prepared by different procedures. *Journal of Membrane Biology*, **15**, 15–28.

GARLAND, P. B., CLEGG, R. A., BOXER, D., DOWNIE, J. A. & HADDOCK, B. A. (1975). Proton-translocating nitrate reductase of *Escherichia coli*. In *Electron Transfer Chains and Oxidative Phosphorylation*, ed. E. Quagliariello, S. Papa, F. Palmieri, E. C. Slater & N. Siliprandi, pp. 351–8. Amsterdam: North-Holland Publishing Company.

GEL'MAN, N. S., LUKOYANOVA, M. A. & OSTROVSKII, D. N. (1975). Bacterial membranes and the respiratory chain. *Biomembranes*, **6**, 129–209. New York & London: Plenum Press.

GIBSON, F. & COX, G. B. (1973). The use of mutants of *Escherichia coli* K12 in studying electron transport and oxidative phosphorylation. *Essays in Biochemistry*, **9**, 1–29.

GRAY, C. T., WIMPENNY, J. W. T., HUGHES, D. E. & RANLETT, M. (1963). A soluble *c*-type cytochrome from anaerobically grown *Escherichia coli* and various Enterobacteriaceae. *Biochimica et Biophysica Acta*, **67**, 157–60.

HADDOCK, B. A. & DOWNIE, J. A. (1974). The reconstitution of functional respiratory chains in membranes from electron-transport deficient mutants of *Escherichia coli* as demonstrated by quenching of atebrin fluorescence. *Biochemical Journal*, **142**, 703–6.

HADDOCK, B. A., DOWNIE, J. A. & GARLAND, P. B. (1976). Kinetic characterization of the membrane-bound cytochromes of *Escherichia coli* grown under a variety of conditions by using a stopped-flow dual wavelength spectrophotometer. *Biochemical Journal*, **154**, 285–94.

HADDOCK, B. A. & JONES, C. W. (1977). Bacterial respiration. *Bacteriological Reviews*, in press.

HADDOCK, B. A. & SCHAIRER, H. U. (1973). Electron transport chains of

Escherichia coli: reconstitution of respiration in a 5-aminolaevulinic acid-requiring mutant. *European Journal of Biochemistry*, **35**, 34–45.

HAROLD, F. M. (1972). Conservation and transformation of energy by bacterial membranes. *Bacteriological Reviews*, **36**, 172–230.

HAROLD, F. M. (1976). Membranes and energy transduction in bacteria. *Current Topics in Bioenergetics*, in press.

HARRISON, D. E. F. (1972). Physiological effects of dissolved oxygen tension and redox potential on growing populations of microorganisms. *Journal of Applied Chemistry and Biotechnology*, **22**, 417–40.

HARRISON, D. E. F. (1973). A study of the effect of growth conditions on the cytochromes of chemostat-grown *Klebsiella aerogenes* and kinetic changes of a 500 nm absorption band. *Biochimica et Biophysica Acta*, **275**, 83–92.

HARRISON, D. E. F. (1976). The regulation of respiration rate in growing bacteria. *Advances in Microbial Physiology*, in press.

HARRISON, D. E. F. & LOVELESS, J. E. (1971). The effect of growth conditions on respiratory activity and growth efficiency in facultative anaerobes grown in chemostat culture. *Journal of General Microbiology*, **68**, 35–43.

HEMPFLING, W. P. (1970a). Studies on the efficiency of oxidative phosphorylation in intact *Escherichia coli*. *Biochimica et Biophysica Acta*, **205**, 169–82.

HEMPFLING, W. P. (1970b). Repression of oxidative phosphorylation in *Escherichia coli* B by growth in glucose and other carbohydrates. *Biochemical and Biophysical Research Communications*, **41**, 9–15.

HEMPFLING, W. P. & BEEMAN, D. K. (1971). Release of glucose repression of oxidative phosphorylation in *Escherichia coli* B by cyclic adenosine 3′,5′-monophosphate. *Biochemical and Biophysical Research Communications*, **45**, 924–30.

HEMPFLING, W. P. & MAINZER, S. E. (1975). Effects of varying the carbon source limiting growth on yield and maintenance characteristics of *Escherichia coli* in continuous culture. *Journal of Bacteriology*, **123**, 1076–87.

HENDLER, R. W., TOWNE, D. W. & SHRAGER, R. I. (1975). Redox properties of *b*-type cytochromes in *Escherichia coli* and rat liver mitochondria and techniques for their analysis. *Biochimica et Biophysica Acta*, **376**, 42–62.

HORIO, T. & KAMEN, M. D. (1970). Bacterial cytochromes; II. Functional aspects. *Annual Reviews of Microbiology*, **24**, 399–428.

HOUGHTON, R. L., FISHER, R. J. & SANADI, D. R. (1975). Energy-linked and energy-independent transhydrogenase activities in *Escherichia coli* vesicles. *Biochimica et Biophysica Acta*, **396**, 17–23.

IMAI, K., ASANO, A. & SATO, R. (1967). Oxidative phosphorylation in *Micrococcus denitrificans*. I. Preparation and properties of phosphorylating membrane fragments. *Biochimica et Biophysica Acta*, **143**, 462–76.

ISHAQUE, M. & ALEEM, M. I. H. (1970). Energy coupling in *Hydrogenomonas eutropha*. *Biochimica et Biophysica Acta*, **223**, 388–97.

ISHAQUE, M., DONAWA, A. & ALEEM, M. I. H. (1971). Oxidative phosphorylation in *Pseudomonas saccharophila* under autotrophic and heterotrophic growth conditions. *Biochemical and Biophysical Research Communications*, **44**, 245–51.

JOHN, P. & HAMILTON, W. A. (1970). Respiratory control in membrane particles from *Micrococcus denitrificans*. *FEBS Letters*, **10**, 246–8.

JOHN, P. & HAMILTON, W. A. (1971). Release of respiratory control in particles from *Micrococcus denitrificans* by ion-translocating antibiotics. *European Journal of Biochemistry*, **23**, 528–32.

JOHN, P. & WHATLEY, F. R. (1970). Oxidative phosphorylation coupled to oxygen uptake and nitrate reduction in *Micrococcus denitrificans*. *Biochimica et Biophysica Acta*, **216**, 342–52.

JOHN, P. & WHATLEY, F. R. (1975). *Paracoccus denitrificans* and the evolutionary origin of the mitochondrion. *Nature, London*, **254**, 495–8.

JONES, C. W. (1973). The inhibition of *Azotobacter vinelandii* terminal oxidases by cyanide. *FEBS Letters*, **36**, 347–50.

JONES, C. W., ACKRELL, B. A. C. & ERICKSON, S. K. (1971). Respiratory control in *Azotobacter vinelandii* membranes. *Biochimica et Biophysica Acta*, **245**, 54–62.

JONES, C. W., BRICE, J. M., DOWNS, A. J. & DROZD, J. W. (1975). Bacterial respiration-linked proton translocation and its relationship to respiratory chain composition. *European Journal of Biochemistry*, **52**, 265–71.

JONES, C. W., BRICE, J. M., WRIGHT, V. & ACKRELL, B. A. C. (1973). Respiratory protection of nitrogenase in *Azotobacter vinelandii*. *FEBS Letters*, **29**, 77–81.

JONES, C. W. & MEYER, D. J. (1976). The distribution of cytochromes in bacteria. In *Handbook of Microbiology*, ed. H. LeChevalier, vol. II. Cleveland: Cleveland Rubber Company Press. (In press.)

JONES, C. W. & REDFEARN, E. R. (1967). The cytochrome system of *Azotobacter vinelandii*. *Biochimica et Biophysica Acta*, **143**, 340–53.

JURTSHUK, P., MUELLER, T. J. & ACORD, W. C. (1975). Bacterial terminal oxidases. *Cleveland Rubber Company Critical Reviews in Microbiology*, **3**, 399–468.

KAUFFMAN, H. F. & VAN GELDER, B. F. (1973). The respiratory chain of *Azotobacter vinelandii*. II. The effect of cyanide on cytochrome *d*. *Biochimica et Biophysica Acta*, **314**, 276–83.

KAUFFMAN, H. F. & VAN GELDER, B. F. (1974). The respiratory chain of *Azotobacter vinelandii*. III. The effect of cyanide in the presence of substrates. *Biochimica et Biophysica Acta*, **333**, 218–27.

KLINGENBERG, M. (1970). Metabolite transport in mitochondria; an example for intracellular membrane function. *Essays in Biochemistry*, **6**, 119–59.

KNOBLOCK, K., ISHAQUE, M. & ALEEM, M. I. H. (1971). Oxidative phosphorylation in *M. denitrificans* under autotrophic growth conditions. *Archiv für Mikrobiologie*, **76**, 114–25.

KNOWLES, C. J. & SMITH, L. (1970). Measurements of ATP level of intact *Azotobacter vinelandii* under different conditions. *Biochimica et Biophysica Acta*, **197**, 152–60.

KRÖGER, A. & DADÁK, V. (1969). On the role of quinones in bacterial electron transport; the respiratory system of *Bacillus megaterium*. *European Journal of Biochemistry*, **11**, 328–40.

KRÖGER, A., DADÁK, V., KLINGENBERG, M. & DIEMER, F. (1971). On the role of quinones in bacterial electron transport; differential roles of ubiquinone and menaquinone in *Proteus rettgeri*. *European Journal of Biochemistry*, **21**, 322–33.

KRÖGER, A. & KLINGENBERG, M. (1970). Quinones and nicotinamide nucleotides associated with electron transfer. *Vitamins and Hormones*, **28**, 533–74.

LAM, Y. & NICHOLAS, D. J. D. (1969). Aerobic and anaerobic respiration in *Micrococcus denitrificans*. *Biochimica et Biophysica Acta*, **172**, 450–61.

LAWFORD, H. G., COX, J. C., GARLAND, P. B. & HADDOCK, B. A. (1976). Electron transport in aerobically grown *Paracoccus denitrificans*: kinetic characterisation of the membrane bound cytochromes and the stoichiometry of proton translocation. *FEBS Letters*, **64**, 369–74.

LAWFORD, H. G. & HADDOCK, B. A. (1973). Respiration-driven proton translocation in *Escherichia coli*. *Biochemical Journal*, **136**, 217–20.

LEMBERG, R. & BARRETT, J. (1973). Bacterial cytochromes and cytochrome oxidases. In *Cytochromes*, pp. 217–36. New York & London: Academic Press.

LINTON, J. D., HARRISON, D. E. F. & BULL, A. T. (1975). Molar growth yields, respiration and cytochrome patterns of *Beneckea natriegens* when grown at different medium dissolved-oxygen tensions. *Journal of General Microbiology*, **90**, 237–46.

LINTON, J. D., HARRISON, D. E. F. & BULL, A. T. (1976). The effect of rate of respiration on sensitivity to cyanide and carbon monoxide in *Beneckea natriegens* grown in batch and continuous culture. *FEBS Letters*, **64**, 358–63.

McFETERS, G. A., WILSON, D. F. & STROBEL, G. A. (1970). Cytochromes in a cyanide-resistant strain of *Bacillus cereus*. *Canadian Journal of Microbiology*, **16**, 1221–6.

MAINZER, S. E. & HEMPFLING, W. P. (1976). Effects of growth temperature on yield and maintenance during glucose-limited continuous culture of *Escherichia coli*. *Journal of Bacteriology*, **126**, 251–6.

MEYER, D. J. & JONES, C. W. (1973a). Distribution of cytochromes in bacteria: relationship to general physiology. *International Journal of Systematic Bacteriology*, **23**, 459–67.

MEYER, D. J. & JONES, C. W. (1973b). Reactivity with oxygen of bacterial cytochrome oxidases a_1, aa_3 o. *FEBS Letters*, **33**, 101–5.

MITCHELL, P. (1966). Chemiosmotic coupling in oxidative and photosynthetic phosphorylation. *Biological Reviews*, **41**, 445–502.

MITCHELL, P. (1968). *Chemiosmotic Coupling and Energy Transduction*. Bodmin, Cornwall: Glynn Research Ltd.

MITCHELL, P. (1975a). Protonmotive redox mechanism of the cytochrome b–c_1 complex in the respiratory chain: protonmotive ubiquinone cycle. *FEBS Letters*, **56**, 1–6.

MITCHELL, P. (1975b). The protonmotive Q cycle: a general formulation. *FEBS Letters*, **59**, 137–9.

MOYLE, J. & MITCHELL, P. (1973a). Proton translocation quotient for the adenosine triphosphatase of rat liver mitochondria. *FEBS Letters*, **30**, 317–20.

MOYLE, J. & MITCHELL, P. (1973b). The proton-translocating nicotinamide adenine dinucleotide (phosphate) transhydrogenase of rat liver mitochondria. *Biochemical Journal*, **132**, 571–85.

MURTHY, P. S. & BRODIE, A. F. (1964). Oxidative phosphorylation in fractionated bacterial systems. XV. Reduced nicotinamide adenine dinucleotide phosphate-linked phosphorylation. *Journal of Biological Chemistry*, **239**, 4292–7.

NEIJSSEL, O. M. & TEMPEST, D. W. (1975). The regulation of carbohydrate metabolism in *Klebsiella aerogenes* NCTC 418 organisms, growing in chemostat culture. *Archives of Microbiology*, **106**, 251–8.

NEIJSSEL, O. M. & TEMPEST, D. W. (1976). Bioenergetic aspects of aerobic growth of *Klebsiella aerogenes* NCTC 418 in carbon-limited and carbon-sufficient chemostat culture. *Archives of Microbiology*, **107**, 215–21.

NIEUWENHUIS, F. J. R. M., KANNER, B. I., GUTNICK, D. L., POSTMA, P. W. & VAN DAM, K. (1973). Energy conservation in membranes of mutants of *Escherichia coli* defective in oxidative phosphorylation. *Biochimica et Biophysica Acta*, **325**, 62–71.

NIVEN, D. F., COLLINS, P. A. & KNOWLES, C. J. (1975). The respiratory system of *Chromobacterium violaceum* grown under conditions of high and low cyanide evolution. *Journal of General Microbiology*, **90**, 271–85.

OKA, T. & ARIMA, K. (1965). Cyanide resistance in *Achromobacter*. II. Mechanism of cyanide resistance. *Journal of Bacteriology*, **90**, 744–7.

PATERSON, C. A. (1970). Effect of culture conditions on the cyanide sensitivity of respiration in *Pseudomonas aeruginosa*. *Journal of Microbiology*, **73**, xxviii.

PIRT, S. J. (1965). The maintenance energy of bacteria in growing cultures. *Proceedings of the Royal Society of London, Series B*, **163**, 224–31.

POOLE, R. K. & HADDOCK, B. A. (1974). Energy-linked reduction of nicotinamide-adenine dinucleotide in membranes derived from normal and various respiratory-deficient mutant strains of *Escherichia coli* K12. *Biochemical Journal*, **144**, 77–85.

PUDEK, M. R. & BRAGG, P. D. (1974). Inhibition by cyanide of the respiratory chain oxidases of *Escherichia coli*. *Archives of Biochemistry and Biophysics*, **164**, 682–93.

PUDEK, M. R. & BRAGG, P. D. (1976). Redox potentials of the cytochromes in the respiratory chain of aerobically grown *Escherichia coli*. *Archives of Biochemistry and Biophysics*, **174**, 546–52.

RACKER, E. (1970). The two faces of the inner mitochondrial membrane. *Essays in Biochemistry*, **6**, 1–22.

RAGLAND, T. E., KAWASAKI, T. & LOWENSTEIN, J. M. (1966). Comparative aspects of some bacterial dehydrogenases and transhydrogenases. *Journal of Bacteriology*, **91**, 236–44.

RAVEN, J. A. & SMITH, F. A. (1976). The evolution of chemiosmotic energy coupling. *Journal of Theoretical Biology*, **57**, 301–12.

REEVES, J. P., LOMBARDI, F. J. & KABACK, H. R. (1972). Mechanisms of active transport in isolated bacterial membrane vesicles. VII. Fluorescence of 1-anilino-8-naphthalene sulphonate during D-lactate oxidation by membrane vesicles from *Escherichia coli*. *Journal of Biological Chemistry*, **247**, 6204–11.

REVSIN, B. & BRODIE, A. F. (1969). Carbon monoxide binding pigments of

Mycobacterium phlei and *Escherichia coli*. *Journal of Biological Chemistry*, **244**, 3101–4.

ROSEN, B. P. (1973). β-Galactoside transport and proton movements in an adenosine triphosphate-deficient mutant of *Escherichia coli*. *Biochemical and Biophysical Research Communications*, **53**, 1289–96.

SALTON, M. J. R. (1974). Membrane associated enzymes in bacteria. *Advances in Microbial Physiology*, **11**, 213–83.

SAPSHEAD, L. M. & WIMPENNY, J. W. T. (1972). The influence of oxygen and nitrate on the formation of the cytochrome pigments of the aerobic and anaerobic respiratory chain of *Micrococcus denitrificans*. *Biochimica et Biophysica Acta*, **267**, 388–97.

SCHOLES, P. B. & MITCHELL, P. (1970). Respiration-driven proton translocation in *Micrococcus denitrificans*. *Journal of Bioenergetics*, **1**, 309–23.

SCHOLES, P. B. & SMITH, L. (1968). Composition and properties of the membrane-bound respiratory chain system of *Micrococcus denitrificans*. *Biochimica et Biophysica Acta*, **153**, 363–73.

SHIPP, W. S. (1972*a*). Cytochromes of *Escherichia coli*. *Archives of Biochemistry and Biophysics*, **150**, 459–72.

SHIPP, W. S. (1972*b*). Absorption bands of multiple *b* and *c* cytochromes in bacteria detected by numerical analysis of absorption spectra. *Archives of Biochemistry and Biophysics*, **150**, 482–8.

SIMONI, R. D. & POSTMA, P. W. (1975). The energetics of bacterial active transport. *Annual Reviews of Biochemistry*, **44**, 523–54.

SINCLAIR, P. & WHITE, D. C. (1970). Effects of nitrate, fumarate and oxygen on the formation of the membrane-bound electron transport system of *Haemophilus parainfluenzae*. *Journal of Bacteriology*, **101**, 365–72.

SMITH, L. (1968). The respiratory chain system of bacteria. In *Biological Oxidations*, ed. T. P. Singer, pp. 55–122. New York: Interscience Publishers Inc.

SMITH, L., WHITE, D. C., SINCLAIR, P. & CHANCE, B. (1970). Rapid reactions of cytochromes in *Haemophilus parainfluenzae* on addition of substrates or oxygen. *Journal of Biological Chemistry*, **245**, 5096–100.

STOUTHAMER, A. H. (1969). Determination and significance of molar growth yields. In *Methods in Microbiology*, ed. J. R. Norris & D. W. Ribbons, vol. 1, pp. 629–63. London & New York: Academic Press.

STOUTHAMER, A. H. (1973). A theoretical study on the amount of ATP required for synthesis of microbial cell material. *Antonie van Leeuwenhoek*, **39**, 545–65.

STOUTHAMER, A. H. & BETTENHAUSSEN, C. W. (1973). Utilisation of energy for growth and maintenance in continuous and batch cultures of microorganisms. A re-evaluation of the method for the determination of ATP production by measuring molar growth yields. *Biochimica et Biophysica Acta*, **301**, 53–70.

STOUTHAMER, A. H. & BETTENHAUSSEN, C. W. (1975). Determination of the efficiency of oxidative phosphorylation in continuous cultures of *Aerobacter aerogenes*. *Archives of Microbiology*, **102**, 187–92.

SWEETMAN, A. J. & GRIFFITHS, D. E. (1971*a*). Studies on energy-linked reactions; energy-linked reduction of oxidised nicotinamide adenine dinucleotide by succinate in *Escherichia coli*. *Biochemical Journal*, **121**, 117–24.

SWEETMAN, A. J. & GRIFFITHS, D. E. (1971*b*). Studies on energy-linked reactions; energy-linked transhydrogenase reaction in *Escherichia coli*. *Biochemical Journal*, **121**, 125–30.

TIKHONOVA, G. V. (1974). Some properties of $\Delta\mu_H+$ generators in *Micrococcus lysodeikticus* membranes. *Biochemical Society Special Publication*, **4**, 131–43.

VAN DER BEEK, E. G. & STOUTHAMER, A. H. (1973). Oxidative phosphorylation in intact bacteria. *Archiv für Mikrobiologie*, **89**, 327–39.

VAN DE STADT, R. J., NIEUWENHUIS, F. J. R. M. & VAN DAM, K. (1971). On the reversibility of the energy-linked transhydrogenase. *Biochimica et Biophysica Acta*, **234**, 173–6.

VAN VERSEVELD, H. W. & STOUTHAMER, A. H. (1976). Oxidative-phosphorylation in *Micrococcus denitrificans*. Calculation of the P/O ratio in growing cells. *Archives of Microbiology*, **107**, 241–7.

WEST, I. C. & MITCHELL, P. (1972). Proton-coupled β-galactoside translocation in non-metabolising *Escherichia coli*. *Journal of Bioenergetics*, **3**, 445–62.

WEST, I. C. & MITCHELL, P. (1974). The proton-translocating adenosine triphosphatase of *Escherichia coli*. *FEBS Letters*, **40**, 1–4.

WESTON, J. A., COLLINS, P. A. & KNOWLES, C. J. (1974). The respiratory system of the marine bacterium *Beneckea natriegens*. II. Terminal branching of respiration to oxygen and resistance to inhibition by cyanide. *Biochimica et Biophysica Acta*, **368**, 148–57.

WESTON, J. A. & KNOWLES, C. J. (1973). A soluble CO-binding *c*-type cytochrome from the marine bacterium *Beneckea natriegens*. *Biochimica et Biophysica Acta*, **305**, 11–18.

WESTON, J. A. & KNOWLES, C. J. (1974). The respiratory system of the marine bacterium *Beneckea natriegens*. I. Cytochrome composition. *Biochimica et Biophysica Acta*, **333**, 228–36.

WHISTANCE, G. R. & THRELFALL, D. R. (1968). Effect of anaerobiosis on the concentrations of demethylmenaquinone, menaquinone and ubiquinone in *Escherichia freundii*, *Proteus mirabilis* and *Aeromonas punctata*. *Biochemical Journal*, **108**, 505–7.

WHITE, D. C. (1962). Cytochrome and catalase patterns during growth of *Haemophilus parainfluenzae*. *Journal of Bacteriology*, **83**, 851–9.

WHITE, D. C. (1963). Factors affecting the affinity for oxygen of cytochrome oxidases in *Haemophilus parainfluenzae*. *Journal of Biological Chemistry*, **238**, 3757–61.

WHITE, D. C. & SINCLAIR, P. R. (1971). Branched electron transport systems in bacteria. *Advances in Microbial Physiology*, **5**, 173–211.

WHITE, D. C. & SMITH, L. (1962). Haematin enzymes of *Haemophilus parainfluenzae*. *Journal of Biological Chemistry*, **237**, 1332–6.

WORCEL, A., GOLDMAN, D. S. & CLELAND, W. W. (1965). An allosteric reduced nicotinamide adenine dinucleotide oxidase from *Mycobacterium tuberculosis*. *Journal of Biological Chemistry*, **240**, 3399–407.

YATES, M. G. & JONES, C. W. (1974). Respiration and nitrogen-fixation in *Azotobacter*. *Advances in Microbial Physiology*, **11**, 97–135.

PHOSPHORYLATIVE ELECTRON TRANSPORT WITH FUMARATE AND NITRATE AS TERMINAL HYDROGEN ACCEPTORS

ACHIM KRÖGER

Institut für Physiologische Chemie,
Ludwig-Maximilians-Universität München,
BRD-8000 München 2, Pettenkoferstrasse 14a, West Germany

INTRODUCTION

Certain bacteria growing under anaerobic conditions can use nitrate, fumarate, sulphate or carbonate as terminal acceptors of reducing equivalents. The reactions of these acceptors with molecular hydrogen are given in (1) to (4) below, together with their standard free energy differences at pH 7 ($\Delta G'_0$).

$$NO_3^- + H_2 \rightarrow NO_2^- + H_2O \qquad -38.5 \text{ kcal/2e}^- \qquad (1)$$

$$\text{Fumarate} + H_2 \rightarrow \text{succinate} \qquad -20.6 \text{ kcal/2e}^- \qquad (2)$$

$$SO_4^{2-} + 4H_2 + 2H^+ \rightarrow H_2S + 4H_2O \qquad -9.2 \text{ kcal/2e}^- \qquad (3)$$

$$HCO_3^- + 4H_2 + H^+ \rightarrow CH_4 + 3H_2O \qquad -8.1 \text{ kcal/2e}^- \qquad (4)$$

The energy metabolism associated with these acceptors involves energy-conserving mechanisms which are different from the known substrate-level phosphorylation reactions, and appear to be similar to mitochondrial oxidative phosphorylation. It is well established that electron-transport-dependent phosphorylation is coupled to the reduction of nitrate and fumarate. These cases will be discussed in this article with particular reference to the mechanisms of electron transport and energy transduction. The systems of sulphate and carbonate reduction are less well understood, and have been extensively reviewed (Peck, 1968; Decker, Jungermann & Thauer, 1970; Wolfe, 1971; Barker, 1972; Barton, LeGall & Peck, 1972; Quayle, 1972; LeGall & Postgate, 1973; Siegel, 1975; Thauer, Jungermann & Decker, 1976); they will therefore not be discussed here. The transport of amino acids and sugars which is driven by fumarate and nitrate reduction has also been omitted, because recent reviews are available (Harold, 1972; Hamilton, 1975; Konings & Boonstra, 1976).

Table 1. *Standard redox potentials at pH 7 (E'_0) and free energy changes of the reactions with fumarate of various donors which were found to reduce fumarate in the membrane fraction of bacteria*

Donor (SH$_2$)/acceptor (S)	E'_0 (mV)	$\Delta G'_0$ of reaction with fumarate (kcal mol^{-1})
H$_2$/2H$^+$	−420	−20.6
Formate/HCO$_3^-$	−416	−20.4
NADH/NAD$^+$	−320	−16.0
Lactate/pyruvate	−197	−10.3
Glycerol-1-phosphate/di-hydroxyacetone-phosphate	−190	−10.0
Malate/oxaloacetate	−172	−9.2

From Decker *et al.* (1970).

FUMARATE REDUCTION

Energetics

The standard redox potential of the fumarate/succinate couple at pH 7 ($E'_0 = +30$ mV) is about 200 mV greater than that of most of the other redox couples of metabolism. This causes fumarate to be a suitable hydrogen acceptor for anaerobic metabolism by bacteria. The membrane fraction of various bacteria has been shown to catalyse the reduction of fumarate by the donors (SH$_2$) listed in Table 1 according to reaction (5):

$$\text{Fumarate} + \text{SH}_2 \rightarrow \text{succinate} + \text{S.} \tag{5}$$

The free energies of the reactions of hydrogen, formate and NADH with fumarate are sufficiently negative to allow the formation of 1 mol ATP per mol fumarate reduced. These reactions are energetically equivalent to those of the mitochondrial respiratory chain, where about 17 kcal per 2 electrons transported are available for the synthesis of 1 mol ATP. The formation of ATP with the last three donors of Table 1 should either require the support of additional exergonic reactions associated with fumarate reduction, or yield less than 1 mol ATP per mol fumarate reduced.

Occurrence

The Gram-negative bacteria have been systematically investigated by Mannheim & Holländer's group for their ability to reduce fumarate (for references see Table 2). The various species were divided into seven different categories according to their quinone content and their growth ability in the presence and absence of oxygen (Tables 2 and 3). The

Table 2. *Occurrence of fumarate reduction and quinones in some chemo-organotrophic Gram-negative bacteria (representative strains of category a–g)*

Category	Species/strain	Acceptor during growth	Cell density (mg dry wt l⁻¹)	Quinone content (μmol per g protein)[a]			Rate of NADH oxidation (μmol min⁻¹ per g protein) by:	
				MK	DMK	Q	Fumarate	O₂
a	Bacteroides fragilis H. Schwabacher WPMH1 (NCTC 8560)	O₂	n.g.[b]	—	—	—	—	—
		—	110	0.63	0	0	38.4	58.4
		Fumarate	160	0.14	0	0	109.8	57.6
b	Flavobacterium meningosepticum, serovar A E. O. King 14 (NCTC 10016)	O₂	2500	2.73	0	0	50.5	16.9
		—	20	1.74	0	0	10.8	8.1
		Fumarate	570	4.50	0	0	279.4	22.6
c	Cytophaga aurantiaca H. Bortels (ATCC 12208)	O₂	2500	1.0	0	0	0	16.5
		—	n.g.	—	—	—	—	—
		Fumarate	n.g.	—	—	—	—	—
d	Edwardsiella tarda D. L. Adler 'King' (Bartholomew) (NCTC 10398)	O₂	920	0.18	0	0.19	2.5	104.3
		—	110	0.52	0	0.29	16.1	43.5
		Fumarate	250	0.60	0	0.27	50.2	32.7
e	Haemophilus influenzae H. J. Bensted RAMC 18 (NCTC 4560)	O₂	500	0	1.06	0	19.0	34.7
		—	30	—	—	—	—	—
		Fumarate	80	0	1.44	0	82.2	91.4
f	Pasteurella multocida H. Schütze HS (NCTC 3195)	O₂	340	0	0.53	1.21	96.3	257.6
		—	60	—	—	—	—	—
		Fumarate	140	0	0.68	0.12	112.5	221.1
g	Moraxella urethralis K. Bøvre A200/71 (NCTC 11008)	O₂	230	0	0	1.59	0	119.5
		—	n.g.	—	—	—	—	—
		Fumarate	n.g.	—	—	—	—	—

References: Fischer (1972); Jediss (1973); Holländer (1974); Mannheim, Jediss & Zabel (1974a); Mannheim, Jediss, Zabel & Holländer (1974b); Holländer & Mannheim (1975); Holländer (1976).
[a] MK, menaquinone; DMK, desmethylmenaquinone; Q, ubiquinone.
[b] No growth.

Table 3. *Occurrence of fumarate reduction and quinones in some chemoorganotrophic Gram-negative bacteria. Extension of Table 2. (The strains characterised in Table 2 are representative of categories a–g below.)*

Category	Quinones present	Growth $+O_2$	Growth $-O_2$	Fumarate reduction	Species in category
a	MK	−	+	+	Several species of Bacteroides including *B. fragilis* and *B. melaninogenicus*
b	MK	+	+	+	*Flavobacterium meningosepticum* (all serovars) and related strains (IIb group), *F. tirrenicum*, *F. uliginosum*, *Flexibacter elegans*, *Cytophaga johnsonii* and several unnamed species
c	MK	+	−	−	*Flavobacterium* (*breve*, *pectinovorum*), several strains of *Flavobacterium–Cytophaga*, *Cytophaga* (*aurantiaca*, *fermentans*)
d	MK+Q	+	+	+	Enterobacteriaceae: *Salmonella* (*choleraesuis*, *typhimurium*), *Enterobacter* (*aerogenes*, *clocacae*), *Erwinia carotovora*, *Escherichia coli*, *Citrobacter freundii*, *Hafnia alvei*, *Serratia marcescens*, *Shigella dysenteriae* (serovar 1), *Klebsiella* (*pneumoniae*, *aerogenes*, *oxytoca*), *Edwardsiella tarda*, *Proteus* (*vulgaris*, *mirabilis*, *rettgeri*, *morganii*, *inconstans*), *Yersinia* (*pseudotuberculosis*, *enterocolitica*), *Beneckea* (*campbelli*, *pelagia*)
e	DMK	+	+	+	Leading species of *Haemophilus* (*aegytius*, *agni*, *aphrophilus*, *ducreyi*, *influenzae*, *parainfluenza* (in part), *paraphrophilus*, *parahaemolyticus*, *somnus*), *Actinobacillus* (*actinomycetumcomitans*, *equuli* (in part), *suis*), *Pasteurella* (*bettii*, *pneumotropica*)
f	DMK+Q	+	+	+	Leading species of *Actinobacillus* (*equuli* (in part), *liquieresii*, *seminis*), leading species of *Pasteurella* (*gallinarum*, *haemolytica* biovar A and biovar T, *mastitidis*, *multocida*, *piscicida*, *urae*, and several unnamed species), *Haemophilus* (*haemoglobinophilus*, *paragallinarum*, *parainfluenzae* (in part), *parasuis*), *Vibrio* (*cholerae*, *fischeri*, *parahaemolyticus*), *Aeromonas* (*hydrophilia*, *shigelloides*)
g	Q	+	−	−	*Acetobacter aceti*, *Acinetobacter* (*calcoaceticus*, *haemolyticus*, *metalcaligenes*), *Agrobacterium tumefaciens*, *Alcaligenes faecalis*, *Azotobacter vinelandii*, *Beggiatoa* sp., *Bordetella bronchiseptica*, *Brucella abortus*, *Caulobacter crescentus*, *Chromobacterium violaceum*, *Comamonas percolans*, *Eikenella corrodens*, *Flavobacterium* (*aquatile*, *capsulatum*, *devorans*, *halmephilum*), *Francisella tularensis*, *Haemophilus* (*vaginalis*, *piscium*), *Hyphomicrotium neptunium*, *Moraxella* (*nonliquefaciens*, *osloensis*, *urethralis*), *Neisseria menigitidis*, *Paracoccus denitrificans*, *Pseudomonas* (*fluorescens*, *mallei*, *maltophilia*) *Rhizobium lupini*, *Sphaerotilus natans*, *Spirillum serpens*, *Vitreoscilla* sp., *Xanthomonas phaseoli*, *Zoogloea ramigera*

bacteria were grown on proteose/peptone medium either anaerobically without addition of acceptors or with oxygen or fumarate present. The cell yields in the early stationary phase, and the rates of NADH oxidation with fumarate and oxygen by sonicated cell suspensions were measured. The data from strains representative of each category are shown in Table 2. In Table 3 the species belonging to each category are given.

The bacteria of five categories (*a*, *b*, *d*, *e* and *f*) of the seven were found to be capable of reducing fumarate. The activity of fumarate reduction is of the same order of magnitude as the respiratory activity. The capacity to reduce fumarate is associated with the presence of either menaquinone (MK) or desmethylmenaquinone (DMK) in the bacteria and an increased cell density under anaerobic conditions with fumarate. In bacteria of categories *a*, *b* and *d*, fumarate reduction is detectable following anaerobic growth in the absence of added fumarate, but a marked stimulation of the activity is observed in the presence of fumarate. The anaerobic growth of the bacteria in categories *e* and *f* seems to be absolutely dependent on the presence of fumarate.

To summarise, fumarate reduction appears to be a wide-spread property among Gram-negative bacteria, and seems to play an important role in anaerobic metabolism. Utilisation of fumarate as an electron acceptor, as indicated by an increased cell yield caused by the presence of fumarate in anaerobic growth and by the presence of NADH-fumarate reductase, was also observed with some facultatively aerobic Gram-positive bacteria (strains of *Bacillus* and *Staphylococcus*, *Leptotrichia dentium* and *Kurthia zopfii*; W. Mannheim, unpublished observations). These bacteria contain mostly MK and never Q (ubiquinone). In general, the specific activities of fumarate reduction are lower and the effect of fumarate on the cell yield is less pronounced than with Gram-negative bacteria.

Metabolic significance

The increased cell yield of many bacteria produced by addition of fumarate during anaerobic growth on complex media (Table 2) may indicate that ATP is formed by electron-transport-dependent phosphorylation coupled to fumarate reduction. Alternatively fumarate may serve merely as a high-potential sink for reducing equivalents. The function of this sink would be to drive substrate-level phosphorylation reactions which would not occur in the absence of fumarate. For instance, fumarate can accept the reducing equivalents of glucose fermentation instead of pyruvate in *Streptococcus faecalis*. This leads to the formation of acetate instead of lactate and is associated with an additional site for substrate-level phosphorylation (Deibel & Kvetkas, 1964).

Escherichia coli (Macy, Kulla & Gottschalk, 1976) and *Vibrio succinogenes* (Wolin, Wolin & Jacobs, 1961) grow at the expense of reaction (2). As the oxidation of hydrogen by fumarate cannot involve substrate-level phosphorylation reactions, this shows that the ATP required for growth is provided by fumarate reduction.

V. succinogenes can also grow on formate and fumarate according to reaction (6):

$$\text{Formate} + \text{fumarate} + H_2O \rightarrow HCO_3^- + \text{succinate}. \qquad (6)$$

Bicarbonate and succinate are the only products. From the growth yield (about 4 g dry wt per mol fumarate) it may be concluded that less than 1 mol ATP is obtained from the reduction of 1 mol fumarate.

Proteus rettgeri may serve as an example of a bacterium which makes use of the reduction of fumarate, although other pathways of ATP synthesis are available. This bacterium is closely related to *E. coli*. It can grow either aerobically or anaerobically on glucose or other substrates, and contains Q and MK together with the types of cytochromes present in *E. coli*. In contrast to *E. coli*, which requires certain donors (glucose, glycerol, lactate or hydrogen) in addition to fumarate for anaerobic growth, *P. rettgeri* can grow with fumarate as the only source of energy, carbon and hydrogen. Furthermore, the fumarate reductase of *P. rettgeri* is induced by growth on complex media even in the absence of added fumarate, whereas this enzyme is present only in *E. coli* cells which are grown with fumarate.

The anaerobic growth of *P. rettgeri* on fumarate proceeds at the expense of reaction (7) (Kröger, Schimkat & Niedermaier, 1974):

$$7 \text{ fumarate} + 8H_2O \rightarrow 6 \text{ succinate} + 4HCO_3^-$$
$$+ 2H^+ - 106 \text{ kcal}. \qquad (7)$$

The metabolic pathway for the fermentation of fumarate is given in Fig. 1. Two mol of fumarate are converted to 4 mol bicarbonate and 1 mol succinate. The reducing equivalents liberated in this reaction are used for the reduction of an additional 5 mol of fumarate. The metabolic pathway is confirmed by the induction of fumarase, malic enzyme, malate dehydrogenase, citrate synthase, aconitase and isocitrate dehydrogenase during growth on fumarate. The growth yield is 5.5 g cells per mol succinate, and this indicates that about 1 mol ATP is synthesised per mole of succinate formed. As 5 mol succinate are

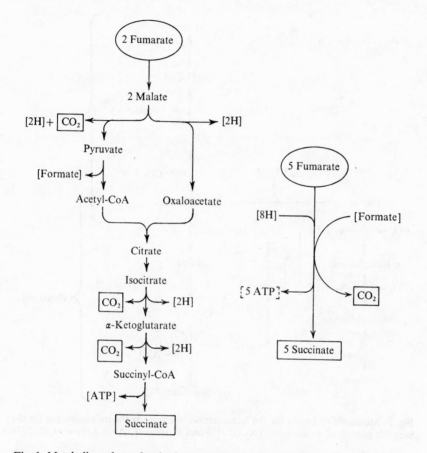

Fig. 1. Metabolic pathway for the fermentation of fumarate as the only substrate for the anaerobic growth of *P. rettgeri*. (After Kröger *et al.*, 1974.)

formed by fumarate reduction and only 1 mol from citrate (Fig. 1), 83% of the ATP is formed by fumarate reduction. The preference of *P. rettgeri* for using fumarate reduction is evident from the fact that the acetyl-CoA formed as an intermediate is not converted to acetate, although this pathway is used in the anaerobic growth on pyruvate (Kröger *et al.*, 1974). *Klebsiella aerogenes* metabolises citrate via pyruvate (Fig. 2). This pathway does not involve any hydrogen transfer and ATP is formed only from acetyl-CoA. In contrast, half of the ATP of the citrate metabolism of *P. rettgeri* is gained from fumarate reduction (Kröger *et al.*, 1974). The hydrogen required for this reaction is generated by the oxidation of citrate to succinate. One mole ATP per mole citrate is gained in both pathways.

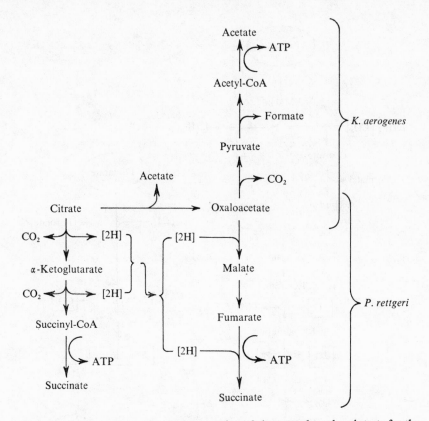

Fig. 2. Metabolic pathways for the fermentation of citrate as the only substrate for the anaerobic growth of *K. aerogenes* (NCIB 418) and *P. rettgeri*. (After Kröger *et al.*, 1974.)

The anaerobic propionibacteria are known to contain membrane-bound fumarate reductase, with NADH, lactate and glycerol-1-phosphate as donors, together with MK and *b* cytochromes (Sone, 1972; DeVries, van Wyck-Kapteyn & Stouthamer, 1973; Schwartz & Sporkenbach, 1975). The succinate formed by the reduction of fumarate is subsequently converted to propionate (Allen, Kellermeyer, Stjernholm & Wood, 1964). *Propionibacterium freudenreichii* and *P. pentosaceum* were shown to grow on lactate and glycerol as the only sources of energy, according to reactions (8) and (9) respectively (Bauchop & Elsden, 1960; DeVries *et al.*, 1973).

$$3 \text{ Lactate} \rightarrow 2 \text{ propionate} + \text{acetate} + HCO_3^- + H^+ - 41 \text{ kcal.} \qquad (8)$$

$$\text{Glycerol} \rightarrow \text{propionate} + H^+ + H_2O - 36.2 \text{ kcal.} \qquad (9)$$

The metabolic pathway of reaction (8) suggests the formation of 1 mol ATP per mol lactate, of which two-thirds would be formed by

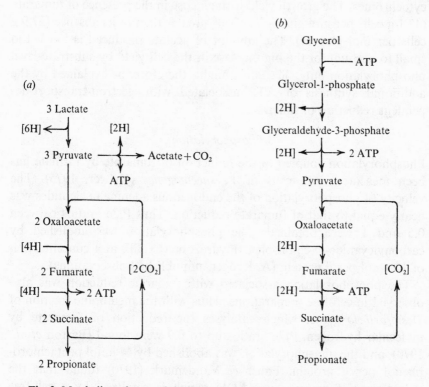

Fig. 3. Metabolic pathways of the fermentation of lactate (a) and glycerol (b) in propionibacteria. (According to Wood & Utter, 1965 and DeVries *et al.*, 1973.)

fumarate reduction and one-third by substrate-level phosphorylation (Fig. 3a). The cell yield following growth on lactate (8.1–12.9 g cells per mol lactate) was interpreted as indicating the formation of only 2 mol ATP per 3 mol lactate (DeVries *et al.*, 1973); however, it may also be consistent with a yield of 1 mol ATP per mol lactate. The growth yield with glycerol (24.1–27.0 g cells per mol glycerol) indicates the formation of 2 mol ATP per mol glycerol in agreement with the free energy change of reaction (9) and the metabolic pathway (Fig. 3b), according to which one-third of the ATP can be formed by fumarate reduction.

S. faecalis was found to produce acetate and succinate when grown on glucose plus fumarate, whereas only lactate is formed from glucose alone (Deibel & Kvetkas, 1964). NADH-fumarate reductase was found to be present in the membrane fraction of the bacterium (Aue & Deibel, 1967) which is known to contain DMK (Baum & Dolin, 1965), but no cytochromes. The growth yield with glucose in the presence of fumarate (47.1 g cells per mol glucose) is 70% greater than in its absence (27.9 g cells per mol glucose). The amount of acetate produced is much too small to account for the big increase in the cell yield by substrate-level phosphorylation only. The effect might, therefore, be explained by the additional synthesis of ATP associated with electron-transport-dependent reduction of fumarate.

Phosphorylation

Phosphorylation coupled to the reduction of fumarate by formate has been measured with cells of *V. succinogenes* (Kröger, 1975). The velocity of phosphorylation of the endogenous adenine nucleotides was nearly equal to that of fumarate reduction. Thus $P/2e^-$ ratios between 0.5 and 1 were obtained. The phosphorylation was abolished by carbonylcyanide-m-chlorophenylhydrazone (CCCP) at a concentration of 10 μmol per g protein (A. Kröger, unpublished observations).

The phosphorylation associated with fumarate reduction was also observed in cell-free preparations. Thus with the membrane fraction of *Desulfovibrio gigas*, which catalyses the reduction of fumarate by molecular hydrogen, $P/2e^-$ ratios up to 0.9 were found (Barton *et al.*, 1970), and the phosphorylation was abolished by 44 μmol pentachlorophenol per g protein. Faust & Vandemark (1970) found that the reduction of fumarate by NADH which is catalysed by a cell-free preparation of *S. faecalis* stimulates the esterification of inorganic phosphate. The $P/2e^-$ ratio was 0.19; phosphorylation was not inhibited by carbonyl-cyanide-p-trifluoro-phenylhydrazone in an amount (8–24 μmol per g protein) which is about 50 times that required for uncoupling mitochondrial oxidative phosphorylation. The membrane fraction of *E. coli* grown anaerobically on glycerol and fumarate catalyses fumarate reduction by glycerol-1-phosphate (Miki & Lin, 1975a). This reaction is associated with phosphate esterification. The $P/2e^-$ ratio (0.1) is decreased by only 60% in the presence of 15 μmol CCCP per g protein.

Proton extrusion

The electrogenic extrusion of protons was found to be coupled to formate–fumarate reduction in intact cells of *V. succinogenes* (Kröger, 1975). With a suspension of *V. succinogenes* in the presence of carbonic anhydrase and formate, the addition of limiting amounts of fumarate causes the appearance of protons in the medium, which disappear in a second phase after the exhaustion of the fumarate. Proton extrusion is completely abolished by CCCP (10 μmol per g protein). The reaction depends on the presence of potassium in the medium, but is not significantly stimulated by valinomycin. The ratio $H^+/2e^-$, obtained by extrapolation of the phase of proton uptake to the moment of the addition of formate, is between 1 and 2.

The electron transport chain

The electron transport systems which catalyse the reduction of fumarate with the various donors listed in Table 1 are localised in the cytoplasmic membrane of the bacteria (Kröger & Dadák, 1969; Kröger, Dadák Klingenberg & Diemer, 1971). On fractionation of *V. succinogenes* using either ampicillin spheroplasts or cells treated with EDTA and lysozyme which are then lysed by osmotic shock, the formate–fumarate reductase is found to be present only in the membrane fraction (Kröger & Innerhofer, 1976c). MK, cytochrome *b* and covalently bound FAD are exclusively localised in the membrane fraction, whereas cytochrome *c*, acid-extractable FAD and non-haem iron are present also in the soluble fraction. FMN is only present in the soluble fraction.

Quinones

As the concomitant occurrence of either MK or DMK with the ability to reduce fumarate in various bacteria (Table 2) suggests, these quinones seem to be essential mediators of electron transport with fumarate as the acceptor. The reduction of fumarate by NADH, malate and glycerol-1-phosphate in *Bacillus megaterium*, which contains only MK, was shown to depend on the presence of MK (Kröger & Dadák, 1969). In *P. rettgeri*, which contains both Q and MK, fumarate reduction by NADH and formate could be reactivated by MK, whereas Q preferentially restored the respiratory activity (Kröger et al., 1971). Fumarate reduction in *D. gigas* (Hatchikian, 1974) and *Propionibacterium arabinosum* (Sone, 1974) was also shown to depend on MK. Experiments with a mutant of *E. coli*, deficient in MK synthesis, indicated that MK is required for the oxidation of dihydroorotate by fumarate under anaerobic conditions (Newton, Cox & Gibson, 1971). It appears very probable, moreover, that MK is also involved in the metabolically significant reduction of fumarate during the growth of *E. coli* on fumarate together with either glucose, hydrogen, glycerol or lactate.

The formate–fumarate reductase of *V. succinogenes* is dependent on MK (Kröger, 1975). Extraction of 95% of the MK produces an almos complete inhibition of overall electron transport, and, by reincorporation of MK into the membrane, 25% of the activity is restored. Formate oxidase and formate–ferricyanide reductase are also dependent on MK. In contrast, formate dehydrogenase, as measured with benzylviologen as acceptor, and fumarate reductase, which is measured in terms of succinate–ferricyanide reduction, do not require MK for activity. This is also true for fumarate reduction by reduced benzylviologen.

The function of DMK, which is found in several bacteria capable of reducing fumarate, appears to be similar to that of MK (Holländer, 1976). This was shown by an extraction and reincorporation experiment with *Haemophilus influenzae*. Extraction of the DMK inhibited the activity of NADH–fumarate reduction and, on reincorporation of DMK, full reactivation was obtained. MK restored fumarate reduction, whereas with Q only negligible stimulation was found (Holländer, 1976); neither quinone is synthesised by this particular strain of *H. influenzae*. Each of the three quinones was capable of restoring oxygen-dependent respiratory activity with NADH.

The failure of Q to reactivate fumarate reduction (see also Kröger *et al.*, 1971), is probably a consequence of its redox potential. From the standard potentials at pH 7 (Schnorf, 1966; R. Holländer, personal communication), it is evident that reduced MK ($E'_0 = -74$ mV) and DMK ($E'_0 = +36$ mV) can serve as donors for fumarate, whereas Q ($E'_0 = +113$ mV) cannot. Also the observation that all three quinones can serve as mediators in oxygen-dependent NADH oxidation is consistent with their redox potentials. Furthermore, Q and DMK, though not MK, are capable of mediating oxygen-dependent succinate oxidation (Holländer, 1976).

The specificity of quinone interaction is governed not only by the redox potentials, but also by the structure of the quinone. Thus in *B. megaterium*, which contains only MK, oxygen-dependent respiratory activities with NADH, malate and glycerol-1-phosphate can be reactivated only by MK, and not by Q (Kröger & Dadák, 1969). The oxygen-dependent respiration of aerobically grown *P. rettgeri* with NADH and formate exhibits a distinct specificity for Q. However, MK was also found to reactivate these activities in *P. rettgeri* grown anaerobically (Kröger *et al.*, 1971). This variation in specificity correlates with the fact that Q is preferentially synthesised under aerobic growth conditions and MK under anaerobic conditions.

The redox reactions of the MK present in the membrane of *B. megaterium* (Kröger & Dadák, 1969), *P. rettgeri* (Kröger *et al.*, 1971) and *V. succinogenes* (Kröger, 1975) confirmed the participation of MK in

Table 4. *Inhibition by NQNO and p-CMS of various electron transport activities of the membrane fraction of V. succinogenes*

Inhibitor	Concentration (μmol per g protein)	Formate $\rightarrow$ fumarate	Formate $\rightarrow$ BV[a]	Formate $\rightarrow$ Fe(CN)$_6^{3-}$	Succinate $\rightarrow$ Fe(CN)$_6^{3-}$	Reduced BV[a] $\rightarrow$ fumarate
		(μmol min^{-1} per g protein				
(Control)		940	1010	1100	2080	12 000
		% inhibition				
NQNO	1.5	98	2	70	0	0
p-CMS	15	97	4	2	98	0

The titres of NQNO (2(n-nonyl)-4-hydroxyquinoline-N-oxide) and p-CMS (p-chloromercuriphenylsulphonate) were 0.81 and 10.3 μmol per g protein respectively. (Kröger & Innerhofer, 1976c.)
[a] BV, benzylviologen.

electron transport with fumarate as terminal acceptor. MK is fully reducible by each of the donors, and is partially oxidised in the steady state when fumarate is being reduced. The degree of reduction of MK in the steady state is a function of the activity of both the dehydrogenase and fumarate reductase.

Fumarate reductase

Of all the bacterial membrane-bound fumarate reductases, only that of *V. succinogenes* has been isolated. This enzyme catalyses the reduction of fumarate by reduced benzylviologen, and can be separated from formate dehydrogenase, which catalyses the reduction of benzylviologen by formate, by means of ion-exchange chromatography. Covalently bound FAD is the prosthetic group of fumarate reductase (Kröger & Innerhofer, 1976c) and the FAD is linked at position 8-α to the N-3 of histidine (W. C. Kenney & T. P. Singer, personal communication). Thus the prosthetic group of the fumarate reductase is similar to that of the mitochondrial succinate dehydrogenase.

The succinate–ferricyanide reductase, which is purified together with the reduced benzylviologen–fumarate reductase, involves an iron–sulphur protein in addition to the fumarate reductase. This is deduced from the observations that the succinate–ferricyanide reductase, in contrast to the fumarate reductase, is inhibited by p-chloromercuriphenylsulphonate (p-CMS) as shown in Table 4 and that the titre of p-CMS inhibition is equivalent to the iron–sulphur protein content of the membrane fraction. The involvement in overall electron transport of both fumarate reductase and the iron–sulphur protein is shown by the identical p-CMS inhibition curves of the succinate–ferricyanide and the formate–fumarate reductase activities (Kröger, 1975). The iron–sulphur protein seems to be the component which directly reacts with the fumarate reductase, since the cytochromes and MK interact on the formate side of the iron–sulphur protein.

Dehydrogenases

Depending on the organism and the growth conditions, the membrane fractions from bacteria which catalyse fumarate reduction contain dehydrogenases which are specific for the donors listed in Table 1. Although some of the dehydrogenases have been isolated, the nature of the prosthetic groups is not known in most cases. The hydrogenase of *D. gigas* (LeGall *et al.*, 1971) is an iron–sulphur protein and the glycerol-1-phosphate dehydrogenase of *E. coli* appears to be a flavoprotein (Kistler & Lin, 1972). Both the lactate dehydrogenase and the malate dehydrogenase are independent of NAD$^+$.

The formate dehydrogenase of *V. succinogenes*, as measured with benzylviologen as acceptor, was purified by ion-exchange chromatography and found to be a molybdoprotein (A. Kröger & A. Innerhofer, unpublished observations). The purest preparation contained 6 μmol Mo per g protein, with no flavin, c cytochrome, or selenium. An amount of cytochrome b corresponding to that of molybdenum was not separated by the purification procedure. This shows the spectral properties of a low-potential b cytochrome (see Fig. 5) and is reduced by formate. However, it is not yet clear whether or not this reduction is mediated by MK. It is unlikely that the enzyme contains iron–sulphur protein, because at least 90% of the iron–sulphur protein of the membrane fraction is associated with the fumarate reductase.

The formate dehydrogenase of *E. coli* grown anaerobically on nitrate (Enoch & Lester, 1975) is similar to that of *V. succinogenes* with respect to its content of molybdenum and cytochrome b, but differs from it in its content of selenium and iron–sulphur protein and its acceptor specificity.

Carbon dioxide and not bicarbonate is the product of the formate dehydrogenase reaction. This is demonstrated by measuring the pH changes caused by the addition of limiting amounts of formate to a suspension of cells of *V. succinogenes* containing CCCP and fumarate in excess. At pH < 6 (Fig. 4a), where the hydration of carbon dioxide does not occur, $1H^+$ per formate disappears according to reaction (10):

$$\text{Formate} + \text{fumarate} + H^+ \to CO_2 + \text{succinate}. \tag{10}$$

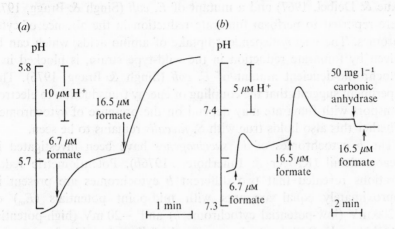

Fig. 4. pH movements associated with the oxidation of formate by fumarate catalysed by *V. succinogenes*. Cells were anaerobically incubated in 0.15 M KCl, 5 mM MgCl$_2$, 0.33 mM fumarate and 10 μmol CCCP per g protein. (After Kröger, 1975.)

In Fig. 4b the same experiment is done at pH > 7. The initial decrease in pH is followed by a slower acidification resulting from carbon dioxide hydration according to reaction (11):

$$CO_2 + H_2O \rightarrow H^+ + HCO_3^-. \tag{11}$$

This is shown by the fact that the whole cyclic pH movement is abolished by the presence of carbonic anhydrase.

Therefore, the formate dehydrogenase acts by removing H$^-$ from formate. It is not known whether the hydride is split into proton and electrons by the formate dehydrogenase alone or whether the low-potential cytochrome *b* is required in addition. Experiments with artificial molybdenum complexes led Stiefel (1973) to propose the first mechanism for aldehyde dehydrogenases.

Cytochromes

Most of the bacteria which are capable of reducing fumarate, ncluding *D. gigas* (Hatchikian & LeGall, 1972), were found to contain *b* cytochromes. Part of the *b* cytochromes of *B. megaterium* (Kröger & Dadák, 1969), *P. rettgeri* (Kröger *et al.*, 1971), *Propionibacterium freudenreichii* (DeVries *et al.*, 1973), *P. shermanii* (Schwartz & Sporkenbach, 1975), *P. arabinosum* (Sone, 1974) and *V. succinogenes* (Jacobs & Wolin, 1963a, b) were shown to be oxidised by fumarate in the steady state. This may indicate, but does not prove, that *b* cytochromes are involved in electron transport to fumarate. On the other hand *S. faecalis*

(Aue & Deibel, 1967) and a mutant of *E. coli* (Singh & Bragg, 1975) were reported to perform fumarate reduction in the absence of cytochromes. The energy-dependent uptake of amino acids, which can be driven by fumarate reduction in the wild-type strain, is blocked in a cytochrome-deficient mutant of *E. coli* (Singh & Bragg, 1976). This experiment suggests that the coupling of energy transduction to electron transport with fumarate may depend on the presence of cytochromes. Whether this also holds true with *S. faecalis* remains to be seen.

The *b* cytochromes of *V. succinogenes* have been investigated in greater detail (Kröger & Innerhofer, 1976*b*). Potentiometric redox titrations revealed that two different *b* cytochromes are present in approximately equal amounts with mid-point potentials (E_m) of −200 mV (low-potential cytochrome *b*) and −20 mV (high-potential cytochrome *b*). Both *b* cytochromes can be fully reduced by formate and fully reoxidised by fumarate under anaerobic conditions. The two *b* cytochromes can also be differentiated by low-temperature spectroscopy. Most of the high-potential cytochrome *b* is reduced by succinate. The corresponding low-temperature difference spectrum has a maximum at 561 nm. The reduction is independent of the presence of MK, and is abolished by *p*-CMS. This indicates that the high-potential cytochrome *b* is reduced via the fumarate reductase, and that it is situated between the iron–sulphur protein and MK (Fig. 5). After the extraction of MK from the membrane fraction, only about half of the total cytochrome *b* is oxidised in the presence of formate and fumarate. The corresponding difference spectrum has a maximum at 559 nm, which is attributed to the low-potential cytochrome *b*. The same maximum is observed with the *b* cytochrome isolated with the formate dehydrogenase. A low-temperature spectrum with a maximum at 559 nm is also observed with the membrane fraction in the presence of 2(n-nonyl)-4-hydroxyquinoline-*N*-oxide (NQNO), formate and fumarate.

From these observations the position of the *b* cytochromes in the formate–fumarate reductase chain has been deduced, as shown in Fig. 5. The low-potential cytochrome *b* is situated on the formate side of MK, and is probably the acceptor for electrons from formate dehydrogenase. The inhibition by NQNO is probably due to the reaction of this inhibitor with the low-potential cytochrome *b*. This is suggested by the finding that the amount of NQNO required for full inhibition is equal to the content of this cytochrome and that the mid-point potential of the low-potential cytochrome *b* is shifted by −30 mV in the presence of NQNO, whereas that of the high-potential cytochrome *b* remains unaffected.

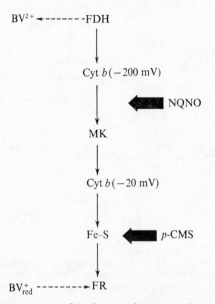

Fig. 5. Sequence of the components of the formate–fumarate reductase chain of *V. succino-genes*. FDH, formate dehydrogenase; FR, fumarate reductase; Fe–S, iron–sulphur protein; BV^{2+}, benzylviologen; BV^+_{red}, reduced benzylviologen. (After Kröger & Innerhofer, 1976*a*.)

The redox reactions of the two *c* cytochromes (mid-point potentials −160 and +70 mV) which are also present in *V. succinogenes* do not suggest a functional role in electron transport from formate to fumarate.

Mechanism of generation of the electrochemical proton potential

According to the chemiosmotic hypothesis, the coupling of electron transport to phosphorylation and other energy-linked reactions is performed by the electrogenic translocation of protons across the membrane (Mitchell, 1966). Although it is not yet decided whether proton translocation is an essential prerequisite of phosphorylation, the ability to create an electrochemical proton potential appears to be common to all systems of phosphorylative electron transport.

The mechanism of proton transport by coupling to a redox reaction as postulated by Lundegardh (1945), Robertson (1960) and Mitchell (1966) has not as yet been proved experimentally. The known sequence, of electron transport carriers from formate to fumarate in *V. succino-genes* may help in the understanding of the mechanism of proton generation. The experimental evidence so far available permits only suggestions for possible mechanisms; experimental support is still required.

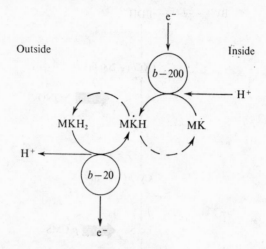

Fig. 6. Possible role of MK and the *b* cytochromes in the electrogenic translocation of protons in *V. succinogenes*. MKH, neutral radical of MK; *b*–200 low-potential cytochrome *b* ($E_m = -200$ mV); *b*–20, high-potential cytochrome *b* ($E_m = -20$ mV).

The proposed mechanisms of proton translocation (Lundegardh, 1945; Robertson, 1960; Mitchell, 1966) consist in the reduction of a hydrogen carrier on the inside of the membrane and its oxidation on the outside by electron carriers acting as electron donors and acceptors respectively. This situation would be met by the sequence cytochrome *b* (−200 mV) → MK → cytochrome *b* (−20 mV) of the formate–fumarate chain (Figs. 5 and 6), provided that the low-potential cytochrome *b* is situated on the inside of the membrane, with the high-potential cytochrome *b* on the outside. This would lead to the transport of $1H^+$ per electron on the basis that either the neutral radical (MKH) is formed as the intermediate (Fig. 6), or that no radical is involved. Alternatively $2H^+$ per electron could be translocated with the radical anion (MK^-) as the intermediate (Kröger, 1976). Experimental evidence either for or against proton translocation mediated by quinones is lacking so far.

Another mechanism for generating an electrochemical proton potential is shown in Fig. 7. This scheme is based on the assumption that protons are liberated on the outside of the membrane by the reaction of formate dehydrogenase with formate, and that protons are taken up on the inside by the reduction of fumarate. Thus the chain connecting the two enzymes serves only to transport electrons, and not hydrogen or protons. This mechanism is in agreement with the proposal (Fig. 5) that both

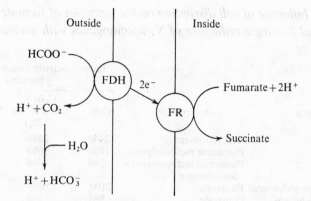

Fig. 7. Possible mechanism of generating an electrochemical proton potential across the bacterial membrane by electron transport from formate to fumarate. FDH, formate dehydrogenase; FR, fumarate reductase.

formate dehydrogenase and fumarate reductase react with electron rather than with hydrogen carriers *in situ*. Furthermore, both enzymes react directly with the artificial electron carrier benzylviologen. The liberation of 2 protons on the outside of the membrane per 2 electrons transported is consistent with measurements of proton extrusion.

The experimental results shown in Tables 5 and 6 indicate that formate dehydrogenase is accessible from the outside of the bacterium, whereas fumarate reductase is not. The activities of the enzymes with charged dyes are measured with cells in the absence and in the presence of a detergent (0.2% triton X-100; Table 5). For comparison the activities measured after sonic disruption of the cells are also given. The activity of formate reduction of benzylviologen, which is a divalent cation, is the same with cells in the absence and the presence of the detergent, and represents a maximum rate as seen by comparison with the sonic homogenate. Also the activity measured with the membrane fraction prepared by osmotic lysis of spheroplasts is not significantly greater (not shown). In contrast, the reduction of fumarate by reduced benzylviologen as a monovalent cation is six times faster in the presence of detergent than in its absence. Assuming that the velocities of penetration of oxidised and reduced benzylviologen are slow when compared with the activities of their redox reactions, these results mean that formate dehydrogenase is accessible to benzylviologen from the outside and fumarate reductase only from the inside of membrane. The results on formate dehydrogenase are confirmed with phenazine methosulphate and its sulphonic acid derivative as acceptors. In intact cells, fumarate

Table 5. *Influence of cell disruption on the activities of formate dehydrogenase and fumarate reductase of* V. succinogenes *with various charged dyes*

		Enzyme activity (μmol min^{-1} per g protein)		
Donor	Acceptor	Cells	Cells+ triton X-100 (0.2%)	Sonic homogenate
Formate	Benzylviologen	2440	2440	2900
	Phenazine methosulphate	1800	1600	1400
	Phenazine methosulphate-2-sulphonate	240	360	440
Reduced benzylviologen	Fumarate	2100	12 600	11 000
Anthra-9,10-hydro-quinone-2-sulphonate	Fumarate	380	380	1630
Succinate	Fe(CN)$_6^{3-}$	203	2110	1340

Table 6. *Influence of cell disruption on the Michaelis constants* (K_M) *for formate of the formate dehydrogenase and for fumarate of the fumarate reductase of* V. succinogenes

		K_M value (mM)		
Activity measured	K_M for:	Cells	Cells+ triton X-100 (0.2%)	Sonic homogenate
Formate–benzylviologen reductase	Formate	0.36	0.42	0.36
Reduced benzylviologen fumarate reductase	Fumarate	0.63	0.07	0.12

reductase is also not accessible to anthra-9,10-hydroquinone-2-sulphonate and ferricyanide. Anthra-9, 10-hydroquinone-2-sulphonate, like reduced benzylviologen, reacts directly with the fumarate reductase, whereas ferricyanide reacts with the iron–sulphur protein which is not accessible to *p*-CMS in intact cells (not shown). For a reason still unknown the maximum activity of fumarate reduction with anthra-9,10-hydroquinone-2-sulphonate is only seen after sonication or with the membrane fraction; no stimulation is obtained with the detergent.

The accessibility of formate dehydrogenase to formate and of fumarate reductase to fumarate can be studied by measuring the respective K_M values of the reactions with benzylviologen (Table 6). The same value is obtained for formate with cells, in both the absence and the presence of detergent, and with the membrane fraction. Furthermore, the K_M obtained with intact cells is not influenced by variations of pH between 7.0 and 7.9 nor by the presence of CCCP (not shown). These results are consistent with the view that formate dehydrogenase reacts with formate on the outside, and argue against the possibility that formate has to pass through the membrane (as formic acid) in order to reach the dehydrogenase. In contrast, the K_M for fumarate of fumarate reduction is more than five times less when measured with detergent present or with the membrane fraction than with intact cells. This may indicate that with intact cells the K_M of the fumarate transport protein is measured rather than that of the fumarate reductase. This would mean that fumarate needs to be transported before reaction with fumarate reductase can occur. It is likely that fumarate is transported in exchange for succinate, but no experimental evidence is available to confirm this view. In summary, the results support the sidedness of the active sites of formate dehydrogenase and fumarate reductase as proposed in Fig. 7, and are also consistent with the possibility that the proton exchange is coupled with the action of the enzymes as suggested above.

The hydrogenase of *D. gigas* is localised, together with a low-potential cytochrome *c* as the electron acceptor of the hydrogenase, in the periplasmic space of the bacterium (Bell, LeGall & Peck, 1974). This suggests that the hydrogenase enzyme, involved in the phosphorylating hydrogen–fumarate reductase, may also liberate protons on the outside of the cytoplasmic membrane. Thus the hydrogen–fumarate reductase may be constructed in a similar fashion to the formate–fumarate reductase, as proposed in Fig. 7.

NITRATE REDUCTION

The bacterial nitrate reductase, its occurrence and metabolic function have recently been reviewed (White & Sinclair, 1971; Payne, 1973; Pichinoty, 1973; Stouthamer, 1976; Konings & Boonstra, 1976; Thauer *et al.*, 1976; Haddock & Jones, 1977). In this section only some aspects of electron transport and energy transduction with nitrate as the terminal acceptor will therefore be discussed.

Phosphorylation

In Table 7 the donors (SH_2) for the membrane-bound reduction of nitrate are listed, together with the free energy at pH 7 of the reaction with nitrate according to reaction (12):

$$NO_3^- + SH_2 \rightarrow NO_2^- + S + H_2O. \qquad (12)$$

The reduction of nitrate with almost all the donors is sufficiently exergonic to allow the formation of 2 mol ATP per mol nitrate, due to the

Table 7. *Free energies of the reduction of nitrate to nitrite with various donors*

Donor (SH_2)	$\Delta G'_0$ (kcal/$2e^-$)
H_2	38.5
Formate	38.3
NADH	34
Lactate	28.5
Glycerol-1-phosphate	28.2
Succinate	20.7

From Decker *et al.* (1970).

relatively high redox potential of the NO_3^-/NO_2^- couple ($E'_0 = +420$ mV). Only the reduction of nitrate by succinate can yield only 1 mol ATP per mol nitrate.

The molar growth yields of various bacteria under anaerobic conditions are considerably increased by nitrate, but are always smaller with nitrate than with oxygen as acceptor (Stouthamer, 1976). With Clostridia, which are unable to perform oxidative phosphorylation, the greater cell yield with nitrate is caused merely by enhanced substrate-level phosphorylation (Hasan & Hall, 1975; Stouthamer, 1976). From a detailed analysis of the cell yields with other bacteria, Stouthamer (1976) calculated a P/$2e^-$ ratio of nearly 2 with NADH as the donor and of 1 with lactate.

With a cell-free preparation of *Paracoccus denitrificans* the P/$2e^-$ ratio was found to be 1 with NADH (John & Whatley, 1970), and the phosphorylation was fully inhibited by dinitrophenol. This ratio may be regarded as a maximum value, because it is much greater than all the other values obtained with cell-free extracts (Ohnishi & Mori, 1960; Yamanaka, Ota & Okunuki, 1962; Ohnishi, 1963; Ota, Yamanaka & Okunuki, 1964; Miki & Lin, 1975b) and because the preparation exhibited a distinct control of electron transport by the phosphorylation reaction. On the other hand, John & Whatley (1970) did not find phosphorylation with succinate as the donor, in contrast to Ohnishi (1963) who reported a P/$2e^-$ ratio of 0.25 with *Pseudomonas denitrificans*. Thus the efficiency of the phosphorylation coupled to nitrate reduction is not yet known with certainty.

The electron transport chain

The formate–nitrate reductase of *E. coli* is the best known system of electron transport with nitrate as acceptor. The formate dehydrogenase and the nitrate reductase of this system have been isolated and the overall electron transport chain reconstituted from the two enzymes.

Nitrate reductase

Nitrate reductase has been isolated from various bacteria, and found to contain molybdenum (Forget, 1971; Rosso, Forget & Pichinoty, 1973; Forget, 1974; MacGregor, Schnaitman, Normansell & Hodgins, 1974; Enoch & Lester, 1974; MacGregor, 1975; van t'Riet, Van Ee, Wever & Planta, 1975; Clegg, 1976). The *E. coli* enzyme consists of two or three different subunits. The subunit composition and the molecular weight is not yet known with certainty. In addition Enoch & Lester (1974) and Clegg (1976) found a *b* cytochrome to be associated with the enzyme which can be dissociated without affecting nitrate reductase activity with reduced benzylviologen as the donor. The molybdenum was found to undergo redox reactions (Forget & Der Vartanian, 1972; DerVartanian & Forget, 1975; Bray *et al.*, 1976) which are similar to those of other well known molybdenum-containing enzymes (Bray, 1976). The reduction of the iron–sulphur protein also present in the enzyme gives rise to two different electron paramagnetic resonance signals which respond to oxidation with nitrate (Forget & DerVartanian, 1972; DerVartanian & Forget, 1975).

Formate dehydrogenase of E. coli

Formate dehydrogenase (Enoch & Lester, 1975) is composed of two or three polypeptides, and the molecular weight is about 600×10^3 daltons. The enzyme contains molybdenum, cytochrome *b* and selenium in a molar ratio of $1:1:1$, and iron–sulphur protein in a fifteen- to twenty-fold excess. Flavins and quinones are not present in the preparation. The cytochrome *b* is reducible by formate. The redox reactions of the molybdenum, the selenium and the iron–sulphur protein have not yet been studied. It is likely, by analogy with other aldehyde dehydrogenases (Stiefel, 1973; Bray, 1976), that molybdenum is the first acceptor of reducing equivalents from formate. The enzyme is similar to that isolated from *V. succinogenes* with respect to its specific activity with phenazine methosulphate as the acceptor and its content of molybdenum and cytochrome *b*; it differs in its low specific activity with benzylviologen as the acceptor, and in containing selenium and iron–sulphur protein. It is possible that the apparently greater complexity of the enzyme from *E. coli* is related to the fact that this bacterium can induce formate-hydrogenlyase (Ruiz-Herrera & Alvarez, 1972; Stouthamer, 1976), in contrast to *V. succinogenes* (Jacobs & Wolin, 1963*a*).

Cytochromes

The reduction of nitrate by physiological donors (including formate) is inhibited in haem-deficient mutants of various bacteria (de Groot & Stouthamer, 1970; MacGregor, 1975; Kemp, Haddock & Garland, 1975), and can be restored by adding 5-aminolaevulinate to the growth medium. This indicates that cytochromes are necessary for nitrate reduction and is in agreement with the finding that functionally active *b* cytochromes are associated with formate dehydrogenase and with nitrate reductase (see above). The cytochrome *b* of formate dehydrogenase is characterised by an α-band at 559 nm and that of the nitrate reductase by an α-band at 558 nm (Enoch & Lester, 1974) in room temperature difference spectra. The cytochromes have not yet been characterised by potentiometric redox titration. The two cytochromes differ in their redox responses to formate and nitrate, the cytochrome *b* of formate dehydrogenase being fully reduced by formate, whereas that of nitrate reductase requires, in addition, the presence of formate dehydrogenase and Q for reduction. On the other hand, the cytochrome *b* of the nitrate reductase can be oxidised by nitrate while that of formate dehydrogenase cannot. The oxidation of the formate dehydrogenase cytochrome *b* by nitrate, in the presence of nitrate reductase and Q, has not yet been demonstrated. The view that the formate–nitrate reductase complex of *E. coli* contains two different *b* cytochromes is further supported by the observation that the oxidation of half of the *b* cytochromes by nitrate is inhibited by 2-heptyl-4-hydroxy-quinoline-*N*-oxide (Ruiz-Herrera & DeMoss, 1969). All of cytochrome *b* can be fully reduced by formate and fully oxidised by nitrate in the absence of the inhibitor.

Quinones

Quinones are necessary redox carriers for electron transport with nitrate as the acceptor. This was shown for the formate–nitrate reductase complex of *E. coli* (Itagaki, 1964) and the NADH–nitrate reductase complex of *K. aerogenes* (Knook & Planta, 1973) by extraction of the quinones and subsequent restoration of the inhibited activity with quinones. The formate dehydrogenase and the nitrate reductase preparations of Enoch & Lester (1974) have been shown to restore an overall formate–nitrate reductase activity when either Q or MK are present as mediators. As Q is more efficient, both in the reconstitution and the reactivation after extraction, it was concluded that Q rather than MK is involved in formate–nitrate reductase. The NADH-nitrate reductase of

K. aerogenes could only be reactivated by Q since MK was inefficient. As judged from their mid-point redox potentials, both Q and MK should mediate electron transport to nitrate with all the donors except succinate, which requires Q. It is likely that in Gram-positive bacteria, which contain only MK, electron transport to nitrate is mediated by this quinone. Thus a mutant of *Staphylococcus aureus* deficient in MK was found to lack electron-transport-dependent nitrate reduction (Săsărman, Purvis & Portelance, 1974).

The electron transport chain suggested by the reconstitution of formate–nitrate reductase activity (Ruiz-Herrera & DeMoss, 1969; Enoch & Lester, 1974) is shown below.

$$FDH - Cyt\ b_{FDH} \rightarrow Q \rightarrow Cyt\ b_{NR} - NR$$
$$\text{(Mo, Se, Fe–S)} \qquad\qquad\qquad \text{(Fe–S, Mo)}$$

The molybdenum of both formate dehydrogenase (FDH) and nitrate reductase (NR) is probably directly involved in the exchange of reducing equivalents with the respective substrates (see above). By analogy with the formate–fumarate reductase it is suggested that the *b* cytochromes of formate dehydrogenase and nitrate reductase serve as the donor and acceptor of Q.

Mechanism of energy transduction

In view of the similarities of the electron transport sequences of the formate–nitrate reductase complex of *E. coli* and the formate–fumarate reductase complex of *V. succinogenes* the speculation may be made that the mechanisms of energy transduction are also similar. In this respect it is interesting that the uptake of amino acids by *E. coli* can be driven both by the formate–nitrate and the formate–fumarate reductase (Boonstra, Huttunen, Konings & Kaback, 1975) and also that the quenching of the atebrin fluorescence, which is thought to indicate capacity for energy transduction of the membrane, is observed with both systems (Haddock & Kendall-Tobias, 1975).

Garland, Downie & Haddock (1975) found that $2H^+$ per NO_3^- are liberated into the medium by spheroplasts of *E. coli* with formate as reductant in a reaction which is sensitive to uncoupling agents. This ratio is equivalent to that measured with *V. succinogenes* for formate and fumarate. On the premise that a ratio of $2H^+$ per $2e^-$ is equivalent to the capacity for synthesising 1 mol ATP per $2e^-$, this indicates that 1 mol ATP per mol nitrate is synthesised by *E. coli*. This would be in agreement with the ATP yield of the formate–nitrate reduction of *V. succinogenes*, which was estimated from the molar growth yield.

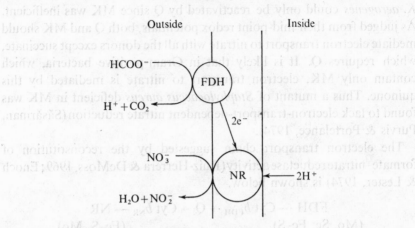

Fig. 8. Possible mechanism of generating an electrochemical proton potential across the bacterial membrane by the electron transport from formate to nitrate. FDH, formate dehydrogenase; NR, nitrate reductase. According to Garland et al. (1975) the nitrate reductase translocates protons.

Garland et al. (1975) suggested that the protons liberated by formate-nitrate reduction in E. coli are translocated by the quinone mechanism, essentially as proposed in Fig. 6. In addition these authors assumed that the nitrate reductase acts by conducting protons from the inside to the outside of the membrane. These protons are required for the reduction of nitrate that occurs on the outside of the membrane (see Fig. 8). The capacity of the nitrate reductase to translocate protons is thought to be indicated by the finding that the hydrogen carrier reduced FMN serves as a donor of nitrate reductase only when accessible to the inside of the membrane, whereas reduced benzylviologen acts as a donor in both broken and intact cells (Garland et al., 1975). Alternatively, the protons required for nitrate reduction may be provided by a quinone mechanism which translocates $2H^+$ per electron (Kröger, 1976).

In these considerations Garland et al. (1975) assumed that the oxidation of formate by formate dehydrogenase occurs on the inside of the membrane. The oxidation of formate on the outside (Fig. 8) would provide two additional protons. In this case the proton-translocating quinone mechanism would not be required in order to account for the H^+ per NO^-_3 ratio of 2, if the protons necessary for nitrate reduction are translocated by the nitrate reductase as proposed in Fig. 8.

CONCLUSIONS

The reduction of fumarate and nitrate by various donor substrates is catalysed by enzyme systems which resemble the mitochondrial respiratory chain. The systems of both fumarate and nitrate reduction are membrane-bound and require a quinone as redox mediator. This clearly identifies both systems as electron transport chains. At least one b cytochrome is required for formate-dependent nitrate reduction, whereas the requirement for b cytochromes in fumarate reduction is not yet clearly established. Cytochromes of the a-, c- and d-type appear not to be involved in either system.

The phosphorylation reactions driven by fumarate and nitrate reduction are clearly different from those of substrate-level phosphorylation and show most of the properties of oxidative phosphorylation. As shown especially with cell-free preparations the phosphorylation is associated with the membrane and is abolished by uncoupling agents. Control of electron transport by the phosphorylation reaction was observed in one case with nitrate as the acceptor, but appears to be less pronounced in fumarate reduction. Formate and nitrate reduction are associated with the generation of an electrochemical proton potential across the membrane which is abolished by uncoupling agents. Finally fumarate and nitrate reduction can provide the energy required for the active uptake of amino acids by the bacteria.

The author is indebted to Drs B. A. Haddock, C. W. Jones, W. N. Konings, A. H. Stouthamer and R. K. Thauer for making available their reviews before publication and to Dr G. R. N. Jones for his help in preparing the manuscript.

REFERENCES

ALLEN, S. H. G., KELLERMEYER, R. W., STJERNHOLM, R. L. & WOOD, H. G. (1964). Purification and properties of enzymes involved in the propionic acid fermentation. *Journal of Bacteriology*, **87**, 171–87.

AUE, B. J. & DEIBEL, R. H. (1967). Fumarate reductase activity of *Streptococus faecalis*. *Journal of Bacteriology*, **93**, 1770–6.

BARKER, H. A. (1972). ATP formation by anaerobic bacteria. In *Horizons of Bioenergetics*, ed. A. San Pietro & H. Gest, pp. 7–31. New York & London: Academic Press.

BARTON, L. L., LeGALL, J. & PECK, H. D. (1970). Phosphorylation coupled to oxidation of hydrogen with fumarate in extracts of the sulfate reducing bacterium, *Desulfovibrio gigas*. *Biochemical and Biophysical Research Communications*, **41**, 1036–42.

4

BARTON, L. L., LEGALL, J. & PECK, H. D. (1972). Oxidative phosphorylation in the obligate anaerobe, *Desulfovibrio gigas*. In *Horizons of Bioenergetics*. ed. A. San Pietro & H. Gest, pp. 35–51. New York & London: Academic Press.

BAUCHOP, T. & ELSDEN, S. R. (1960). The growth of microorganisms in relation to their energy supply. *Journal of General Microbiology*, 23, 457–69.

BAUM, R. H. & DOLIN, M. I. (1965). Isolation of 2-solanesyl-1,4-naphthoquinone from *Streptococcus faecalis*, 10C1. *Journal of Biological Chemistry*, 240, 3425–33.

BELL, G. R., LEGALL, J. & PECK, H. D. (1974). Evidence for the periplasmic location of hydrogenase in *Desulfovibrio gigas*. *Journal of Bacteriology*, 120, 994–7.

BOONSTRA, J., HUTTUNEN, M. T., KONINGS, W. N. & KABACK, H. R. (1975). Anaerobic transport in *Escherichia coli* membrane vesicles. *Journal of Biological Chemistry*, 250, 6792–8.

BRAY, R. C. (1976). Molybdenum iron–sulfur flavin hydroxylases and related enzymes. *The Enzymes*, 12, 299–419.

BRAY, R. C., VINCENT, S. P., LOWE, D. J., CLEGG, R. A. & GARLAND, P. B. (1976). Electron-paramagnetic-resonance-studies on the molybdenum of nitrate reductase from *Escherichia coli* K12. *Biochemical Journal*, 155, 201–3.

CALLIES, E. (1974). Auftrennung des Flavobacterium-Cytophaga-Komplexes aufgrund der Atmungschinongehalte. V. *Tagung der Deutschen Gesellschaft für Hygiene und Mikrobiologie, Mainz, Germany*.

CLEGG, R. A. (1976). Purification and some properties of nitrate reductase from *Escherichia coli* K12. *Biochemical Journal*, 153, 533–41.

DECKER, K., JUNGERMANN, K. & THAUER, R. K. (1970). Energy production in anaerobic organism. *Angewandte Chemie (International Edition)*, 9, 138–58.

DEGROOT, G. N. & STOUTHAMER, A. H. (1970). Regulation of reductase formation in *Proteus mirabilis*. II. Influence of growth with azide and of heme deficiency on nitrate reductase formation. *Biochimica et Biophysica Acta*, 208, 414–27.

DEIBEL, R. H. & KVETKAS, M. J. (1964). Fumarate reduction and its role in the diversion of glucose fermentation by *Streptococcus faecalis*. *Journal of Bacteriology*, 88, 858–64.

DERVARTANIAN, D. V. & FORGET, P. (1975). The bacterial nitrate reductase, EPR studies on the enzyme A of *Escherichia coli* K12. *Biochimica et Biophysica Acta*, 379, 74–80.

DEVRIES, W., VAN WYCK-KAPTEYN, W. M. C. & STOUTHAMER, A. H. (1973). Generation of ATP during cytochrome-linked anaerobic electron transport in propionic acid bacteria. *Journal of General Microbiology*, 76, 31–41.

ENOCH, H. G. & LESTER, R. L. (1974). The role of a novel cytochrome *b*-containing nitrate reductase and quinone in the *in vitro* reconstruction of formate–nitrate reductase activity of *E. coli*. *Biochemical and Biophysical Research Communications*, 61, 1234–41.

ENOCH, H. G. & LESTER, R. L. (1975). The purification and properties of formate dehydrogenase and nitrate reductase from *Escherichia coli*. *Journal of Biological Chemistry*, **250**, 6693–705.

FAUST, P. J. & VANDEMARK, P. J. (1970). Phosphorylation coupled to NADH oxidation with fumarate in *Streptococcus faecalis* 10C1. *Archives of Biochemistry and Biophysics*, **137**, 392–8.

FISCHER, P. (1972). Die Dynamik der Atmungskette bei *Achromobacter mucosus*. MD thesis, University of Marburg, Germany.

FORGET, P. (1971). Les nitrate-réductases bactériennes. Solubilisation, purification et propriétés de l'enzyme A de *Micrococcus denitrificans*. *European Journal of Biochemistry*, **18**, 442–50.

FORGET, P. (1974). The bacterial nitrate reductases. Solubilization, purification and properties of the enzyme A of *Escherichia coli* K12. *European Journal of Biochemistry*, **42**, 325–32.

FORGET, P. & DERVARTANIAN, D. V. (1972). The bacterial nitrate reductases: EPR studies on nitrate reductase A from *Micrococcus denitrificans*. *Biochimica et Biophysica Acta*, **256**, 600–6.

GARLAND, P. B., DOWNIE, J. A. & HADDOCK, B. A. (1975). Proton translocation and respiratory nitrate reductase of *Escherichia coli*. *Biochemical Journal*, **152**, 547–59.

HADDOCK, B. A. & JONES, C. W. (1977). Bacterial respiration. *Bacteriological Reviews*, in press.

HADDOCK, B. A. & KENDALL-TOBIAS, M. W. (1975). Functional anaerobic electron transport linked to the reduction of nitrate and fumarate in membranes from *Escherichia coli* as demonstrated by quenching of atebrin fluorescence. *Biochemical Journal*, **152**, 655–9.

HAMILTON, W. A. (1975). Energy coupling in microbial transport. *Advances in Microbial Physiology*, **12**, 1–53.

HAROLD, F. M. (1972). Conservation and transformation of energy by bacterial membranes. *Bacteriological Reviews*, **36**, 172–230.

HASAN, S. M. & HALL, J. B. (1975). The physiological function of nitrate reduction in *Clostridium perfringens*. *Journal of General Microbiology*, **87**, 120–8.

HATCHIKIAN, E. C. (1974). On the role of menaquinone-6 in the electron transport of hydrogen: fumarate reductase system in the strict anaerobe *Desulfovibrio gigas*. *Journal of General Microbiology*, **81**, 261–6.

HATCHIKIAN, E. C. & LEGALL, J. (1972). Evidence for the presence of a *b*-type cytochrome in the sulfate reducing bacterium *Desulfovibrio gigas* and its role in the reduction of fumarate by molecular hydrogen. *Biochimica et Biophysica Acta*, **267**, 479–84.

HOLLÄNDER, R. (1974). Respiratorische Chinone der Bakterien der *Haemophilus*-Gruppe. *V. Tagung der Deutschen Gesellschaft für Hygiene und Mikrobiologie, Mainz, Germany*.

HOLLÄNDER, R. (1976). Physiologie und Physiotaxonomie einiger Vertreter der Gattung *Haemophilus* Winslow *et al.* PhD thesis, University of Marburg, Germany.

HOLLÄNDER, R. & MANNHEIM, W. (1975). Characterization of hemophilic and related bacteria by their respiratory quinones and cytochromes. *International Journal of Systematic Bacteriology*, **25**, 102–7.

ITAGAKI, E. (1964). The role of lipophilic quinones in the electron transport system of *Escherichia coli*. *Journal of Biochemistry, Tokyo*, **55**, 432–45.

JACOBS, N. J. & WOLIN, M. J. (1963*a*). Electron-transport system in *Vibrio succinogenes*. I. Enzymes and cytochromes of the electron-transport system. *Biochimica et Biophysica Acta*, **69**, 18–28.

JACOBS, N. J. & WOLIN, M. J. (1963*b*). Electron-transport system of *Vibrio succinogenes*. II. Inhibition of electron transport by 2-heptyl-4-hydroxy-quinoline-*N*-oxide. *Biochimica et Biophysica Acta*, **69**, 29–39.

JEDISS, R. (1973). Cytochrome und respiratorische Chinone der Enterobacteriaceae. MD thesis, Universität Marburg, Germany.

JOHN, P. & WHATLEY, F. R. (1970). Oxidative phosphorylation coupled to oxygen uptake and nitrate reduction in *Micrococcus denitrificans*. *Biochimica et Biophysica Acta*, **216**, 342–52.

KEMP, M. B., HADDOCK, B. A. & GARLAND, P. B. (1975). Synthesis and sidedness of membrane-bound respiratory nitrate reductase (EC 1.7.99.4) in *Escherichia coli* lacking cytochromes. *Biochemical Journal*, **148**, 329–33.

KISTLER, W. S. & LIN, E. C. C. (1972). Purification and properties of the flavine-stimulated anaerobic L,α-glycerophosphate dehydrogenase of *Escherichia coli*. *Journal of Bacteriology*, **112**, 539–47.

KNOOK, D. L. & PLANTA, R. J. (1973). The function of ubiquinone in *Klebsiella aerogenes*. *Archiv für Mikrobiologie*, **93**, 13–22.

KONINGS, W. N. & BOONSTRA, J. (1976). Anaerobic electron transfer and active transport in bacteria. *Current Topics in Membranes and Transport*, in press.

KRÖGER, A. (1975). The electron transport-coupled phosphorylation of the anaerobic bacterium *Vibrio succinogenes*. In *Electron Transfer Chains and Oxidative Phosphorylation*, ed. E. Quagliariello, S. Papa, F. Palmieri, E. C. Slater & N. Siliprandi, pp. 265–70. Amsterdam: North-Holland Publishing Company.

KRÖGER, A. (1976). The interaction of the radicals of ubiquinone in mitochondrial electron transport. *FEBS Letters*, **65**, 278–80.

KRÖGER, A. & DADÁK, V. (1969). On the role of quinones in bacterial electron transport. The respiratory system of *Bacillus megaterium*. *European Journal of Biochemistry*, **11**, 328–40.

KRÖGER, A., DADÁK, V., KLINGENBERG, M. & DIEMER, F. (1971). On the role of quinones in bacterial electron transport. Differential roles of ubiquinone and menaquinone in *Proteus rettgeri*. *European Journal of Biochemistry*, **21**, 322–33.

KRÖGER, A. & INNERHOFER, A. (1976*a*). The electron transport of the anaerobic *Vibrio succinogenes*. *Zeitschrift für Physiologische Chemie*, **357**, 270.

KRÖGER, A. & INNERHOFER, A. (1976*b*). The function of the *b*-cytochromes in the electron transport from formate to fumarate of *Vibrio succinogenes*. *European Journal of Biochemistry*, in press.

KRÖGER, A. & INNERHOFER, A. (1976*c*). The function of menaquinone, covalently bound FAD and iron–sulfur protein in the electron transport from formate to fumarate of *Vibrio succinogenes*. *European Journal of Biochemistry*, in press.

KRÖGER, A., SCHIMKAT, M. & NIEDERMAIER, S. (1974). Electron-transport phosphorylation coupled to fumarate reduction in anaerobically grown *Proteus rettgeri*. *Biochimica et Biophysica Acta*, **347**, 273–89.

LEGALL, J., DERVARTANIAN, D. V., SPILKER, E., LEE, J. P. & PECK, H. D. (1971). Evidence for the involvement of non-heme iron in the active site of hydrogenase from *Desulfovibrio vulgaris*. *Biochimica et Biophysica Acta*, **234**, 525–30.

LEGALL, J. & POSTGATE, J. R. (1973). The physiology of sulfate-reducing bacteria. *Advances in Microbial Physiology*, **10**, 81–133.

LUNDEGARDH, H. (1945). Absorption, transport and exudation of inorganic ions by the roots. *Arkiv für Botanik*, **32A**, 1–139.

MACGREGOR, C. H. (1975). Solubilization of *Escherichia coli* nitrate reductase by a membrane-bound protease. *Journal of Bacteriology*, **121**, 1102–10.

MACGREGOR, C. H., SCHNAITMAN, C. A., NORMANSELL, D. E. & HODGINS, M. G. (1974). Purification and properties of nitrate reductase from *Escherichia coli* K12. *Journal of Biological Chemistry*, **249**, 5321–7.

MACY, J., KULLA, H. & GOTTSCHALK, G. (1976). H_2-dependent anaerobic growth of *Escherichia coli* on L-malate: succinate formation. *Journal of Bacteriology*, **125**, 423–8.

MANNHEIM, W., JEDISS, R. & ZABEL, R. (1974a). Die Atmungschinone der chemoorganotrophen, fermentativen gramnegativen Bakterien. *V. Tagung der Deutschen Gesellschaft für Hygiene und Mikrobiologie, Mainz, Germany*.

MANNHEIM, W., JEDISS, R., ZABEL, R. & HOLLÄNDER, R. (1974b). Taxonomic significance of respiratory quinones in chemoorganotrophic Gram-negative, and fermentative bacteria. *XI Conference on the Taxonomy of Bacteria, Brno, Czechoslovakia*.

MIKI, K. & LIN, E. C. C. (1975a). Anaerobic energy-yielding reaction associated with transhydrogenation from glycerol-3-phosphate to fumarate by an *Escherichia coli* system. *Journal of Bacteriology*, **124**, 1282–7.

MIKI, K. & LIN, E. C. C. (1975b). Electron transport chain from glycerol-3-phosphate to nitrate in *Escherichia coli*. *Journal of Bacteriology*, **124**, 1288–94.

MITCHELL, P. (1966). Chemiosmotic coupling in oxidative and photosynthetic phosphorylation. *Biological Reviews*, **41**, 445–502.

NEWTON, N. A., COX, G. B. & GIBSON, F. (1971). The function of menaquinone (vitamin K_2) in *Escherichia coli* K-12. *Biochimica et Biophysica Acta*, **244**, 155–66.

OHNISHI, T. (1963). Oxidative phosphorylation coupled with nitrate respiration with cell free extracts of *Pseudomonas denitrificans*. *Journal of Biochemistry, Tokyo*, **53**, 71–9.

OHNISHI, T. & MORI, T. (1960). Oxidative phosphorylation coupled with denitrification in intact cell systems. *Journal of Biochemistry, Tokyo*, **48**, 406–11.

OTA, A., YAMANAKA, T. & OKUNUKI, K. (1964). Oxidative phosphorylation coupled with nitrate respiration. II. Phosphorylation coupled with anaerobic nitrate reduction in a cell-free extract of *Escherichia coli*. *Journal of Biochemistry, Tokyo*, **55**, 131–5.

PAYNE, W. J. (1973). Reduction of nitrogenous oxides by microorganisms. *Bacteriological Reviews*, **37**, 409–52.

PECK, H. D. (1968). Energy-coupling mechanisms in chemolithotrophic bacteria. *Annual Review of Microbiology*, **22**, 489–518.

PICHINOTY, F. (1973). La réduction bactérienne des composés oxygénés minéraux de l'azote. *Bulletin de l'Institut Pasteur*, **71**, 317–95.

QUAYLE, J. R. (1972). The metabolism of one-carbon compounds by microorganisms. *Advances in Microbial Physiology*, **7**, 119–203.

ROBERTSON, R. N. (1960). Ion transport and respiration. *Biological Reviews*, **35**, 231–64.

ROSSO, J. P., FORGET, P. & PICHINOTY, F. (1973). Les nitrate-reductases bactériennes, solubilisation, purification et propriétés de l'enzyme A de *Micrococcus halodenitrificans*. *Biochimica et Biophysica Acta*, **321**, 443–55.

RUIZ-HERRERA, J. & ALVAREZ, A. (1972). A physiological study of formate dehydrogenase, formate oxidase and hydrogenlyase from *Escherichia coli* K12. *Antonie van Leeuwenhoek*, **38**, 479–91.

RUIZ-HERRERA, J. & DeMOSS, J. A. (1969). Nitrate reductase complex of *Escherichia coli* K12: participation of specific formate dehydrogenase and cytochrome b_1 components in nitrate reduction. *Journal of Bacteriology*, **99**, 720–9.

SĂSĂRMAN, A., PURVIS, P. & PORTELANCE, V. (1974). Role of menaquinone in nitrate respiration in *Staphylococcus aureus*. *Journal of Bacteriology*, **117**, 911–13.

SCHNORF, U. (1966). Der Einfluss von Substituenten auf Redoxpotential und Wuchsstoffeigenschaften von Chinonen. Thesis No. 3871, Eidgenössische technische Hochschule, Zürich, Switzerland.

SCHWARTZ, A. C. & SPORKENBACH, J. (1975). The electron transport system of the anaerobic *Propionibacterium shermanii*, cytochrome and inhibitor studies. *Archiv für Mikrobiologie*, **102**, 261–73.

SIEGEL, L. M. (1975). Biochemistry of the sulfur cycle. *Metabolic Pathways*, **7**, 217–86.

SINGH, A. P. & BRAGG, P. D. (1975). Reduced nicotinamide adenine dinucleotide dependent reduction of fumarate coupled to membrane energization in a cytochrome deficient mutant of *Escherichia coli* K12. *Biochimica et Biophysica Acta*, **396**, 229–41.

SINGH, A. P. & BRAGG, P. D. (1976). Anaerobic transport of amino acids coupled to the glycerol-3-phosphate–fumarate oxidoreductase system in a cytochrome-deficient mutant of *Escherichia coli*. *Biochimica et Biophysica Acta*, **423**, 450–61.

SONE, N. (1972). The redox reactions in propionic acid fermentation. I. Occurrence and nature of an electron transfer system in *Propionibacterium arabinosum*. *Journal of Biochemistry, Tokyo*, **71**, 931–40.

SONE, N. (1974). The redox reactions in propionic acid fermentation. IV. Participation of menaquinone in the electron transfer system in *Pro-*

pionibacterium arabinosum. Journal of Biochemistry, Tokyo, **76**, 137–45.

STIEFEL, E. I. (1973). Proposed molecular mechanism for the action of molybdenum in enzymes: coupled proton and electron transfer. *Proceedings of the National Academy of Sciences, USA*, **70**, 988–92.

STOUTHAMER, A. H. (1976). Biochemistry and genetics of nitrate reductase in bacteria. *Advances in Microbial Physiology*, **14**, 315–75.

THAUER, R. K., JUNGERMANN, J. & DECKER, K. (1976). Energy conservation in chemotrophic anaerobic bacteria. *Bacteriological Reviews*, in press.

VAN'T RIET, J., VAN EE, J. H., WEVER, R., VAN GELDER, B. F. & PLANTA, R. J. (1975). Characterization of the respiratory nitrate reductase of *Klebsiella aerogenes* as a molybdenum-containing iron–sulfur enzyme. *Biochimica et Biophysica Acta*, **405**, 306–17.

WHITE, D. A. & SINCLAIR, P. R. (1971). Branched electron-transport systems in bacteria. *Advances in Microbial Physiology*, **5**, 173–211.

WOLFE, R. S. (1971). Microbial formation of methane. *Advances in Microbial Physiology*, **6**, 107–46.

WOLIN, M. J., WOLIN, E. A. & JACOBS, N. J. (1961). Cytochrome-producing anaerobic vibrio, *Vibrio succinogenes* sp.n. *Journal of Bacteriology*, **81**, 911–17.

WOOD, H. G. & UTTER, M. F. (1965). The role of CO_2 fixation in metabolism. *Essays in Biochemistry*, **1**, 1–27.

YAMANAKA, T., OTA, A. & OKUNUKI, K. (1962). Oxidative phosphorylation. I. Evidence for phosphorylation coupled with nitrate reduction in a cell free extract of *Pseudomonas aeruginosa. Journal of Biochemistry, Tokyo*, **51**, 253–8.

THE ISOLATION OF PHENOTYPIC AND GENOTYPIC VARIANTS FOR THE FUNCTIONAL CHARACTERISATION OF BACTERIAL OXIDATIVE PHOSPHORYLATION

B. A. HADDOCK

Department of Biochemistry, Medical Sciences Institute, University of Dundee, Dundee DD1 4HN, UK

INTRODUCTION

The ability of the bacterial cell to alter the components responsible for oxidative phosphorylation in response to manipulation by the investigator, and thereby adapt to different environmental situations, makes possible a variety of experimental opportunities for the study of the mechanism of energy conservation that are not readily available in equivalent studies with mitochondria. Clearly yeast and other fungi share some of these potential advantages; however, the interpretation of the results obtained is often complicated by the fact that certain mitochondrial enzyme complexes contain polypeptides coded for by the mitochondrial, as well as the nuclear, DNA.

Three general approaches can be used to exploit this adaptability of bacteria in the study of oxidative phosphorylation. The first approach involves a comparative study of different bacteria that have adapted to life in particular ecosystems; for example, (*a*) photolithotrophs, chemolithotrophs and chemo-organotrophs; (*b*) obligate anaerobes, obligate aerobes and facultative anaerobes; and (*c*) acidophiles and alkalophiles. A second general approach involves the study of those phenotypic changes that accompany the adaptation of a particular organism to a new environmental situation and include: (*a*) changes in the energy source available for growth; for example, transitions from phototrophic to chemotrophic growth conditions by photosynthetic bacteria or, alternatively, the effects of catabolite repression; (*b*) changes in the availability of different terminal electron acceptors for respiration; for example, during anaerobic to aerobic growth transitions; (*c*) changes in the availability of other nutrients that are required, after suitable processing by the cell, for the functional activity of particular proteins; for example, the effect of iron-limited growth conditions on the cytochrome content of cells; (d) the presence of specific inhibitors of electron-transport-dependent ATP synthesis in the growth medium; and (e) those changes that accompany gross alterations to cellular morphology and function in response to other environmental pressures, for example during sporulation. The third general approach is the isolation of mutants of a particular organism defective in their ability to catalyse electron-transport-dependent ATP synthesis and the study of the effects of these genetic lesions on, for example: (a) bacterial metabolism under different growth conditions; and (b) the synthesis and assembly of both indivdual proteins (from apoproteins and prosthetic groups) and complete multimeric enzyme complexes into functionally active constituents of the cytoplasmic membrane.

These latter two approaches are considered here with particular emphasis on the experimental methods used to attain phenotypic and genotypic modifications to normal function. A brief outline of their potential use and application to the understanding of the functional organisation and the control mechanisms regulating the synthesis and activity of those membrane-bound components that are responsible for electron-transport-dependent ATP synthesis is also given.

ISOLATION OF MUTANTS FOR THE STUDY OF OXIDATIVE PHOSPHORYLATION IN BACTERIA

It is hardly surprising that the organism of choice for many workers attempting to isolate mutants deficient in electron-transport-dependent ATP synthesis has been *Escherichia coli* (Gibson & Cox, 1973; Cox & Gibson, 1974; Haddock & Jones, 1977). Indeed this organism has many potential advantages since not only is it amenable to sophisticated genetic analysis and manipulation but also it is a facultative anaerobe capable of deriving energy for growth by a variety of different mechanisms. However, relevant mutants of other bacteria have also been isolated and extensively characterised. Although in many cases the isolation procedures used and types of mutant obtained are similar to those for *E. coli*, in certain cases significant differences can be found, both in the type and properties of the isolated mutants, which reflect the wide variety of functional components and metabolic activities found in different bacteria.

Many experimental approaches have been used for the isolation of bacterial mutants with defects in their ability to generate ATP by electron transport. These approaches are best considered as different examples of the following general methods for mutant isolation.

Differential ability to utilise alternative carbon sources for growth

This is perhaps the most widely used method for mutant screening and depends upon the ability of bacteria like *E. coli* to grow both fermentatively and oxidatively. For growth, the bacterial cell needs to produce ATP for its biosynthetic reactions and this can be derived either from substrate level phosphorylation or alternatively by electron-transport-dependent ATP synthesis. In addition, the bacterial cell must be able to establish and maintain a protonmotive force across the cytoplasmic membrane for the accumulation of various solutes. This latter requirement can be satisfied either by a functional proton translocating respiratory chain or by ATP hydrolysis via the proton translocating adenosine triphosphatase [EC 3.6.1.3; ATPase]. Naturally if an organism is to grow on a non-fermentable carbon source, such as succinate, then it must be able to synthesise all those components that are required for electron-transport-dependent ATP synthesis, since this is the only mechanism for generating intracellular ATP. However, a mutant lacking either a functional electron transport chain or a functional ATPase will still be able to grow on a fermentable carbon source, such as glucose, since ATP can be produced via substrate level phosphorylation, and, provided that either the respiratory chain or the ATPase is still able to translocate H^+, this can be used to establish and maintain the required protonmotive force across the cytoplasmic membrane. Naturally a double mutant with lesions in both electron transport and ATPase activity would be unable to grow under any conditions since it has no mechanism for generating a transmembranous protonmotive force. Many laboratories have isolated mutants that are able to grow on glucose but not on non-fermentable carbon sources, such as succinate, D-lactate, glycerol, malate, and acetate, and subsequent analysis has shown that this method can be used to obtain a wide variety of mutants with defects in their ability to synthesise ubiquinone (*ubi*), menaquinone (*men*), components of the iron transport system (*ent, fep, fes*), functional ATPase activity (*unc*), and several dehydrogenase activities (*dld, sdh, glp*D) as summarised in Table 1. In order to increase the frequency of mutants defective in some aspect of oxidative phosphorylation over mutants defective either in their ability to transport the non-fermentable carbon source used in the screen or in enzymes of the tricarboxylic acid cycle, it is preferable to select for mutants able to grow on glucose but not on plates containing a mixture of acetate, malate and succinate (Schairer & Haddock, 1972).

It has been suggested that mutants lacking ATPase activity can be

Table 1. *Mutants of* E. coli *with defects in their ability to synthesise components required for electron-transport-dependent ATP synthesis*

Mutation	Gene	Approximate map position (min)	Notes	Reference
Ubiquinone synthesis	ubiA	90	Various structural genes for the enzymes responsible for the biosynthesis of ubiquinone-8 from chorismic acid (no regulatory genes yet described)	Gibson (1973)
	ubiB	84		
	ubiC	90		
	ubiD	84		
	ubiE	84		
	ubiF	15		
	ubiG	48		
	ubiH	62		
Menaquinone synthesis	menA	87	Accumulates 1,4-dihydroxy-2-naphthoic acid	Young (1975)
	menB	?	Accumulates 2-succinylbenzoic acid	
Haem biosynthesis	hemA (popC)	4	5-Aminolaevulinic acid synthetase [EC 2.3.1.37]	Powell *et al.* (1973); Wulff (1967); Săsărman, Surdeanu & Horodniceanu (1968); Haddock & Schairer (1973)
[Note: confusion as to nomenclature; proposals of Săsărman *et al.* (1975) have been adopted with alternative designations, given in the original references, quoted in brackets]	hemA	26	Both are 5-aminolaevulinic acid auxotrophs	
	hemB (popD)	1	Porphobilinogen synthetase [EC 4.2.1.24]	Powell *et al.* (1973)
	hemC (popE)	84	Uroporphyrinogen I synthetase [EC 4.3.1.8]	Powell *et al.* (1973)
	hemD (hemC)	90	Uroporphyrinogen II cosynthetase	McConville & Charles (1975)
	hemE	89	Uroporphyrinogen decarboxylase [EC 4.1.1.37]	Săsărman *et al.* (1975)
	hemF (popB)	17	Coproporphyrinogen III oxidase [EC 1.3.3.3]	Powell *et al.* (1973)
	hemG (popA)	11	Ferrochelatase [EC 4.99.1.1]	Cox & Charles (1973)
	?similar to hem⁻			Săsărman *et al.* (1968b); Beljanski & Beljanski (1957)
Iron limitation	ent A–G	13	Lesions in the synthesis of enterochelin from chorismic acid	Young *et al.* (1971); Luke & Gibson (1971); Woodrow, Young & Gibson (1975)
	fep	13	Lesion in the transport of ferric-enterochelin complex	Cox *et al.* (1970)
	fes B	13	Unable to hydrolyse ferric-enterochelin complex	Langman *et al.* (1972)
	ton B	27	Affects all known active transport systems for iron	e.g. Frost & Rosenberg (1975)

	Gene	Map	Characteristics	References
Anaerobic electron-transport-dependent nitrate reduction	chlA	17	Pleiotropic mutants lacking several enzyme activities. Locus possibly divisible into two complementation groups. Suggested primary lesion in ability to synthesise a Mo-cofactor	Venables (1972); MacGregor (1975); MacGregor & Schnaitman (1973)
	chlB	85	Pleiotropic, possibly three complementation groups. Suggested primary lesion in synthesis of an association factor (Fa) required for attachment or insertion of Mo-cofactor into nitrate reductase [EC 1.7.99.4]	Venables (1972); MacGregor (1975); MacGregor & Schnaitman (1973); Rivière et al. (1975)
	chlC	27	Structural gene for nitrate reductase, possibly α-subunit	MacGregor (1975); Guest (1969)
	chlD	17	Pleiotropic mutants whose phenotype is restored to that of the wild-type by addition of excess Mo to growth medium. Suggested lesion in the processing of Mo	Glaser & DeMoss (1971)
	chlE	18	Pleiotropic mutants, possibly two complementation groups. Has been proposed as a structural gene for nitrate reductase, possibly γ-subunit, but requires further investigation	MacGregor (1975); MacGregor & Schnaitman (1971)
	chlF	26	Suggested as structural gene for formate dehydrogenase	Glaser & DeMoss (1972)
	chlG	0	Not known	Glaser & DeMoss (1972)
	dld	?	D-lactate dehydrogenase [EC 1.1.2.4]	Simoni & Shallenberger (1972); Hong & Kaback (1972)
Dehydrogenase activity	sdh	16	Succinate dehydrogenase [EC 1.3.99.1]	Spencer & Guest (1974)
	frd	93	Fumarate reductase [EC 1.3.99.-]	Spencer & Guest (1973)
	glpA	48	L-α-glycerophosphate dehydrogenase [EC 1.1.99.-]. Soluble enzyme required for anaerobic growth with fumarate as electron acceptor	Kistler & Lin (1971)
	glpD	74	L-α-glycerophosphate dehydrogenase [EC 1.1.99.5]. Particulate enzyme required for growth aerobically and used preferentially during anaerobic growth with NO₃⁻ as electron acceptor	Kistler & Lin (1971)
	nut	?	Energy-linked transhydrogenase [EC 1.6.1.1]	Cited in Gibson & Cox (1973)
	ndh	23	NADH dehydrogenase [EC 1.6.99.-]	Young & Wallace (1976)

Table 1. Continued

Mutation	Gene	Approximate map position (mins)	Notes	Reference
Additional mutants in non-energy-conserving electron transport reactions	hyd	57	Hydrogenase [EC 1.12.7.1] a component of formate hydrogen lyase activity	Pascal et al. (1975)
	$nirA$ $nirB$	26 61–83}	Nitrite reductase [EC 1.6.6.4]	{Cole & Ward (1973) Kavanagh & Cole (1976)
Proton translocating adenosine triphosphatase activity [EC 3.6.1.3]	$uncA$	83	Mutants lacking ATPase activity. Wide variety described probably reflecting point mutations in the structural genes for the catalytically active polypeptides of the F_1-subunit of the enzyme. Require further characterisation	See reviews of Simoni & Postma (1975); Haddock & Jones (1977)
	$uncB$	83	Mutants exhibiting ATPase which is not, however, functionally organised in the membrane and capable of catalysing those energy-dependent reactions that require ATP hydrolysis. Wide variety of mutants described reflecting point mutants in the structural genes for the polypeptides of the Fo-subunit of the enzyme and for those polypeptides that attach the catalytically active F_1 polypeptides to the Fo-subunit, e.g. in mutant etc-15 (Bragg, Davies & Hou, 1973). Further characterisation required	See reviews of Simoni & Postma (1975); Haddock & Jones (1977)
	$uncD$	83	Mutants with decreased ATPase activity exhibiting altered divalent metal ion specificity	Thipayathasana (1975)
Additional but uncharacterised mutants with defects in energy conservation	Strain AN295	86	High levels of ATPase activity; not known if this is the direct result of a mutation in a regulator gene or the indirect effect of a mutation elsewhere	Cox et al. (1973)
	Strain 15b⁻ Strain JSH4 (ecf^{ts})	? 61–65	Uncharacterised defect in energy metabolism Temperature-sensitive for the coupling of metabolic energy to the transport of a variety of amino acids and sugars	Turnock et al. (1972) Lieberman & Hong (1974, 1976)

Gene designations and approximate map positions are in accordance with Bachmann, Low & Taylor (1976).

distinguished from those deficient in electron transport by their differential ability to grow on glycerol (Schairer & Haddock, 1972). The pathway of glycerol metabolism in *E. coli* can be summarised as:

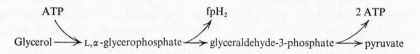

It is apparent that, although this pathway results in the net synthesis of ATP by substrate level phosphorylation, a functional respiratory chain is required for the reoxidation of the flavin component of L,α-glycerophosphate dehydrogenase and is, therefore, essential for growth on glycerol. It was expected, therefore, that ATPase-deficient mutants, but not electron-transport-deficient mutants, should grow on glycerol and this was experimentally demonstrated in certain strains of *E. coli* (Schairer & Haddock, 1972; Simoni & Shallenberger, 1972). However, subsequent investigations have shown that other ATPase-deficient mutants were unable to grow on glycerol and, in addition, a variety of sugars (Daniel, Roisin, Burstein & Kepes, 1975). The common feature shared by these nutrients was their metabolism by inducible pathways, so it was suggested that, in the absence of ATPase activity, the intracellular ATP concentration and hence cyclic AMP level was lowered, which in turn resulted in the impaired induction of the necessary catabolic enzymes. Clearly these proposals require experimental verification but the different results reported by different laboratories suggest that the observed phenotype of a particular ATPase mutant may be dependent on the genetic background of the parental strain from which it is derived.

It is now well established that cyclic AMP interacts with a specific binding protein called the cyclic AMP receptor protein (CRP protein) to stimulate transcription from many operons of *E. coli*. Mutants with defects in their ability to produce adenylate cyclase [EC 4.6.1.1] (*cya*) and the CRP protein (*crp*) are known (e.g. Brickman, Soll & Beckwith, 1973; Sabourin & Beckwith, 1975) and it is worth remembering that these mutants exhibit superficially a similar phenotype to mutants with defects in oxidative phosphorylation. However, *cya* and *crp* mutants are capable of electron-transport-dependent ATP synthesis, with decreased effectiveness in comparison with wild-type strains, but they cannot grow on non-fermentable carbon sources since they are unable to induce the necessary transport systems and/or catabolic enzymes to utilise these substrates. Thus it has been shown that a *cya* mutant when grown aerobically contains reduced cytochrome levels and shows decreased electron transport activity; the addition of cyclic AMP to the growth medium stimulated cytochrome synthesis and restored growth on glucose to the normal rate observed with the parental strain (Broman, Dobrogosz & White, 1974). In addition Hempfling (1970) has demonstrated that the efficiency of oxidative phosphorylation in *E. coli* B was decreased during aerobic growth in the presence of glucose but that the effect could be reversed by the addition of cyclic AMP to the growth medium (Hempfling & Beeman, 1971). These data certainly suggest a role for cyclic AMP and/or CRP protein in regulating the functional activity of certain proteins required for oxidative phosphorylation, but further work is clearly required to identify differences in the synthesis of specific proteins (Daoud & Haddock, 1976), to distinguish between effects on apoprotein and prosthetic group synthesis, and to determine if these alterations are a primary or secondary consequence of cyclic AMP and/or CRP protein deficiency.

Differential ability to utilise alternative terminal electron acceptors for growth

E. coli can derive energy for growth from electron-transport-dependent ATP synthesis using oxygen, or under anaerobic conditions, either fumarate or nitrate, as terminal electron acceptor. Methods have been described for the isolation of mutants defective in these anaerobic electron transport pathways, based on the selection of colonies able to grow on plates containing non-fermentable carbon sources incubated aerobically but not on similar plates incubated anaerobically and containing, in addition, either fumarate (Spencer & Guest, 1973) or nitrate (Venables & Guest, 1968). Interestingly, it has been shown (Lambden & Guest, 1976) that mutants selected as unable to grow on lactate with fumarate anaerobically, but capable of aerobic growth with lactate, exhibit a similar phenotype to those pleiotropic nitrate reductase mutants isolated as chlorate-resistant colonies (see later). This suggests that the two anaerobic electron transport chains utilising nitrate and fumarate share common component(s) that are not required for aerobic electron transport. The reverse approach for the isolation of mutants deficient in aerobic electron transport, based upon a selection of strains able to grow on glycerol with nitrate anaerobically, for example, but unable to grow on glycerol aerobically, has not been reported to my knowledge: indeed the approach may not be feasible since *E. coli* can synthesise different redox carriers that provide alternative routes for electron transport in the presence of oxygen (Haddock & Jones, 1977).

Inhibitor resistance

The addition of inhibitors of normal electron-transport-dependent ATP synthesis allows for the forward selection of inhibitor-resistant mutants. Two classes of mutant are to be expected from this approach: (i) mutants showing an increased permeability barrier to the inhibitor in question; and (ii) mutants synthesising modified proteins that are still functionally competent but which demonstrate decreased affinity for the inhibitor. Mutants of the first general class are not relevant to this discussion but in practice many inhibitor-resistant mutants come in this category. An example of the second class of mutant is the dicyclohexyl-carbodiimide-resistant mutant isolated and characterised by Fillingame (1975), which synthesises a functional proton-translocating ATPase that is insensitive to this inhibitor.

A well documented variation in this approach is the use of chlorate resistance to select for mutants with defects in anaerobic nitrate-dependent electron transport. The described experimental procedure is to screen for mutants that are able to grow on glucose anaerobically in the presence of chlorate (e.g. Glaser & DeMoss, 1972). The generally accepted sequence of events is that chlorate serves as an inducer of nitrate reductase and is converted by the enzyme to chlorite which in turn is toxic to the cell; therefore mutants unable to make chlorite, because of a defect in electron-transport-dependent nitrate reduction, are able to grow using the energy derived from the fermentation of glucose. In a comparative study of nitrate reductase mutants of *E. coli* selected by alternative procedures, Glaser & DeMoss (1972) clearly showed that mutants obtained by this method were pleiotropically deficient in a variety of enzyme activities including formate dehydrogenase, nitrate reductase and formate hydrogen lyase.

The pleiotropic defects exhibited by some of these mutants (*chl*A, *chl*B and *chl*D) are readily explained in terms of the proposals summarised in Table 1, since it appears that these mutants are unable to synthesise various proteins required for the incorporation of molybdenum into both apo-formate dehydrogenase and apo-nitrate reductase and, therefore, to form the respective functional holoenzymes. However, the pleiotropic defects demonstrated by other mutants (e.g. *chl*C and *chl*E) are not so easily understood in terms of the proposed primary lesions suggested in Table 1 and require more intensive investigation. A variety of methods has been used to assess the gene products defective in chlorate-resistant mutants and interpretation of the data provided by these different methods is complicated by: (*a*) the possibility that individual *chl* loci may be genetically subdivisible (Venables, 1972); (*b*) the fact that biochemical investigations have usually only been performed with a single mutant isolate thought typical of a particular *chl* locus and in many cases such analyses have been cursory and far from specific for the precise identification of the altered enzyme activity; (*c*) the need, when considering the synthesis and functional assembly of complex membrane-bound proteins from prosthetic groups and apoproteins (which in turn may be composed of two or more polypeptide subunits), to distinguish between the primary effects of the genetic lesion on the rate of synthesis of a particular component and secondary effects that may follow; for example, an increased rate of degradation of apoprotein (or indeed an altered subcellular localisation of the apoprotein) in the absence of prosthetic group synthesis; and (*d*) the uncertainties surrounding the physiological mechanism(s) determining sensitivity and resistance to chlorate. The sequence of events described above, in which chlorate induces nitrate reductase and is converted by the enzyme to chlorite, clearly explains why wild-type strains cannot grow anaerobically in the presence of chlorate and yet grow normally with chlorate under aerobic conditions, since it is known that nitrate reductase is only produced in the absence of oxygen. However, these proposals are not in accordance with two additional observations: (*a*) that certain mutants, selected for their inability to grow on lactate with fumarate anaerobically in the absence of added nitrate were shown in subsequent growth studies to be chlorate-resistant anaerobically (Lambden & Guest, 1976); and (*b*) that an electron-transport-deficient mutant, strain S2–21, was shown to be sensitive to chlorate under aerobic growth conditions in the absence of added nitrate, whereas the parental strain was chlorate resistant, as expected, under these conditions (Simoni & Shallenberger, 1972). We have confirmed this latter observation and shown that a ubiquinone-deficient mutant and a haem-deficient mutant were both sensitive to chlorate when grown aerobically; if the haem-deficient mutant was allowed to synthesise functional cytochromes, by supplementation of the growth medium with 5-aminolaevulinic acid, then aerobic growth was chlorate resistant (J. A. Downie & B. A. Haddock, unpublished observations). The experimental basis for chlorate sensitivity and resistance is, therefore, not as simple as was originally thought and the most plausible explanation is that some other redox carrier, either as well as or instead of nitrate reductase, serves to reduce chlorate to the toxic chlorite. The identity of this carrier is not known but a strong candidate would be menaquinol, since in the experimental situations described above, there is a clear correlation between the presence of high levels of

Table 2. *Examples of mutants with defects in electron transport processes available in bacteria other than E. coli*

Organism	Deficiency	References
Staphylococcus aureus	Haem synthesis	Tien & White (1968); Yegian, Gallo & Toll (1959)
	Menaquinone synthesis	Săsărman et al. (1968a, 1971); Goldenbaum, Keyser & White (1975)
	Haem synthesis	Anderson & Ivanovics (1967); Berek, Miczek & Ivanovics (1974)
Bacillus subtilis	Cytochrome a synthesis	Taber (1974); Taber & Freese (1974)
	Menaquinone synthesis	Farrand & Taber (1973a,b)
	Riboflavin synthesis	Bacher, Eggers & Lingens (1973)
Pseudomonas aeruginosa	Haem synthesis	van Hartingsveldt & Stouthamer (1974)
Salmonella typhimurium	Specifically lacks ATP-dependent transhydrogenase [EC 1.6.1.1] activity of all energy-linked reactions studied	Kay & Bragg (1975)
	Haem synthesis	Săsărman et al. (1970, 1976)
	Tetrathionate reductase activity	Casse, Pascal & Chippaux (1972)
	Formate dehydrogenase activity	Pascal et al. (1973)
	Hydrogenase activity	Pascal et al. (1975)

A variety of mutants from many bacteria have been isolated with defects in anaerobic electron-transport-dependent nitrate reduction as reviewed by Stouthamer (1976).

menaquinone in the cytoplasmic membrane, and chlorate-sensitive growth of the organism.

Antibiotic resistance

Selection for resistant growth on glucose in the presence of various aminoglycosidic antibiotics such as neomycin, streptomycin and kanamycin has been widely used for the isolation of mutants deficient in electron-transport-dependent ATP synthesis. Using this approach *E. coli* mutants deficient in the synthesis of haem (Săsărman et al., 1968b, 1975) and functional ATPase activity (Kanner & Gutnick, 1972; Rosen, 1973; Yamamoto, Mével-Ninio & Valentine, 1973) have been obtained. This technique has also been applied in the isolation of related mutants from other organisms, including those deficient in their ability to synthesise various quinones (see Table 2).

The experimental basis behind the use of these antibiotics for the positive selection of mutants with defects in energy conservation remains obscure since it is generally accepted that these compounds act on the 70 S ribosome to produce misreading of the genetic code and hence inhibit protein synthesis. It has been suggested that respiratory-deficient cells accumulated less antibiotic than normal cells (Săsărman, Sanderson, Surdeanu & Sonea, 1970); however, since both electron-transport-deficient and ATPase-deficient strains can obviously still establish and maintain normal concentration gradients of those solutes that are essential for growth, this proposal requires more rigorous evaluation.

Auxotrophic mutants

This approach has limited application and in general is restricted to the study of mutants with defects in their ability to synthesise redox carriers, such as quinones, or the prosthetic groups of redox carriers, such as haem or riboflavin. In addition, there are certain experimental restrictions to the potential use of this approach since many of the various intermediates in the biosynthetic pathways of these components are either commercially unavailable (or only available at high cost) or alternatively they are not readily accumulated by cells. For example, although certain bacteria, such as *Staphylococcus* and *Bacillus*, are freely permeable to compounds like riboflavin and haem, and, therefore, auxotrophic mutants can be isolated directly (see Table 2), other bacteria, notably *E. coli*, are apparently not permeable to these compounds. One experimental solution to this problem is to isolate double mutants which are both permeable to, and have a growth requirement for, a given compound, as illustrated by the 'haemin-permeable' mutants of *E. coli* described by McConville & Charles (1975). An alternative solution is to isolate auxotrophic mutants whose growth requirements can be satisfied by an added compound which is accumulated by cells via an existing transport system for some other cellular metabolite. An example of this is the isolation of 5-aminolaevulinic acid requiring mutants of *E. coli* (Table 1). 5-Aminolaevulinic acid is the biosynthetic precursor of haem, and mutants unable to synthesise functional cytochromes can be isolated by selecting for colonies able to grow on succinate with 5-aminolaevulinic acid but not on succinate alone; from what has been said before it should be obvious that such mutants are able to grow, albeit at decreased rate, on fermentable substrates such as glucose. It is unlikely that the cell produces a specific transport system for 5-aminolaevulinic acid but presumably the compound can be accumulated via the lysine transport system. It is of interest that apparently two genes, mapping at distinct sites on the chromosome, can mutate to give a specific growth requirement for 5-aminolaevulinic acid on non-fermentable substrates (Table 1) but the significance of this observation remains to be determined.

Use of intrinsic and extrinsic chromophores

In screening a large number of potential mutants, isolated by any method other than an antibiotic- or inhibitor-resistance forward selection screen, it is an obvious advantage to be able to identify those mutants in which one is particularly interested, quickly and specifically from the background colonies. Clearly replica plating and the subsequent visual inspection of two or more plates can be tedious, expensive and often inaccurate since many of the relevant mutants, particularly those with defects in electron transport, form small colonies. Many of these problems can be avoided if conditions can be found in which the required mutants produce, or indeed fail to produce, a characteristic colour.

The possibility of utilising intrinsic chromophoric compounds, normally produced by the cell, is somewhat limited. However, Cox & Charles (1973) have described a method for the isolation of *E. coli* mutants defective in their ability to synthesise haem based upon the observation that porphyrin-accumulating mutants are red–brown in colour while wild-type strains and mutants unable to synthesise porphyrin are white. The experimental procedure was initially to select for mutants that could not grow on lactate but which formed red–brown colonies during growth on glucose. All such isolates were found to be protoporphyrin-accumulating mutants defective in the enzyme ferrochelatase (Powell, Cox, McConville & Charles, 1973). The second stage of the procedure was to select for secondary mutants, from one of these red–brown porphyrin-accumulating strains, which were identified as small, white colonies on glucose plates. All the double mutants isolated by this approach were deficient in ferrochelatase and had a second metabolic lesion in one of the other porphyrin biosynthetic enzymes. An analogous procedure, that has not been reported to my knowledge, could be used to obtain mutants defective in their ability to synthesise flavin, based upon the initial isolation of over-producers of flavin, identified by the intensity of their fluorescence under ultraviolet light, and the subsequent selection of secondary mutants exhibiting decreased fluorescence: since, however, flavin is an obligatory component of many enzymes, other than those required for electron transport, attention would have to be given to the growth supplements required to enable a flavin-deficient mutant to grow under conditions where certain membrane-bound dehydrogenases were inactive in the absence of prosthetic group synthesis. The more direct approach for screening large numbers of potential electron-transport-deficient mutants, based upon the visual identification of yeast mutants with altered absorption spectra at the temperature of liquid nitrogen using a Hartree microspectroscope (Sherman, 1967), is yet to be exploited in equivalent bacterial studies.

An alternative approach, with potentially a much wider application, is the use of added redox dyes which possess a characteristic colour in either the reduced or oxidised form, or reagents which give coloured complexes with metabolic intermediates, to monitor the presence or absence of a particular enzymic reaction directly in whole cells. Thus if benzyl viologen is included in glucose plates incubated under anaerobic conditions, mutants of *Salmonella typhimurium* lacking formate dehydrogenase and hydrogenase (Table 2) or of *E. coli* lacking hydrogenase (Table 1) form white colonies whereas competent strains give blue colonies. The use of eosin/methylene blue dye indicator plates for the isolation of *S. typhimurium* mutants lacking tetrathionate reductase (Table 2) and of *E. coli* lacking nitrate reductase (Ruiz-Herrera, Showe & DeMoss, 1969) has been described. A final example of this approach is the method for the isolation of nitrate reductase mutants of *E. coli* based upon a dye-overlay assay using reagents for nitrite determination (Ruiz-Herrera *et al.*, 1969; Glaser & DeMoss, 1972). Further examples of this general approach exist in the literature and, because the technique enables the investigator to screen large numbers of colonies for potential mutants, additional variations will no doubt be described.

Genetic techniques

As summarised in Tables 1 and 2, many mutants with defects in oxidative phosphorylation have now been isolated from a variety of bacteria. Although chromosomal maps are now being constructed for several of these organisms, *E. coli* K12 remains the most suitable bacterium for detailed genetic studies. This assertion is based not only upon the extensive background information as to the genetic constitution of the bacterium but, also, on the ease with which mutant alleles can be transferred between strains. The opportunities arising from the ability to transfer genes between different bacterial strains using phage or episome vectors, and indeed from the specialised properties of some of these bacteriophages themselves, can be exploited for the isolation of certain mutants that cannot be obtained readily using the approaches already described.

For example using P_1 phage transduction we have transferred the *hem*A mutant allele, from a strain unable to synthesise 5-aminolaevulinic acid and hence functional cytochromes, into an *unc*A strain which cannot synthesise functional ATPase activity (H. U. Schairer & B. A.

Haddock, unpublished observations). The resulting double mutant can grow on glucose only if the growth medium is supplemented with 5-aminolaevulinic acid. The phenotypic properties of this double mutant support the premise made earlier that either a functional electron transport chain or a functional ATPase enzyme is a necessary requirement for growth and that simultaneous loss of both activities is lethal to the cell.

The characteristic properties of certain bacteriophages have also been used to obtain mutants with defects in oxidative phosphorylation. For example, Mu phage lysogenises randomly over the *E. coli* chromosome and, when it inserts within a gene, it confers a strong transcriptional polarity effect upon that and any distal genes within the same operon. Therefore, one approach to determine whether or not linked genes are grouped into an operon is to look for the inactivation of more than one gene by a single Mu insertion. Using this approach Woodrow, Young & Gibson (1975) discovered a previously unrecognised gene (*ent*G) coding for a protein required for enterochelin synthesis and provided evidence that three of the *ent* genes (*ent*A,B,G) are part of an operon.

It is worth noting that although a wide variety of genes specifying components required for oxidative phosphorylation in *E. coli* have now been identified and located on the chromosome map relative to other markers, with the exception of certain of the nitrate reductase genes (Adhya, Cleary & Campbell, 1968; Venables, 1972) little attention has been given to the fine structure genetic analysis of these loci. Such an approach would clearly be informative in the identification of additional structural genes and in determining regulatory genes.

Genetic techniques can also be used to construct partial diploid strains, that is heterogenotes containing two copies of certain regions of the bacterial chromosome. Partial diploids are obviously of value in chromosome mapping studies as an alternative to phage transduction analysis, but more importantly they can be used to determine gene dosage effects on cellular metabolism. For example, the cell normally synthesises a fixed amount of functional ATPase enzyme, but in a partial diploid, containing two copies of the ATPase operon, one could investigate the physiological consequences of this additional ATPase activity on the ability of the cells to generate ATP by electron transport. The episome F'16 (F'-*ilv*) has been used to map certain *unc* mutants and could presumably be used in such a study (Yamamoto *et al.*, 1973). Heterogenotes constructed with certain F*gal* and F*lac* episomes showed an increased cytochrome content in comparison with parental strains.

The increases were explained by the proposal that a number of cyto-chrome structural and/or regulatory genes are localised on the *E. coli* chromosome in a 2 min long region between the *pur*E (13 min) and *sup*E (15 min) loci (Shipp, Piowtrowski & Friedman, 1972). These ob-servations are clearly of interest but further work is required to distin-guish possible effects on the regulation of haem synthesis from effects on the synthesis of particular apocytochromes.

Finally it is worth considering that many of the sophisticated genetic skills and techniques, which have been developed for *E. coli*, and used so successfully to provide information on the molecular biology of the gene, have so far found little application in the construction and analysis of mutants deficient in oxidative phosphorylation.

ISOLATION OF PHENOTYPIC VARIANTS FOR THE STUDY OF OXIDATIVE PHOSPHORYLATION IN BACTERIA

In the Introduction a variety of different methods were briefly described for altering the phenotype of an organism by manipulating the growth environment of the cell. Many of these methods are already well docu-mented in the literature and will not be described here, but attention is directed instead to two general methods that have not, as yet, been widely exploited.

Alteration to the growth-limiting nutrient

The particular respiratory components synthesised by many bacteria alter depending upon the carbon source for growth and upon the pres-ence or absence of different terminal electron acceptors. Even a cursory glance at the literature indicates that most investigators define the phenotype of the organism under study by fixing the carbon source in the medium (for example glucose for fermentative growth and succinate for oxidative growth) and the terminal electron acceptor provided (for example oxygen for aerobic growth and nitrate or fumarate for anaero-bic growth). In these studies all other growth requirements are present in the medium in excess and never limit or affect the growth rate of the organism. Clearly a transition from one mode of growth to another, by changing the carbon source or terminal electron acceptor provided, affords an experimental system in which to study the factors regulating the synthesis and functional activity of those components which change under these conditions.

However, the phenotype of a particular organism can also be altered by making some nutrient, other than the carbon source or the terminal electron acceptor, rate-limiting for growth. For example, iron-limited growth of *E. coli* under aerobic conditions resulted in the decreased synthesis of *b*-type cytochromes and iron–sulphur proteins (Rainnie & Bragg, 1973), while under anaerobic conditions, iron limitation resulted in decreased levels of hydrogenase, and, hence, formate hydrogen lyase, activities (Fukuyama & Ordal, 1965). It should be appreciated that in *E. coli* iron limitation can be achieved either by decreasing the iron content of the growth medium or by using mutants blocked in their ability to transport iron (Table 1; Cox *et al.*, 1970); a comparative study of the effects of iron limitation induced by these two approaches would clearly be of interest.

Since iron is a constituent of both cytochromes and iron–sulphur proteins, the effect of sulphate-limited growth conditions on aerobic cultures of *E. coli* has been investigated in an attempt to characterise specifically the role of iron–sulphur proteins in electron transport and energy conservation. Using a low-sulphate-requiring strain of *E. coli* K12, isolated after prolonged growth of a prototroph under sulphate-limited conditions, it was shown that sulphate-limited growth in a chemostat resulted in the decreased synthesis of iron–sulphur proteins, the adaptive loss of proton translocation associated with the NADH dehydrogenase region of the electron transport chain, and the induced synthesis of alternative cytochrome components (Poole & Haddock, 1975). All these effects could be reversed by increasing the sulphate concentration in the growth medium, indicating the phenotypic rather than genotypic nature of the alterations, but, to date, conditions have not been found for reversing these effects in non-growing cells grown previously under conditions of sulphate limitation. More recently, the experiments have been repeated with *E. coli* W, where no loss of proton translocation associated with the NADH dehydrogenase activity could be demonstrated in cells grown under sulphate-limited conditions (Farmer & Jones, 1976). The reason for this discrepancy is not known and, although it may be accounted for on the basis of a strain difference, further work is obviously required. Relevant to this discussion is the observation that the efficiency of oxidative phosphorylation in *Enterobacter aerogenes*, as calculated from growth yield experiments, is similar in glucose- and histidine-limited cultures but decreased in sulphate-limited chemostat cultures (Stouthamer & Bettenhaussen, 1975; Stouthamer, this volume).

Copper is a functional component of the mitochondrial cytochrome $a + a_3$-type terminal oxidase, and copper-limited growth of the yeast *Torulopsis utilis* gives cells whose mitochondria exhibit defective electron

transport (see Downie & Garland, 1973). Analogous studies with
bacteria are more limited and indeed it is not known if bacterial terminal
oxidases of the cytochrome o-, a_1- and d-type require copper for functional activity (Lund & Raynor, 1975); clearly, however, this is an experimental system worth investigation.

As a final example, phenotypic alterations to the anaerobic, nitrate-
dependent, electron transport chain are worth consideration. Thus it has
been shown that molybdenum-limited growth of *E. coli* under anaerobic
conditions results in the decreased synthesis of formate dehydrogenase
and nitrate reductase activities as both enzymes require molybdenum for
functional activity (Lester & DeMoss, 1971). Since *chl*D mutants are
specifically blocked in their ability to transport molybdenum (Glaser &
DeMoss, 1971) they should exhibit a similar phenotype to wild-type
cells grown under conditions of molybdenum limitation. Formate
dehydrogenase also requires selenium for functional activity and selenite-
limited anaerobic growth results in cells having a decreased formate
dehydrogenase activity (Lester & DeMoss, 1971).

Addition of inhibitors of normal function to the growth medium

Compounds which selectively inhibit growth under conditions where
cellular energy is derived from electron-transport-dependent ATP synthesis, rather than from fermentation, can be used in this approach. Thus,
the presence of low concentrations of cyanide in the growth medium
has been shown to result in the induced synthesis of alternate cytochromes in several bacteria incubated under aerobic conditions including *Achromobacter* sp. (Arima & Oka, 1965), *Bacillus cereus* (McFeters,
Wilson & Strobel, 1970) and *E. coli* (Ashcroft & Haddock, 1975). The
reason for the adaptive synthesis of new components is not known, but
a working hypothesis is that these concentrations of cyanide bind to and
inhibit a cytochrome oxidase with a high affinity for cyanide and that
the cell compensates by induced synthesis of an alternative cytochrome
oxidase with a lower affinity for the inhibitor.

An additional example of this approach was reported by Scott &
DeMoss (1976), who made use of the observation that tungsten is a
competitive inhibitor of molybdenum in many biological systems, and
showed that anaerobic growth of *E. coli* in tungstate-containing medium
resulted in the formation of inactive formate dehydrogenase and nitrate
reductase. Subsequent incubation of tungstate-grown cells with molybdate under non-growing conditions led to the rapid activation of both
enzymic activities.

CONCLUSIONS

Considerable attention has been directed towards describing the rationale and experimental methods for isolating bacterial cells which possess specific defects in their ability to derive ATP from electron transport. The subsequent analysis of these modified cells, both in terms of their physiological properties and also as starting material in attempts to restore normal functional activity, can provide a wealth of information concerning the synthesis and functional activity of those components that are required for oxidative phosphorylation. Indeed, the use of functionally characterised lesions provides a powerful means of probing the complexities of the mechanism of energy conservation in biological membranes.

The application of phenotypic and genotypic variants to these problems has been described in detail (Haddock & Jones, 1977). For example, mutants deficient in their ability to synthesise the prosthetic group required for the functional activity of a particular redox carrier can be used to determine: (*a*) the biosynthetic pathway of that prosthetic group; (*b*) whether or not the apoprotein is synthesised and incorporated into the membrane in the absence of prosthetic group synthesis; and (*c*), if the apoprotein is synthesised, the factors regulating the correct insertion of prosthetic group and production of functional holoenzyme. Mutants defective in their ability to produce a component polypeptide of a multimeric enzyme complex can be used to determine if the other polypeptides of the complex are still produced in the presence (point mutants) or absence (deletion mutants) of the defective polypeptide, their intracellular location, and whether or not they still combine to give an inactive enzyme complex; temperature-sensitive mutants should prove useful in this type of study. Following the isolation of the individual polypeptides of the complex, either separately or following disaggregation of the defective enzyme complex, attempts can be made to re-create a functional complex by complementation studies with polypeptides obtained from competent strains and to determine the factors regulating the assembly and insertion of the complex into the membrane.

The ultimate objective is to isolate all the necessary gene loci on phage or episome vectors, to achieve the in-vitro synthesis of all the necessary proteins, to characterise the factors regulating the synthesis, assembly and incorporation of these components into a defined lipid bilayer and to produce, in the test tube, a functional unit capable of electron-transport-dependent ATP synthesis. Without doubt this is a daunting task and will be realised only after the isolation and characterisation of even more bacterial deviants.

I should like to thank the many colleagues who have helped in a variety of ways to make this review possible, and, in particular, Dr H. P. Charles for his clarifying comments on the nomenclature of the *hem* genes. The work of the author was supported by the SRC through grant B/RG/58518.

REFERENCES

ADHYA, S., CLEARY, P. & CAMPBELL, A. (1968). A deletion analysis of prophage lambda and adjacent genetic regions. *Proceedings of the National Academy of Sciences, USA*, **61**, 956–62.

ANDERSON, T. J. & IVANOVICS, G. (1967). Isolation and some characteristics of haemin dependent mutants of *Bacillus subtilis*. *Journal of General Microbiology*, **49**, 31–40.

ARIMA, K. & OKA, T. (1965). Cyanide resistance in *Achromobacter*. I. Induced formation of cytochrome *a* and its role in cyanide resistant respiration. *Journal of Bacteriology*, **90**, 734–43.

ASHCROFT, J. R. & HADDOCK, B. A. (1975). Synthesis of alternative membrane-bound redox carriers during aerobic growth of *Escherichia coli* in the presence of potassium cyanide. *Biochemical Journal*, **148**, 349–52.

BACHER, A., EGGERS, U. & LINGENS, F. (1973). Genetic control of riboflavin synthetase in *Bacillus subtilis*. *Archiv für Mikrobiologie*, **89**, 73–7.

BACHMANN, B. J., LOW, K. B. & TAYLOR, A. L. (1976). Recalibrated linkage map of *Escherichia coli* K-12. *Bacteriological Reviews*, **40**, 116–67.

BELJANSKI, M. & BELJANSKI, M. (1957). Sur la formation d'enzymes respiratoires chez un mutant *Escherichia coli* streptomycino-resistant et auxotrophe pour l'hemine. *Annales de l'Institut Pasteur*, **92**, 396–412.

BEREK, L., MICZAK, A. & IVANOVICS, G. (1974). Mapping the Δ-aminolevulinic acid dehydrase and porphobilinogen deaminase loci in *Bacillus subtilis*. *Molecular and General Genetics*, **132**, 233–9.

BRAGG, P. D., DAVIES, P. L. & HOU, C. (1973). Effect of removal or modification of subunit polypeptides on the coupling factor and hydrolytic activities of the Ca^{2+}- and Mg^{2+}-activated adenosine triphosphatase of *Escherichia coli*. *Archives of Biochemistry and Biophysics*, **159**, 664–70.

BRICKMAN, E., SOLL, L. & BECKWITH, J. (1973). Genetic characterization of mutations which affect catabolite-sensitive operons in *Escherichia coli*, including deletions of the gene for adenyl cyclase. *Journal of Bacteriology*, **116**, 582–7.

BROMAN, R. L., DOBROGOSZ, W. J. & WHITE, D. C. (1974). Stimulation of cytochrome synthesis in *Escherichia coli* by cyclic AMP. *Archives of Biochemistry and Biophysics*, **162**, 595–601.

CASSE, F., PASCAL, M. C. & CHIPPAUX, M. (1972). A mutant of *Salmonella typhimurium* deficient in tetrathionate reductase activity. *Molecular and General Genetics*, **119**, 71–4.

COLE, J. A. & WARD, F. B. (1973). Nitrite reductase-deficient mutants of *Escherichia coli* K12. *Journal of General Microbiology*, **76**, 21–9.

Cox, G. B. & Gibson, F. (1974). Studies on electron transport and energy-linked reactions using mutants of *Escherichia coli*. *Biochimica et Biophysica Acta*, **346**, 1–25.

Cox, G. B., Gibson, F., Luke, R. K. J., Newton, N. A., O'Brien, I. G. & Rosenberg, H. (1970). Mutations affecting iron transport in *Escherichia coli*. *Journal of Bacteriology*, **104**, 219–26.

Cox, G. B., Gibson, F., McCann, L. M., Butlin, J. D. & Crane, F. L. (1973). Reconstitution of the energy-linked transhydrogenase activity in membranes from a mutant strain of *Escherichia coli* K12 lacking magnesium ion- or calcium ion-stimulated adenosine triphosphatase. *Biochemical Journal*, **132**, 689–95.

Cox, R. & Charles, H. P. (1973). Porphyrin-accumulating mutants of *Escherichia coli*. *Journal of Bacteriology*, **113**, 122–32.

Daniel, J., Roisin, M. P., Burstein, C. & Kepes, A. (1975). Mutants of *Escherichia coli* K12 unable to grow on non-fermentable carbon substrates. *Biochimica et Biophysica Acta*, **376**, 195–209.

Daoud, M. S. & Haddock, B. A. (1976). Electron transport in mutants of *Escherichia coli* deficient in their ability to synthesize cyclic AMP and the catabolite gene activator protein. *Biochemical Society Transactions*, **4**, 711–14.

Downie, J. A. & Garland, P. B. (1973). An antimycin-A and cyanide resistant variant of *Candida utilis* arising during copper-limited growth. *Biochemical Journal*, **134**, 1051–61.

Farmer, I. S. & Jones, C. W. (1976). The energetics of *Escherichia coli* during aerobic growth in continuous culture. *European Journal of Biochemistry*, **67**, 115–22.

Farrand, S. K. & Taber, H. W. (1973a). Pleiotropic menaquinone-deficient mutant of *Bacillus subtilis*. *Journal of Bacteriology*, **115**, 1021–34.

Farrand, S. K. & Taber, H. W. (1973b). Physiological effects of menaquinone deficiency in *Bacillus subtilis*. *Journal of Bacteriology*, **115**, 1035–44.

Fillingame, R. H. (1975). Identification of the dicyclohexylcarbodiimide-reactive protein component of the adenosine 5'-triphosphate energy transducing system of *Escherichia coli*. *Journal of Bacteriology*, **124**, 870–83.

Frost, G. E. & Rosenberg, H. (1975). Relationship between the tonB locus and iron transport in *Escherichia coli*. *Journal of Bacteriology*, **124**, 704–12.

Fukuyama, T. & Ordal, E. J. (1965). Induced biosynthesis of formic hydrogen lyase in iron-deficient cells of *Escherichia coli*. *Journal of Bacteriology*, **90**, 673–80.

Gibson, F. (1973). Chemical and genetic studies on the biosynthesis of ubiquinone by *Escherichia coli*. *Biochemical Society Transactions*, **1**, 317–26.

Gibson, F. & Cox, G. B. (1973). The use of mutants of *Escherichia coli* K12 in studying electron transport and oxidative phosphorylation. *Essays in Biochemistry*, **9**, 1–29.

Glaser, J. M. & DeMoss, J. A. (1971). Phenotypic restoration by molybdate of nitrate reductase activity in *chl* D mutants of *Escherichia coli*. *Journal of Bacteriology*, **108**, 854–60.

GLASER, J. H. & DeMOSS, J. A. (1972). Comparison of nitrate reductase mutants of *Escherichia coli* selected by alternative procedures. *Molecular and General Genetics*, **116**, 1–10.

GOLDENBAUM, P. E., KEYSER, P. D. & WHITE, D. C. (1975). Role of vitamin K_2 in the organization and function of *Staphylococcus aureus* membranes. *Journal of Bacteriology*, **121**, 442–9.

GUEST, J. R. (1969). Biochemical and genetic studies with nitrate reductase *C*-gene mutants of *Escherichia coli*. *Molecular and General Genetics*, **105**, 285–97.

HADDOCK, B. A. & JONES, C. W. (1977). Bacterial respiration. *Bacteriological Reviews*, in press.

HADDOCK, B. A. & SCHAIRER, H. U. (1973). Electron transport chains of *Escherichia coli*: reconstitution of respiration in a 5-amino-laevulinic acid requiring mutant. *European Journal of Biochemistry*, **35**, 34–45.

HEMPFLING, W. P. (1970). Repression of oxidative phosphorylation in *Escherichia coli* B by growth in glucose and other carbohydrates. *Biochemical and Biophysical Research Communications*, **41**, 9–15.

HEMPFLING, W. P. & BEEMAN, D. K. (1971). Release of glucose repression of oxidative phosphorylation in *Escherichia coli* B by cyclic adenosine 3′,5′-monophosphate. *Biochemical and Biophysical Research Communications*, **45**, 924–30.

HONG, J. S. & KABACK, H. R. (1972). Mutants of *Salmonella typhimurium* and *Escherichia coli* pleiotropically defective in active transport. *Proceedings of the National Academy of Sciences, USA*, **69**, 3336–40.

KANNER, B. I. & GUTNICK, D. L. (1972). Use of neomycin in the isolation of mutants blocked in energy conservation in *Escherichia coli*. *Journal of Bacteriology*, **111**, 287–9.

KAVANAGH, B. M. & COLE, J. A. (1976). Characterization of mutants of *Escherichia coli* K12 defective in nitrite assimilation. *Proceedings of the Society for General Microbiology*, **3**, 84.

KAY, W. W. & BRAGG, P. D. (1975). *Salmonella typhimurium* HfrA, a mutant in which adenosine triphosphate can drive amino acid transport but not energy-dependent nicotinamide nucleotide transhydrogenation. *Biochemical Journal*, **150**, 21–9.

KISTLER, W. S. & LIN, E. C. C. (1971). Anaerobic L-α-glycerophosphate dehydrogenase of *Escherichia coli*: its genetic locus and its physiological role. *Journal of Bacteriology*, **108**, 1224–34.

LAMBDEN, P. R. & GUEST, J. R. (1976). A novel method for isolating chlorate-resistant mutants of *Escherichia coli* K12 by anaerobic selection on a lactate plus fumarate medium. *Journal of General Microbiology*, **93**, 173–6.

LANGMAN, L., YOUNG, I. G., FROST, G. E., ROSENBERG, H. & GIBSON, F. (1972). Enterochelin system of iron transport in *Escherichia coli*: mutations affecting ferric-enterochelin esterase. *Journal of Bacteriology*, **112**, 1142–9.

LESTER, R. L. & DeMOSS, J. A. (1971). Effects of molybdate and selenite on formate and nitrate metabolism in *Escherichia coli*. *Journal of Bacteriology*, **105**, 1006–14.

LIEBERMAN, M. A. & HONG, J. S. (1974). A mutant of *Escherichia coli* defective in the coupling of metabolic energy to active transport. *Proceedings of the National Academy of Sciences, USA*, **71**, 4395–9.

LIEBERMAN, M. A. & HONG, J. S. (1976). Energization of osmotic shock-sensitive transport systems in *Escherichia coli* requires more than ATP. *Archives of Biochemistry and Biophysics*, **172**, 312–15.

LUKE, R. K. J. & GIBSON, F. (1971). Location of three genes concerned with the conversion of 2,3,dihydroxybenzoate into enterochelin in *Escherichia coli* K12. *Journal of Bacteriology*, **107**, 557–62.

LUND, T. & RAYNOR, J. B. (1975). Electron spin resonance of some bacterial respiratory membranes. *Journal of Bioenergetics*, **7**, 161–6.

MACGREGOR, C. H. (1975). Synthesis of nitrate reductase components in chlorate resistant mutants of *Escherichia coli*. *Journal of Bacteriology*, **121**, 1117–21.

MACGREGOR, C. H. & SCHNAITMAN, C. A. (1971). Alterations in the cytoplasmic membrane proteins of various chlorate resistant mutants of *Escherichia coli*. *Journal of Bacteriology*, **108**, 564–70.

MACGREGOR, C. H. & SCHNAITMAN, C. A. (1973). Reconstitution of nitrate reductase activity and formation of membrane particles from cytoplasmic extracts of chlorate-resistant mutants of *Escherichia coli*. *Journal of Bacteriology*, **114**, 1164–76.

MCCONVILLE, M. & CHARLES, H. P. (1975). Isolation of 'haemin permeable' mutants and their use in the study of the genetics of haem biosynthesis in *Escherichia coli* K12. *Proceedings of the Society for General Microbiology*, **3**, 14–15.

MCFETERS, G. A., WILSON, D. F. & STROBEL, G. A. (1970). Cytochromes in a cyanide-resistant strain of *Bacillus cereus*. *Canadian Journal of Microbiology*, **16**, 1221–6.

PASCAL, M. C., CASSE, F., CHIPPAUX, M. & LEPELLETIER, M. (1973). Genetic analysis of mutants of *Salmonella typhimurium* deficient in formate dehydrogenase activity. *Molecular and General Genetics*, **120**, 337–40.

PASCAL, M. C., CASSE, F., CHIPPAUX, M. & LEPELLETIER, M. (1975). Genetic analysis of mutants of *Escherichia coli* K12 and *Salmonella typhimurium* LT2 deficient in hydrogenase activity. *Molecular and General Genetics*, **141**, 173–9.

POOLE, R. K. & HADDOCK, B. A. (1975). Effects of sulphate-limited growth in continuous culture on the electron transport chain and energy conservation in *Escherichia coli* K12. *Biochemical Journal*, **152**, 537–46.

POWELL, K. A., COX, R., MCCONVILLE, M. & CHARLES, H. P. (1973). Mutations affecting porphyrin biosynthesis in *Escherichia coli*. *Enzyme*, **16**, 65–73.

RAINNIE, D. J. & BRAGG, P. D. (1973). The effect of iron deficiency on respiration and energy-coupling in *Escherichia coli*. *Journal of General Microbiology*, **77**, 339–49.

RIVIÈRE, C., GIORDANO, G., POMMIER, J. & AZOULAY, E. (1975). Membrane reconstitution in *chl*-r mutants of *Escherichia coli* K12. VIII. Purification and properties of the F_A factor, the product of the *chl*B gene. *Biochimica et Biophysica Acta*, **389**, 219–35.

ROSEN, B. P. (1973). Restoration of active transport in an Mg^{2+}-adenosine triphosphatase-deficient mutant of *Escherichia coli*. *Journal of Bacteriology*, **116**, 1124–9.

RUIZ-HERRERA, J., SHOWE, M. K. & DeMOSS, J. A. (1969). Nitrate reductase complex of *Escherichia coli* K12: isolation and characterization of mutants unable to reduce nitrate. *Journal of Bacteriology*, **97**, 1291–7.

SABOURIN, D. & BECKWITH, J. (1975). Deletion of the *Escherichia coli crp* gene. *Journal of Bacteriology*, **122**, 338–40.

SĂSĂRMAN, A., CHARTRAND, P., PROSCHER, R., DESROCHERS, M., TARDIF, D. & LAPOINTE, C. (1975). Uroporphyrin-accumulating mutant of *Escherichia coli* K12. *Journal of Bacteriology*, **124**, 1205–12.

SĂSĂRMAN, A., DESROCHERS, M., SONEA, S., SANDERSON, K. E. & SURDEANU, M. (1976). Porphobilinogen-accumulating mutants of *Salmonella typhimurium* LT2. *Journal of General Microbiology*, **94**, 359–66.

SĂSĂRMAN, A., SANDERSON, K. E., SURDEANU, M. & SONEA, S. (1970). Hemin-deficient mutants of *Salmonella typhimurium*. *Journal of Bacteriology*, **102**, 531–6.

SĂSĂRMAN, A., SURDEANU, M. & HORODNICEANU, T. (1968). Locus determining the synthesis of δ-aminolevulinic acid in *Escherichia coli*. *Journal of Bacteriology*, **96**, 1882–4.

SĂSĂRMAN, A., SURDEANU, M., PORTELANCE, V., DOBARDZIC, R. & SONEA, S. (1971). Classification of vitamin K deficient mutants of *Staphylococcus aureus*. *Journal of General Microbiology*, **65**, 125–30.

SĂSĂRMAN, A., SURDEANU, M., SZABADOS, J., GRECEANU, V. & HORODNICEANU, T. (1968a). Menaphthone-requiring mutants of *Staphylococcus aureus*. *Revue Canadienne de Biologie*, **27**, 333–9.

SĂSĂRMAN, A., SURDEANU, M., SZÉGLI, G., HORODNICEANU, T., GRECEANU, V. & DUMISTRESCU, A. (1968b). Hemin-deficient mutants of *Escherichia coli* K12. *Journal of Bacteriology*, **96**, 570–2.

SCHAIRER, H. U. & HADDOCK, B. A. (1972). β-Galactoside accumulation in a Mg^{2+}–Ca^{2+}-activated ATPase deficient mutant of *E. coli*. *Biochemical and Biophysical Research Communications*, **48**, 544–51.

SCOTT, R. H. & DeMOSS, J. A. (1976). Formation of the formate–nitrate electron transport pathway from inactive components in *Escherichia coli*. *Journal of Bacteriology*, **126**, 478–86.

SHERMAN, F. (1967). The preparation of cytochrome-deficient mutants of yeast. *Methods in Enzymology*, **10**, 610–16.

SHIPP, W. S., PIOTROWSKI, M. & FRIEDMAN, A. E. (1972). Apparent cytochrome gene dose effects in F-*lac* and F-*gal* heterogenotes of *Escherichia coli*. *Archives of Biochemistry and Biophysics*, **150**, 473–81.

SIMONI, R. D. & POSTMA, R. W. (1975). The energetics of bacterial active transport. *Annual Reviews of Biochemistry*, **44**, 523–54.

SIMONI, R. D. & SHALLENBERGER, M. K. (1972). Coupling of energy to active transport of amino acids in *Escherichia coli*. *Proceedings of the National Academy of Sciences, USA*, **69**, 2663–7.

SPENCER, M. E. & GUEST, J. R. (1973). Isolation and properties of fumarate reductase mutants of *Escherichia coli*. *Journal of Bacteriology*, **114**, 563–70.

SPENCER, M. E. & GUEST, J. R. (1974). Proteins of the inner membrane of *Escherichia coli*: identification of succinate dehydrogenase by polyacrylamide gel electrophoresis with *sdh* amber mutants. *Journal of Bacteriology*, **117**, 947–53.

STOUTHAMER, A. H. (1976). Biochemistry and genetics of nitrate reductase in bacteria. *Advances in Microbial Physiology*, **14**, 315–75.

STOUTHAMER, A. H. & BETTENHAUSSEN, C. (1975). Determination of the efficiency of oxidative phosphorylation in continuous culture of *Aerobacter aerogenes*. *Archives of Microbiology*, **102**, 187–92.

TABER, H. (1974). Isolation and properties of cytochrome *a* deficient mutants of *Bacillus subtilis*. *Journal of General Microbiology*, **81**, 435–44.

TABER, H. & FREESE, E. (1974). Sporulation properties of a cytochrome *a*-deficient mutant of *Bacillus subtilis*. *Journal of Bacteriology*, **120**, 1004–11.

THIPAYATHASANA, P. (1975). Isolation and properties of *Escherichia coli*. ATPase mutants with altered divalent metal specificity for ATP hydrolysis. *Biochimica et Biophysica Acta*, **408**, 47–57.

TIEN, W. & WHITE, D. C. (1968). Linear sequential arrangement of genes for the biosynthetic pathway of protoheme in *Staphylococcus aureus*. *Proceedings of the National Academy of Sciences, USA*, **61**, 1392–8.

TURNOCK, G., ERICKSON, S. K., ACKRELL, B. A. C. & BIRCH, B. (1972). A mutant of *Escherichia coli* with a defect in energy metabolism. *Journal of General Microbiology*, **70**, 507–15.

VAN HARTINGSVELDT, J. & STOUTHAMER, A. H. (1974). Properties of a mutant of *Pseudomonas aeruginosa* affected in aerobic growth. *Journal of General Microbiology*, **83**, 303–10.

VENABLES, W. A. (1972). Genetic studies with nitrate reductase-less mutants of *Escherichia coli*. I. Fine structure analysis of the *nar*A, *nar*B and *nar*E loci. *Molecular and General Genetics*, **114**, 223–31.

VENABLES, W. A. & GUEST, J. R. (1968). Transduction of nitrate reductase loci of *Escherichia coli* by phages P1 and λ. *Molecular and General Genetics*, **103**, 127–40.

WILSON, A. C. & PARDEE, A. B. (1962). Regulation of flavin synthesis in *Escherichia coli*. *Journal of General Microbiology*, **28**, 283–303.

WOODROW, G. C., YOUNG, I. G. & GIBSON, F. (1975). Mu-induced polarity in the *Escherichia coli* K-12 *ent* gene cluster: evidence for a gene (*ent*G) involved in the biosynthesis of enterochelin. *Journal of Bacteriology*, **124**, 1–6.

WULFF, D. L. (1967). δ-aminolevulinic acid-requiring mutant from *Escherichia coli*. *Journal of Bacteriology*, **93**, 1473–4.

YAMAMOTO, T. H., MÉVEL-NINIO, M. & VALENTINE, R. C. (1973). Essential role of membrane ATPase or coupling factor for anaerobic growth and anaerobic active transport in *Escherichia coli*. *Biochimica et Biophysica Acta*, **314**, 267–75.

YEGIAN, D., GALLO, G. & TOLL, M. W. (1959). Kanamycin resistant *Staphylococcus* mutant requiring heme for growth. *Journal of Bacteriology*, **78**, 10–12.

YOUNG, I. G. (1975). Biosynthesis of bacterial menaquinones. Menaquinone mutants of *Escherichia coli*. *Biochemistry*, **14**, 399–406.

YOUNG, I. G., LANGMAN, L., LUKE, R. K. J. & GIBSON, F. (1971). Biosynthesis of the iron-transport compound enterochelin: mutants of *Escherichia coli* unable to synthesize 2,3-dihydroxybenzoate. *Journal of Bacteriology*, **106**, 51–7.

YOUNG, I. G. & WALLACE, B. J. (1976). Mutations affecting the reduced nicotinamide adenine dinucleotide dehydrogenase complex of *Escherichia coli*. *Proceedings of the Australian Biochemical Society*, **9**, 66.

AUTOTROPHY: A CONCEPTUAL PHOENIX

R. WHITTENBURY AND D. P. KELLY

Department of Biological Sciences and Department of Environmental Sciences, University of Warwick, Coventry CV4 7AL, UK

A word means what I want it to mean
(Lewis Carroll, *Alice in Wonderland*)

INTRODUCTION

Autotrophy in microbes was last the subject of a Symposium of the Society for General Microbiology in 1954. At that time the concept was seen to be useful and the topic sufficiently evocative to occupy the whole symposium. In the intervening 23 years the concept of autotrophy has paradoxically become blurred by the more detailed understanding of the biochemistry and the interrelationships of organisms once clearly identifiable as autotrophs and organisms hitherto regarded unquestionably as being heterotrophs. Recent reactions to this state of affairs (e.g. Rittenberg, 1972) have been to consider the concept of obligate autotrophy to be defunct, or to attempt to redefine autotrophy in the present state of knowledge. These latter attempts have led to a proliferation of terms (e.g. *chemolithotrophs, photolithotrophs* or plain *lithotrophs*) often redefined (if they were resurrected terms) or defined (if new) in such a way that later information has necessitated amendments to these definitions. The upshot is a confused image of autotrophy and its fundamental features and frequent suggestions that the concept has outlived its usefulness.

Our aim here is to define as clearly as is possible what appears to be the essential nature of autotrophy and then to compare 'autotrophic' organisms with others we consider to have important characteristics in common with them. As will emerge, we consider a number of microbial species which utilise C_1-compounds other than carbon dioxide to be similar in certain important aspects to the autotrophs; this similarity we think sufficiently fundamental to the updating of the 'autotroph' concept that we have broadened our view of autotrophy after trimming it in other directions. The process of reaching such a viewpoint is detailed below and an overview of organisms which might in future be included under the umbrella term 'autotrophs' is summarised in the description of three categories of microbe which assimilate C_1 compounds. One category is the 'ribulose bisphosphate pathway' group, which includes the organisms now considered to be autotrophs; another is of C_1 compound utilisers, such as methane and methanol utilising aerobes, which are included in the 'ribulose monophosphate pathway' group; and the final category consists of other C_1 compound utilisers, including methane and methanol utilisers, included in the 'serine pathway' group. It is our opinion that the underlying biochemical similarities of the 'ribulose bisphosphate' and the 'ribulose monophosphate' groups point to both the essential and irrelevant features now embodied in present notions of autotrophy, and that the comparative study of both types will strengthen and give further life to the concept of autotrophy.

The definition of autotrophy arrived at over the years

The definition of autotrophy used in the early years has always been precise in principle and has related only to *carbon* metabolism. Recently, however, the definition has been refined at the biochemical level, possibly losing the essence of the meaning given above.

According to Schlegel (1975) and Woods & Lascelles (1954) autotrophs are organisms *which are able* to synthesise their cell substances from inorganic carbonate as the main source of carbon. The '*which are able*' has to be emphasised, since the ability to utilise carbon dioxide as the main source of carbon is the determining characteristic; it is (Schlegel, 1975) of minor importance that several autotrophs are facultative and can occasionally grow with organic substrates.

This idea of autotrophy meaning the use of carbon dioxide as the obligatory major carbon source is, apparently, an old one (Schlegel, 1975) and separates conceptually the carbon and energy metabolism of the organisms falling within Winogradsky's original definition of the 'Anorgoxydanten'. That the carbon dioxide fixing process could be driven by light energy or inorganic oxidations was recognised by the terms (Pfeffer, 1897) photosynthesis and chemosynthesis. Today (Kelly, 1971; Schlegel, 1975) autotrophic carbon assimilation is recognised as being driven by photolithotrophic and chemolithotrophic energy generating processes, and in rather more specialised cases by photo-organotrophic and chemo-organotrophic energy metabolism.

The strict definition of a true autrotroph has come to imply the obligatory presence of the key enzymes of the Benson–Calvin cycle in an organism, namely ribulose-1,5-bisphosphate carboxylase and ribulose-5-phosphate kinase (Kelly, 1971), though Schlegel (1975) has warned that such a definition might prove too narrow. He recommended only that an autotroph should 'use carbon dioxide as the main carbon source' (*sensu* Woods & Lascelles, 1954).

In the century of work since the recognition in bacteria of carbon dioxide autotrophy based on inorganic oxidation energy, there has been a frequent tendency for the term autotrophy to be used more in the sense of Winogradsky's 'Anorgoxydant', i.e. to mean organisms obtaining carbon from carbon dioxide *and* energy from inorganic sources. This has tended to obscure the value of the clear original concept, so aptly restated by Schlegel (1975). As we will show later there are parts of the original 'Anorgoxydant' concept that could usefully be reintroduced to overcome the rather complex plethora of physiological categories of autotrophs which has sprung up in the recent literature (Kelly, 1971; Schlegel, 1975).

We feel at this point in the paper it will be useful to describe the major categories of autotrophs referred to in text-books and the research literature and to which we will be referring:

(1) *Obligate autotroph* is generally accepted as describing an organism which fixes carbon dioxide as its prime carbon source via the Benson–Calvin cycle (ribulose bisphosphate pathway) and uses a reduced inorganic chemical as its energy source or, if photosynthetic, as its electron donor. As far as we know, all autotrophs can use organic compounds (e.g. acetate) as secondary sources of carbon. Some may be auxotrophs but should be considered autotrophs in that a requirement for a vitamin, etc., is a very minor element of metabolism.

(2) *Facultative autotroph* describes organisms which have the ability to grow either autotrophically (as described above) or heterotrophically.

(3) *Mixotroph* describes facultative autotrophs growing in a culture where conditions are so arranged that both autotrophic and heterotrophic pathways are functioning simultaneously (e.g. Rittenberg & Goodman, 1969; Stukas & DeCicco, 1970).

(4) *Photoautotrophs* are organisms growing at the expense of light energy, a reduced inorganic chemical as a source of electrons and carbon dioxide as the prime carbon source.

(5) *Chemoautotrophs* are organisms growing at the expense of energy released from the oxidation of reduced inorganic chemicals and on carbon dioxide as the prime carbon source.

These descriptions do not cover all the nutritional categories of organisms which might be considered to be autotrophs. To coin terms which cover them would only be confusing at this stage; but more important our later arguments would lead them to become redundant

Lastly, we should make some comment about the term *lithotroph*. This crops up in various word combinations (e.g. chemolithotroph) which, by itself, implies only that an inorganic reduced chemical is used as an energy or electron source. Whereas *autotroph* may or may not imply the use of an inorganic electron donor depending upon how you wish to define autotroph, *chemolithoautotroph* is very precise; it means a carbon-dioxide-fixing, reduced-chemical-utilising organism.

'WHY *ARE* THERE OBLIGATE "CHEMOLITHOAUTOTROPHS" AND RELATED BEASTIES?' (e.g. OBLIGATE PHOTOTROPHS LIKE *CHLOROBIUM*)

An answer to this question may lie in the process of '*natural selection*', in this instance a consequence of the survival of the biochemically fittest in nature's chemostat!

Consider the known results of competition between similar organisms in continuous flow chemostat culture. A difference of 0.1% in growth rate between two bacterial strains can be sufficient (Cox & Gibson, 1974) to give one sufficient advantage to overgrow the other in the chemostat. Similarly, an organism synthesising 'unnecessary' protein (e.g. constitutive versus inducible synthesis of an enzyme) can be at a selective disadvantage in the chemostat when subjected to pressure through growth rate (=dilution rate) in the chemostat or substrate conversion efficiency (Horiuchi, Tomizawa & Novick, 1962). Consider what this means to the autotroph in that: (*a*) an organism growing as a chemolithotrophic autotroph is using *more* enzymes than a 'common' heterotroph; and (*b*) facultative heterotrophs tend to lose their autotrophic/lithotrophic capacities when cultured heterotrophically.

The autotroph is faced with two choices – It must *either* be an *obligate* lithoautotroph *or* be a very good facultative heterotroph to survive in natural environments. The question is – why? Possible answers considered from two standpoints are suggested below.

(1) If the bacterium is able to grow facultatively as a heterotroph it must be able to stand competition from other heterotrophs. In straight competition, therefore, its ability to use a variety of substrates and to grow as fast as its competitors in nutrient-limited environments could be crucial to its survival. *Failure* to compete successfully could result if: (*a*) autotrophic carbon-dioxide-fixing mechanisms were not repressed during heterotrophic growth, for these would cause the synthesis of 're-dundant' protein and (if the enzymes were not prevented from acting) waste of energy in fixing carbon dioxide; (*b*) lithotrophic energy-linked

and oxidative enzymes were not repressed, as these would also lead to synthesis of disadvantageous protein; (c) the organism did repress systems (a) and (b) but had poor ability to adapt to a wide range of available substrates and grew slowly on those that it could use; (d) the organism displayed *all* the bad features listed above.

Organisms with these characteristics might be expected to be just able to subsist in a natural ecosystem but they would be difficult to detect and could not dominate in mesophilic environments at neutral pH values. Examples adapted to extremes of high temperature or low pH (or high pH) might dominate in rare cases.

(2) Retreating into the 'shell' of obligate autotrophy protects the organism from the need to compete and so ensures survival. The following are the relevant characteristics of such organisms: (a) constitutive Calvin cycle enzymes; (b) constitutive lithotrophic oxidation systems; (c) failure to show remarkable adaptive response of (a) or (b) enzyme levels in response to environmental changes of substrate availability, organic matter or growth rate; (d) continued dependence on carbon dioxide as a major carbon source.

In a 'normal' environment the obligate autotroph will derive energy from reactions not available to the heterotrophs with which it shares the habitat (e.g. oxidation of nitrite, sulphur or other inorganic reductants) and will fix carbon dioxide which is a major waste product of its co-habitants. It will make opportunistic use of any organic matter available to it but will not be in any way dependent on that organic matter for survival.

It is noteworthy that the studies made to date on autotrophs reveal two quite distinct classes of physiological response much as outlined above (see Table 1).

(a) There are 'obligate autotrophs' (such as *Thiobacillus neapolitanus*, some strains of *T. ferrooxidans* and *T. thiooxidans*, some nitrifying bacteria) that depend absolutely on their specific energy sources and carbon dioxide, and in the rare cases where facultative heterotrophy has been claimed, growth tends to be poor or achieved only with special techniques.

(b) There are good facultative autotrophs. Thus the 'Hydrogenomonas' species, e.g. *Pseudomonas saccharophila* or *Thiobacillus* A2 or an organism such as '*Micrococcus*' (*Paracoccus*) *denitrificans*, are remarkably adaptable, being able to grow on inorganic energy substrates or on carbon compounds of all chain lengths from C_1 to aromatic rings. Their growth rates also tend to be high and their complement of autotrophic

Table 1. *Comparative features of organisms in the 'obligate autotroph' and 'heterotroph' physiological camp*

Camp 1	No-man's-land	Camp 2
Obligate autotroph		Heterotroph and successful facultative autotroph
Dependence on:		*Dependence on:*
(*a*) Carbon dioxide		(*a*) Organic carbon for energy and carbon
(*b*) Lithotrophic energy systems	Few, if any, inter-mediate forms that do not attribute their success to some metabolic or ecological anomaly (e.g. *Beggiatoa*; Pringsheim, 1967)	(*b*) Insignificance of Calvin cycle or of lithotrophic systems during heterotrophic growth
Plus:		(*c*) Readily adaptive enzyme system; synthesis and repression in response to environmental change
(*a*) Opportunistic assimilation of organic carbon		
(*b*) Inability to become heterotrophic		
(*c*) Relatively invariable levels of the enzymes of major catabolic and anabolic pathways		
(*d*) Very efficient repair mechanisms for genetic damage, making defection to auxotrophy unusual		

and lithotrophic enzymes is essentially abolished by heterotrophic culture.

There is little convincing evidence for the existence of an organism that falls into the intermediate category of retaining full *potential* for autotrophic growth when growing heterotrophically, although in the hydrogen bacteria and *T. ferrooxidans* some intermediate level repression of autotrophic carbon-dioxide-fixing systems occurs (Stukas & DeCicco, 1970; Tabita & Lundgren, 1971).

We would argue that such an intermediate organism would be at considerable selective disadvantage. In fact it is interesting to note that although the unusual 'chemolithotrophic heterotroph' *Thiobacillus intermedius* and the 'metabolically maimed' *T. perometabolis* are organisms that grow at the expense of organic matter (London, 1963; London & Rittenberg, 1967), they can also employ energy from thiosulphate oxidation to effect good growth on otherwise poorly metabolised organic nutrients. Such organisms clearly have a survival advantage in being able to employ unusual supplementary energy sources, while being deficient or lacking in carbon-dioxide-fixing ability. They could of course compete with 'obligate autotrophs' in nature. Rittenberg's (1972) contention that *T. neapolitanus* (an obligate chemolithotroph) would not compete effectively with a chemolithotrophic heterotroph in any natural environment in which a reduced sulphur compound and organic mat-

ter were available is, however, debatable. The isolation of obligate lithotrophs is not difficult, indicating they survive, but the chemolithotrophic
heterotroph's survival is perhaps in competition with other heterotrophs,
so the balance of advantage of one autotrophic physiological type of
organism over another in such a situation is a fine one and probably
neither *Thiobacillus* type would have a marked advantage. This distinction of advantage enjoyed by 'pure' autotrophs and heterotrophs is
schematised in Table 1.

IS THERE A UNIQUE PROPERTY RESPONSIBLE FOR 'OBLIGATE AUTOTROPHY'?

As we have stated earlier, organisms exist which depend for growth
absolutely on inorganic oxidations and carbon dioxide as a prime
carbon source; these include both photo- and chemosynthetic organisms.
The question asked is – why? Detailed discussions have already appeared
in the literature (Smith, London & Stanier, 1967; Kelly, 1967a, 1971;
Schlegel, 1975; Anthony, 1975), but whilst many reasons have been
advanced, nothing conclusive has emerged which explains obligateness.
In an earlier section we have pointed out the selective advantage that
may be conferred on such organisms living in natural environments in
potential competition with heterotrophs; consequently we would say
that natural selection answers the question 'why obligateness?'. The
real question seems to us – how is obligateness achieved?

Four principal explanations have been proposed in one form or
another from time to time:

(*a*) Metabolic 'lesions': the lack of a key enzyme central to chemo-
organotrophic growth.

(*b*) A regulatory dependence on carbon-dioxide-based metabolism:
balanced biosynthesis from a primary carbon source other than carbon
dioxide is not attainable.

(*c*) Failure to grow on organic matter because of toxicity of the substrate or through production of autoinhibitory materials from the organic substrates.

(*d*) Inability to obtain sufficient energy from heterotrophic substrates.

(*a*) and (*c*) represent distinct hypotheses, while (*b*) could impinge on
both those areas. Concerning (*a*), various workers have sought to find a
single (i.e. one-point, unique) metabolic explanation, common to all
'obligate autotrophs'. If such existed, it would establish something of a
prima facie case for obligate autotrophy as a metabolic mode.

This hypothesis we now examine critically. Two 'lesions' in particular have received much attention. The first, that obligate autotrophs lack NADH oxidase and thus cannot generate energy by its oxidation (Smith et al., 1967), can be dismissed at once as erroneous – see Schlegel (1975). That NADH oxidation cannot be linked to ATP synthesis in the electron transport chain has been suggested, but if anything the evidence is contrary to this view (Peck, 1968; Suzuki, 1974). The second reason could be that the absence of 2-oxoglutarate dehydrogenase converts the Krebs cycle into a purely biosynthetic pathway (Kelly, 1967b; Smith et al., 1967; Schlegel, 1975) and makes it unfit as an energy-generating process. Failure to develop as aerobic heterotrophs for this reason has been accorded to various thiobacilli, nitrifying bacteria and blue-green algae. This, however, is a rather oversimple interpretation of the metabolic situation in the obligate autotrophs, since there is no obligatory reason for the lack of a functional Krebs cycle to impose a failure to achieve heterotrophic growth. Numerous fermentative bacteria (such as lactic acid bacteria) lack a functional Krebs cycle, as does *Acetobacter suboxydans* (Greenfield & Claus, 1972). In the latter, and in anaerobic *Escherichia coli*, an incomplete set of cycle reactions serves a biosynthetic role as in some autotrophs (Kelly, 1971). The absence of 2-oxoglutarate dehydrogenase *in itself* does not impose obligate autotrophy, but obligate autotrophy can result if the 'metabolic environment' into which a 'lesion' such as lack of this enzyme is introduced is such that no alternative metabolic process can be growth-supporting.

To extend this concept, consider first the carbon metabolism in an autotroph. Carbon from the Calvin cycle supplies carbon skeletons into the Krebs cycle as 4C and 2C units, leading around one route of the biosynthetic 'horseshoe' to glutamate and around the other to succinate and its derivatives (Kelly, 1971). The unique features of the obligate autotrophs containing this system appear only to be that carbon dioxide is the prime carbon source and that energy is from lithotrophic sources. In principle it would seem that carbon dioxide could be replaced by a number of compounds derived from it in the Calvin cycle and related primary pathways (e.g. sugars, C_3 and C_4 acids). The lithotrophic energy source could perhaps be replaced by the Embden–Meyerhof reactions providing NADH, substrate-level phosphorylation and, conceivably, oxidative phosphorylation during NADH oxidation. In counter to this hypothesis, no such successful adaptation is easily achieved, and carbon dioxide cannot be readily replaced by organic nutrients even *with* the lithotrophic energy source available. Note, however, that the feeble facultative heterotroph *Nitrobacter* (90 h generation time on acetate, 144 h generation time on formate) can develop better on nitrite with acetate than with carbon dioxide, although its growth is still very slow (M. Eccleston & R. Whittenbury, unpublished observations).

Heterotrophic growth on acetate or any substrate capable of metabolic conversion to acetate would seem a theoretical possibility for any autotroph even lacking 2-oxoglutarate dehydrogenase if it possessed a functional glyoxylate cycle, which in principle could fulfil all the needs for energy and carbon skeletons exhibited for successful growth. For any obligate chemolithotroph one might argue with conviction that lack of 2-oxoglutarate dehydrogenase will be a factor in determining obligate autotrophy *if the organism also lacks* the glyoxylate cycle enzymes.

Neilson, Holm–Hansen & Lewin (1972) described a mutant of *Chlamydomonas* (wild-type normally able to grow heterotrophically on *acetate*) that became obligately autotrophic because of the loss of isocitrate lyase. This special case produced obligate autotrophy even though a functional Krebs cycle was present, as acetate could not be converted to C_4 acids.

In trying to pinpoint a single metabolic locus as a key to explaining obligate autotrophy, one must recognise the significance of other metabolic deficiencies or special features that may act in concert with the specific lesion being defined. This leads one back to three considerations: (*a*) the possibility of unique control mechanisms in some autotrophs, (*b*) the relative invariability of major enzymes in autotrophs other than facultative ones, (*c*) the possibility of metabolite autotoxicity.

Some indication of metabolic variability and adaptability can be obtained from looking at enzyme levels in facultative autotrophs and at labelling patterns in metabolites of [^{14}C]acetate fed to obligate and facultative autotrophs. The absence of a functional tricarboxylic acid (TCA) cycle explains the restricted metabolism of acetate by the obligate chemolithotrophs and may be a contributory cause also of obligate methylotrophy (Kelly, 1971; Taylor & Anthony, 1976). An example of extreme metabolic versatility among autotrophs is given by *Thiobacillus* A2 (Taylor & Hoare, 1969; Peeters, Liu & Aleem, 1970), which is able to grow chemolithotrophically or heterotrophically on a number of organic compounds. When grown autotrophically on thiosulphate or formate it lacks 2-oxoglutarate dehydrogenase and shows preferential incorporation of [^{14}C]acetate into leucine and the glutamate family amino acids (Table 2), but does not show the clear 'autotrophic pattern' of labelling characteristic of the obligate chemolithotrophs. When grown heterotrophically a complete TCA cycle can function and more universal

Table 2. *Distribution of ^{14}C from [U-^{14}C]acetate incorporated into some amino acids in the protein of Thiobacillus A2 cultured heterotrophically or autotrophically*

| | Distribution (%) of ^{14}C among amino acids in the hydrolysed protein | | | | | Comparative results for: | |
| | Heterotrophic on: | | | Autotrophic on: | | Methylococcus capsulatus on CH_4 | Thiobacillus neapolitanus on $S_2O_3^{2-} + CO_2$ |
	Acetate	Sucrose	Succinate	Formate	Thiosulphate		
Glutamate	14.14	28.01	28.23	31.68	28.96	36.77	38.4
Aspartate	10.02	12.72	4.46	7.11	10.17	1.56	0
Leucine	11.68	11.89	16.38	16.32	15.74	24.16	24.2
Proline	6.81	12.21	12.33	13.77	10.97	17.00	17.5
Arginine	7.45	15.53	14.90	13.22	11.29	13.41	19.9
Alanine	10.15	1.34	4.22	1.98	3.80	0.56	0
Phenylalanine	4.50	0.53	2.29	1.11	1.39	1.32	0
Threonine	3.91	4.69	1.29	2.72	3.98	0.36	0

Table 3. *Effect of organic additives on steady-state chemostat biomass of obligate chemolithotrophs*

	Dilution rate $(h^{-1})^a$	Additives	% increase in steady-state biomass
Thiomicrospira pelophila[b]	0.1 ($S_2O_3^{2-}$ L) ⎫ 0.1 (CO_2 L) ⎭	Acetate and succinate	24–26 27–50
Thiobacillus neapolitanus[b]	0.1 ($S_2O_3^{2-}$ L) 0.1 (CO_2 L)	Acetate Acetate	11–19 15–17
Thiobacillus neapolitanus[c]	0.135 ($S_2O_3^{2-}$ L) 0.135 ($S_2O_3^{2-}$ L)	Acetate (1 mM) 1 mmol each of acetate, glycine, glutamate and aspartate	9 43

[a] L indicates growth-limiting substrate.
[b] Kuenen & Veldkamp (1973).
[c] D. P. Kelly (unpublished data).

labelling patterns are produced (Table 2). The introduction of acetate-carbon into the aspartate family amino acids during growth on thiosulphate or formate (during which formate-carbon and carbon dioxide are equivalent and are assimilated by a common autotrophic mechanism; D. P. Kelly, unpublished observations) is presumably due to the presence of the functional glyoxylate cycle in autotrophic cultures. The similarity of labelling pattern in autotrophic or succinate-grown organisms may indicate also equivalence of the Calvin cycle to a C_4 acid generating system. The danger of attempting to explain obligate chemolithotrophy in terms of the simple lack of one or two enzymes is further illustrated by the fact that *Thiobacillus denitrificans*, which lacks 2-oxoglutarate dehydrogenase, possesses a constitutive glyoxylate cycle but is still obligately chemolithotrophic (Peeters *et al.*, 1970).

There is little evidence that obligate autotrophs show much variation in enzyme content in response to their external environment. Taylor & Hoare (1971) showed that culturing *T. denitrificans* on thiosulphate with 20 mM acetate only reduced sulphite oxidase by 31% and ribulose bisphosphate carboxylase by 26%, while these enzymes were reduced 73% and 98% in *Thiobacillus* A2.

Similarly, Kuenen (1972) and Kuenen & Veldkamp (1973) found ribulose bisphosphate carboxylase levels to be little affected by organic additives.

Not unexpectedly steady-state biomass (and yield) levels of obligate chemolithotrophs are raised in continuous flow chemostat cultures growing on limiting carbon dioxide or inorganic energy source that are subsequently supplemented with organic nutrients (Table 3). Such stimulation shows *only* that the organic carbon is entering metabolism via existing pathways in the organisms thus exerting a 'saving effect' on the chemolithotrophic energy metabolism. It does not in any way represent a step towards facultative heterotrophy in the named bacteria. Similar batch culture stimulation of *Chlorobium* by acetate or glucose occurs.

The autotoxicity theory

The idea that organic matter might be inhibitory to autotrophs can be traced right back to Winogradsky's 'Anorgoxydant' discussions (Rittenberg, 1969, 1972; Kelly, 1971) and was developed by Umbreit and his collaborators in the decades 1940–70. This view has in recent years evolved from the 'submarine' hypothesis, by which autotrophs excluded all organic matter, to the later view that organic substances (such as glucose) could be metabolised and might even support some growth, but rapidly became non-metabolisable through the generation of toxic intermediary metabolites (Pan & Umbreit, 1972). To counteract this autotoxicity Pan & Umbreit employed dialysis culture techniques, whereby alleged toxic products were continuously removed. By such means they were able to obtain growth of a number of supposedly obligate autotrophs (Pan & Umbreit, 1972). Although the amounts of growth obtained were small (15–80 μg protein per ml in 5–20 days depending on the organisms) and the growth rates relatively slow, there seems no doubt that growth *was* obtained. The question remains: on what?

It seems unlikely that sufficient oxidisable inorganic substrates would have been available in these dialysis systems but this possibility has not been rigorously excluded. The hypothesis remains relatively unassailed that a toxic product from the glucose supplied is being removed by the dialysis. The nature of this product remains uncertain. A number of candidates have been proposed, including pyruvate or keto acids such as 2-oxoglutarate and *p*-hydroxyphenylpyruvate, all of which are accumulated by autotrophic thiobacilli. As Umbreit has admitted, a common inhibitory substance may not be produced by all autotrophs. Other possible inhibitory materials could be in one of two classes:

(*a*) Materials that are overproduced because of the different regulatory processes possibly prevailing during growth on organic carbon rather than carbon dioxide. (Numerous amino acids in excess are harmful to many autotrophs.)

(*b*) Materials accumulated by pathways only operating to a large extent because of the central metabolism being based on a single organic substrate rather than operating autotrophically.

An example of the latter case forms the basis of a theoretical model for autoinhibition proposed by Higgins & Slater in 1975 (J. H. Slater, personal communication). They postulated that a common toxic metabolite causing obligate autotrophy could be methylglyoxal, which is toxic to many organisms. There are several possibilities for the formation of methylglyoxal in the bacteria (non-enzymatically from trioses; enzymatically from dihydroxyacetone phosphate; from threonine catabolism) particularly during sugar metabolism. The absence of mechanisms for the further catabolism of methylglyoxal (e.g. the glyoxylase system or 2-oxoaldehyde dehydrogenase) could lead to the ready accumulation of methylglyoxal to toxic levels. The obligate methylotroph *Methylosinus* and the phototrophs *Chromatium* and *Anacystis* lacked such systems, while the facultative methylotroph *Pseudomonas extorquens* and photoheterotrophic *Rhodospirillum rubrum* contained them. This is an attractive hypothesis worthy of further detailed study.

There have been reports of the growth of obligate autotrophs (e.g. *Nitrobacter*) being stimulated by the presence of heterotrophic organisms in culture with them. In some cases this has been attributed to the excretion of organic trace (or other) nutrients by the heterotroph. An alternative explanation could be the *consumption* by the heterotroph of auto-toxic compounds secreted by the autotroph (even autotrophic growth was stimulated by dialysis culture in Pan & Umbreit's work). This could explain the occasional proof that cultures of autotrophs may contain a heterotroph (e.g. the unusual *Thiobacillus ferrooxidans/T. acidophilus* combination (Guay & Silver, 1975) in which the latter is presumed to develop on excreted organic matter from the former) and the reputed difficulty of isolating *T. denitrificans* without also isolating a consortium of interdependent heterotrophs (J. G. Kuenen, personal communication).

What is clear, as a result of this discussion, is that 'obligate autotrophy' cannot yet be shown to have resulted from a unique evolutionary event – indeed, the more we learn about autotrophs the more likely it seems that different organisms will be found to be obligate for different reasons. Leaving aside for a while the concept of conventional autotrophy, we now wish to turn to the question of broadening the scope of autotrophy – to see whether in the light of recent microbiological studies the concept needs to be recast and whether other explanations can be advanced to answer the problems discussed earlier.

The key and irreducible features of autotrophs: how such features
lead to the broadening of the scope of autotrophy

In attempting to broaden the idea of what is an autotrophic organism, the relevant and irrelevant components of the widely accepted notion of an autotroph need to be identified. This notion is that autotrophs are organisms which obtain energy from the oxidation of reduced inorganic chemicals and fix carbon dioxide by the ribulose bisphosphate carboxylase pathway as their prime and indispensable carbon source. Both features – energy production and carbon source – are considered below.

Energy source

The first point to make is that there is *no* shared mechanism of inorganic chemical oxidation amongst autotrophs. The different substrates (NO_2^-, NH_4^+, reduced S compounds, H_2, Fe^{2+}) are all oxidised by different enzyme complexes and pathways; for instance, NH_4^+ requires a mixed function oxygenase to incorporate $\frac{1}{2}O_2$ into the compound to form hydroxylamine *en route* to NO_2^-; H_2 reduces NAD^+ and the product, NADH, is oxidised via the electron transport chain; NO_2^- and reduced sulphur compounds release electrons on oxidation at potentials too high to reduce NAD^+, hence NADH for biosynthesis is provided via ATP – dependent reversed electron transport (see Aleem, this volume). The second point is that inorganic chemicals are oxidised by organisms considered to be conventional heterotrophs; e.g. *Desulfovibrio* and *Desulfotomaculum* species oxidise hydrogen, various pseudomonads oxidise thiosulphate to tetrathionate, strains of actinomycetes oxidise NH_4^+ to nitrite. Some of these oxidations clearly result in energy production for the heterotrophs concerned, e.g. *Desulfovibrio* species generate ATP as a result of SO_4^{2-} reduction to H_2S with H_2. We can conclude, therefore, that reduced inorganic compound oxidation is not a single process, that it is not a magical property restricted to autotrophs, and that it is not necessarily a key feature of autotrophy.

Carbon source

The other major feature of autotrophs is the possession of the Benson–Calvin pathway of carbon dioxide fixation. All autotrophs fix carbon dioxide via the one major pathway, whether they be photo- or chemoautotrophs. Other routes of carbon assimilation have been proposed from time to time (e.g. the reversed TCA pathway in *Chlorobium* species) but to date the ribulose bisphosphate pathway is the proven major and *indispensable* mechanism of carbon assimilation. Secondary carbon sources, such as acetate, can be assimilated and be incorporated into cellular components, but only when carbon dioxide is being assimilated as the prime carbon source. Consequences of this mode of carbon assimilation, for instance the incomplete TCA cycle, are discussed later in another context.

The position we now adopt is that the unique property common to all autotrophs is the carbon-dioxide-fixing mechanism. Consequently in widening our view of what are autotrophs, it is this feature which we have in mind. A consideration of all C_1 compound utilising microbes – their properties in general and carbon compound synthesis pathways in particular – leads us to propose that all of them should be involved in a new and wider definition of autotrophy. The present definitions are the result of accidents of knowledge at the time they were made. Recent studies on C_1 compound utilisers indicate quite clearly that the time is right to look at autotrophy in a detached manner and in the light of the new knowledge. Arguments presented below convince us that groupings based on C_1 compound metabolism reinforce the viewpoint stated earlier that carbon compound assimilation pathways override previous considerations in discussions on relationships of autotrophs, as presently defined, with other microbes, and that the concept of autotrophy is still of value in pinpointing major principles of microbial nutrition.

C_1 COMPOUND UTILISING MICROBES

The organisms collected under this heading all utilise C_1 compounds as reductants or anaerobic oxidants for energy generation. They are heterotrophs or photoheterotrophs in the presently accepted sense, with the exception of the carbon dioxide reducers which are autotrophs as presently defined.

The properties of these microbes highlight, in our opinion, the importance of C_1 compound pathways as being the key to autotrophy and also demonstrate that the organic or inorganic nature of the energy source is not of prime interest in defining autotrophy. The different groups are described below and their relevance to this discussion indicated.

Anaerobic carbon dioxide reducers

Two groups of bacteria, the methane producers and the acetic acid producers, are partly composed of certain species which can grow in a totally inorganic medium and reduce carbon dioxide (acting as an oxidant) at the expense of an energy source, hydrogen. The latter group, clostridia able to form acetic acid from carbon dioxide and hydrogen (Wieringa, 1940), are not well documented and will not be considered further, although it is worth mentioning that other clostridia are well described (Andreesen *et al.*, 1973) which reduce carbon dioxide via formate to acetic acid at the expense of NADH formed from glycolysis.

The methane producers growing on hydrogen and carbon dioxide fulfil all the requirements of autotrophy, except in recent definitions where the presence of a Benson–Calvin cycle (ribulose bisphosphate pathway) is stipulated. No evidence is available as yet to say exactly how these organisms synthesise their carbon compounds from carbon dioxide, though there is some circumstantial radioisotope evidence (Taylor, Kelly & Pirt, 1975) indicating that a 'serine pathway' (see Quayle, 1972) may be the route of carbon into the biosynthetic mechanisms of the cell. What is clear is that carbon dioxide is not *directly* condensed with a C_5 or other carbon compound, but is reduced to methane via various C_1 intermediates. The actual carbon source assimilated is probably formaldehyde.

It can be argued, therefore, that methane producers whilst adopting an autotrophic mode of life in one sense (growing on inorganic substrates) are in another sense (using formaldehyde or other reduced intermediate *en route* to methane) heterotrophs in the outward guise of autotrophs. The value of these organisms in this discussion on autotrophy is that they point to the idea that *mechanisms of metabolism rather than the nature (organic or inorganic) of substrates are of significance.* This particular point in respect of the methane producers is emphasised by the following groups of microbes.

Anaerobic methane oxidisers

These bacteria have been isolated for the first time in the laboratory of Professor R. Hanson and his colleagues in the past year at the University of Wisconsin (R. Hanson, personal communication). Different strains appear to have different carbon assimilation pathways; all appear to oxidise methane anaerobically with sulphate as the oxidant, the end-products being carbon dioxide and hydrogen sulphide. One strain at least obtains carbon for biosynthesis from an intermediate, possibly formaldehyde, of methane oxidation. In essence, such bacteria seem to be similar in metabolic principle to the methane producers in that both draw upon a reduced one-carbon compound for biosynthesis – the methane oxidisers from CH_4 being oxidised to CO_2 and the methane producers from CO_2 being reduced to CH_4. Energy formed is by a chemolithotrophic process in one case and by a chemo-organotrophic process in the other. This particular difference, however, seems fundamentally secondary in significance to the C_1 compound assimilation process which may prove identical in both cases. This point about energy sources is brought very clearly into focus in the next group of microbes discussed.

Formate and methanol utilisers which fix carbon dioxide as their prime carbon source

Three types of microbe are considered here: (a) *Rhodopseudomonas* species (Stokes & Hoare, 1969) which use methanol as electron donor in photosynthesis; (b) *Paracoccus denitrificans* (Cox & Quayle, 1975) which uses methanol as an energy source; and (c) *Pseudomonas oxalaticus* (Quayle, 1961, 1972) which uses formate as an energy source. All use other electron donors or energy sources but the significant feature in the cases just mentioned is that carbon for biosynthesis is from carbon dioxide fixed by the Benson–Calvin pathway; the substrates, methanol and formate, are used solely as electron donors (by the photosynthetics) or energy sources.

Other points about these particular organisms are that:

(a) The phototrophs can behave as 'conventional' autotrophs using hydrogen or hydrogen sulphide as electron donors, or as photoheterotrophs.

(b) *Paracoccus denitrificans* is metabolically a very versatile microbe, capable of conventional autotrophic growth with hydrogen as the electron donor and also of heterotrophic growth (Knobloch, Ishaque & Aleem, 1971).

(c) *Pseudomonas oxalaticus* only grows heterotrophically; however, when formate is utilised the organism is effectively an autotroph using the hydrogen component as the electron donor and fixing the resultant carbon dioxide by the ribulose bisphosphate pathway. In other words the formate is treated metabolically as hydrogen and carbon dioxide.

All these examples reinforce the notion that autotrophy has to be redefined in terms of carbon assimilation pathways; in all cases the carbon compound oxidised served only as an energy source, as do inorganic chemicals in 'conventional' autotrophy. This idea is now extended to the last group of bacteria being considered here; they do not possess the ribulose bisphosphate mechanism of carbon dioxide fixation but assimilate their carbon, in the known cases, as formaldehyde.

C_1 compound utilisers which assimilate reduced C_1 compounds as
their carbon source

Many microbes including yeast species and filamentous fungi come into
this category. Some, such as formate and/or methanol-utilising species
of *Bacillus*, *Caulobacter*, *Assticacaulis* and the 'mushroom-shaped
bacterium' (a budding bacterium; Whittenbury & Nicoll (1971)), still
remain unexplored as far as their C_1 metabolism is concerned. However,
a great deal has been learned about the carbon assimilation pathways of
the methane oxidising bacteria and the pseudomonads, hyphomicrobia
and yeasts which oxidise methanol, the *N*-methyl compounds and other
C_1 compounds. As with the previous group of organisms, Professor
Quayle and his colleagues at Sheffield have been the major contributors
to the unravelling of the pathways of carbon assimilation in these cases.

Two major pathways have emerged, the 'ribulose monophosphate
pathway' and the 'serine pathway' (e.g. Quayle, 1972). Both have variants but the essential features, outlined in Fig. 1, are that formaldehyde
formed in the oxidation of the C_1 compounds serves as the prime carbon
source for biosynthesis, and that carbon dioxide, acetate and other
carbon compounds serve as secondary carbon sources.

The 'serine pathway' clearly has no obvious similarity to the Benson–
Calvin pathway and will not be discussed further at this stage. The
'ribulose monophosphate' pathway – starting with a $C_1 + C_5$ condensation (formaldehyde and ribulose monophosphate) – is clearly
analogous to the $C_1 + C_5$ condensation process which initiates the Benson–Calvin cycle. It is found in hyphomicrobia, pseudomonads and
yeasts, but was first detected in the obligate methane oxidisers (Anthony,
1975) and it is these last organisms that we will compare in more detail
with the autotrophs to bring out other similarities which add to the
case of considering organisms with the 'ribulose monophosphate'
pathways to be part of the autotrophic complex of microbes.

The methane oxidisers

The obligate methane oxidisers fall into two distinct groups as judged
by biochemical, physiological, morphological and ultrastructural studies
(Whittenbury, Dalton, Eccleston & Reed, 1975). One group assimilates
formaldehyde resulting from methane oxidation via the 'ribulose monophosphate pathway' and the other via the 'serine pathway'. A sub-
group of the 'ribulose monophosphate' group – *Methylococcus capsul-*
latus – appears to be a link in some ways between the two main groups;
the curious feature is the presence in this organism of enzymes involved

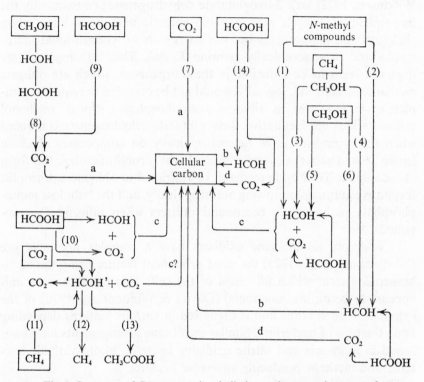

Fig. 1. Summary of C_1 compound assimilation pathways and sources of assimilated carbon in various organisms.

(1) *Diplococcus* PAR; (2) *Bacillus* PM6; (3) *Methylosinus trichosporium*; (4) *Methylomonas methanica*; (5) *Pseudomonas* AM1; (6) *Pseudomonas methylotropha*; (7) *Nitrosomonas europea*; (8) *Paracoccus denitrificans*; (9) *Pseudomonas oxalaticus*; (10) *Pseudomonas* AM1; (11) anaerobic methane oxidiser; (12) anaerobic methane producers; (13) *Clostridium aceticum*; (14) *Bacillus* PM6 (Colby & Zatman, 1975). a, ribulose bisphosphate pathway; b, ribulose monophosphate pathway; c, serine pathway; d, e.g. phosphoenolpyruvate carboxylase (presumed). Boxes represent initial sources of carbon.

in both assimilation pathways (Whittenbury *et al.*, 1975), with the implication (H. L. Reed, unpublished observations) that the 'serine pathway', as well as the 'ribulose monophosphate pathway' may contribute significantly at high growth temperatures to the assimilation of formaldehyde.

All these organisms require carbon dioxide as an essential carbon source *in addition* to formaldehyde.

Similarities of the 'ribulose monophosphate' methane oxidisers to 'conventional' autotrophs are quite marked in a number of respects. For example all the strains of this group tested (Davey, Whittenbury &

Wilkinson, 1972) lack 2-oxoglutarate dehydrogenase; consequently the incomplete TCA cycle acts as a biosynthetic unit, providing carbon skeletons and amino acids but not energy, as in 'conventional' autotrophs and in anaerobically growing *E. coli*. This 'missing' enzyme does not seem to be inducible in these organisms, which are obligate methane utilisers and, logically, would not be expected to require a complete cycle. However, in 'ribulose monophosphate pathway' methanol utilisers which are facultative, 2-oxoglutarate dehydrogenase is induced when the organisms grow heterotrophically on compounds of chain length 2C and above as energy sources. In the aerobic microbes, this form of incomplete TCA cycle seems to be restricted to obligate autotrophs, facultative autotrophs growing autotrophically, and the 'ribulose monophosphate pathway' C_1 compound utilisers when utilising C_1 compounds.

In addition, all methane oxidisers have a complex fine structure (Whittenbury *et al.*, 1975) the most prominent feature being the membraneous system which fills most of the cell. This membraneous unit appears to be an invagination(s) (Davies & Whittenbury, 1970) of the cytoplasmic membrane and is organised in various patterns depending upon the type of bacterium. Similar membrane arrangements have been found in ammonia and nitrite oxidizing bacteria, photosynthetic bacteria, and methane producing anaerobic bacteria.

The interesting point here is that certain autotrophs and the methane oxidising bacteria share this membraneous development. What the function of these membranes is remains unproven. However, it is possible to make an educated guess by considering the physiology of the organisms concerned.

Photosynthetic bacteria (e.g. Holt, Conti, Fuller, 1966; O. T. G. Jones, this volume) have been shown to have a membrane and chlorophyll content in an inverse ratio to the intensity of illumination at which they are grown, i.e. at low light (light-limiting) intensities, chlorophyll and membrane content is high, whereas at saturating light intensities chlorophyll and membrane content are considerably reduced. This implies that the organism is able to adjust – within limits – its rate of ATP synthesis per cell by varying these two components according to the photon intensity, but more importantly, points to the idea that increased membrane content over that of the cytoplasmic membrane is a way of overcoming kinetic problems of one sort or another related to poor and/or very low concentrations of an energy source or electron donor.

The nitrite oxidisers are probably amongst the least favoured of bacteria in respect of the high energy demand for their biosynthesis coupled with low energy yield resulting from the oxidation of the substrate nitrite to nitrate. This situation is well described by Aleem (Aleem, 1970; Aleem, this volume). NADH and ATP requirements for carbon dioxide fixation, the synthesis of NADH at the expense of ATP and reduced cytochrome c and other demands are set against the energy yield from nitrite and, logically, such an organism would seem to need a vast number of nitrite oxidising sites in order to compensate for the low energy yielding oxidations by high turnover rates. As nitrite oxidising and associated ATP synthesising systems are located in the membrane a proliferation of such membranes would be an advantage to such bacteria as compared with a nitrite oxidising cell with a simple cytoplasmic membrane. Turnover rates in the same cell with multiple membranes could be increased twenty-fold or more as judged by the membrane content of nitrite oxidisers in addition to the cytoplasmic membranes. Even so, nitrite oxidisers are slow growers (20+ h generation times) and nitrate accumulation is pronounced before any turbidity is observed in cultures.

Ammonia oxidisers pose similar problems to the nitrite oxidisers as regards energy production and consumption for biosynthesis (e.g. carbon dioxide fixation and NADH synthesis). Even though ammonia oxidation releases more energy per mole than does nitrite oxidation, an additional energy-demanding reaction is necessary in the hydroxylation of ammonia to its presumed first intermediate, hydroxylamine. Assuming this to be a conventional mono-oxygenase type of reaction, NADH is expended in this reaction without any return in the form of ATP synthesis (e.g. Watson, Graham, Remsen & Valois, 1971).

Perhaps the ammonia oxidisers (of all autotrophs) and the 'ribulose monophosphate pathway' methane oxidisers have the closest *apparent* relationship, because in addition to having in common a complex fine structure, an incomplete TCA cycle, the ability to assimilate C_1 compounds by mechanisms similar in principal (a C_5+C_1 condensation), methane oxidisers are also able to oxidise ammonia to nitrite – as do the ammonia oxidisers (Whittenbury, Phillips & Wilkinson, 1970). Although it is not yet clear whether 'useful' energy is released in this oxidation, other 'co-oxidations' of similar type (Whittenbury *et al.*, 1975) do so. The methane mono-oxygenase enzyme complex appears to be the same one which oxidises ammonia (Whittenbury *et al.*, 1970).

Although we have gone to some length to construct this similarity between ammonia oxidisers and the methane oxidisers, we realise that arguments may be advanced to illustrate that these similarities may reflect convergence rather than common ancestry.

However, we feel the methane oxidisers we have described, and the other organisms we have referred to in this same context, show basic similarities of principle, if not of the exact mechanisms of metabolism, with the conventional autotrophs and that it would be unrealistic to ignore this fact in reassessing the concept of autotrophy. Perhaps this can be summarised, in one example. Methane oxidisers can grow on hydrogen and formaldehyde in continuous flow systems (R. Whittenbury & H. L. Reed, unpublished observations) using the 'ribulose monophosphate pathway' for both the assimilation of formaldehyde and the fixation of carbon dioxide resulting from the oxidation of formaldehyde, but not the Benson–Calvin pathway; hydrogenomonads growing on hydrogen and carbon dioxide using the 'ribulose bisphosphate pathway' to assimilate carbon dioxide are already obtaining their energy and carbon in a very similar way to these methane oxidisers. Both must share many biosynthetic mechanisms in common. This latter point, of course, can be made for all microbes in relation to each other and also serves to warn against the temptation to overemphasise differences between groups of microbes. Our tentative conclusion is that in redefining autotrophy, all C_1 compound utilisers should be within the concept.

NEW WINE IN OLD BOTTLES – THE 'OLD' AND THE 'NEW' AUTOTROPHS

Perhaps the best way to begin to illustrate our view of the revised concept of autotrophy is to present the transition of our ideas in a series of tables. All through this discussion we have deliberately refrained from descending into the minutiae of the biochemistry concerned as this has been extensively done in a number of reviews on autotrophy (Rittenberg, 1969; Kelly, 1971; Schlegel, 1975) and methylotrophy (Quayle, 1972; Wolfe, 1972; Anthony, 1975) and also because such a treatment would obscure the points we are trying to make about generalised concepts of C_1 compound metabolism.

Table 4 presents the conventional view of autotrophs as purveyed in text-books and in review literature. Table 5 groups organisms according to their energy sources and carbon assimilation pathways. It is concerned with chemolithotrophy – part of conventional autotrophy – and organic and inorganic carbon assimilation and is designed to show up the incongruities which bedevil the present definitions of autotrophy.

Table 6 is a classification of organisms based upon our view that carbon assimilation pathways, not energy-yielding mechanisms, or the inorganic or organic nature of the energy sources, are of paramount importance in redefining autotrophy.

Table 4. *Classification of the 'old' autotrophs (can use carbon dioxide as sole/main carbon source)*

Energy substrate	Examples	2-oxoglutarate dehydrogenase present (+) or absent (−)	Obligate (O)/ facultative (F) growth potential
Hydrogen (aerobic)	Many bacteria (e.g. Schlegel, 1966; Davis, Doudoroff & Stanier, 1969)	+	F
Sulphur and reduced sulphur compounds	*Thiobacillus thiooxidans*	−	O
	T. thioparus	−	O
	T. neapolitanus	−	O
	T. novellus	+	F
	T. acidophilus	?	F
	T. A2	+/−[a]	F
Ferrous iron	*Thiobacillus ferrooxidans*	+/−[a]	O/F?
	Gallionella sp.		
Ammonia	*Nitrosolobus multiformis*	−	O
Nitrite	*Nitrobacter agilis* (Winogradsky)	+	F
Hydrogen (anaerobic)	*Methanobacterium thermoautotrophicum*	?	O
	Methanobacterium formicicum	?	F

[a] Inducible in those strains growing on organic compounds but negative when growing autotrophically.

Table 5. *Categories of organisms according to the organic or inorganic nature of their energy and carbon sources, and facultative or obligate ability*

Nature of energy and carbon sources	Examples
Inorganic energy and inorganic carbon sources	*Thiobacillus thioparus*
Inorganic energy and inorganic carbon, inorganic energy+organic carbon, organic energy+organic carbon	(a) Most rapid growth on inorganic energy +organic carbon: *Thiobacillus intermedius* *Nitrobacter winogradskyi* (b) Most rapid growth on organic energy+ organic carbon: *Paracoccus (Micrococcus) denitrificans* *Pseudomonas saccharophila*
Inorganic energy+organic carbon, organic energy+organic carbon	(a) Most rapid growth on inorganic energy: *Thiobacillus perometabolis* (b) Most rapid growth on organic energy: *Desulfovibrio* spp.
Organic energy+organic carbon, *but* autotrophic mode of metabolism	*Pseudomonas oxalaticus* (formate) *Paracoccus (Micrococcus) denitrificans* (methanol) *Thiobacillus A2* (formate) *Methanomonas methanica* (methane)
Organic energy+organic carbon *but* mode of metabolism akin to autotrophy Organic energy+organic carbon using 'serine' pathway	*Methylosinus trichosporium* (methane)
Inorganic energy+inorganic carbon, *but* inorganic carbon (carbon dioxide) serves initially as an oxidant	*Methanobacterium* sp. (anaerobic formation of methane from hydrogen and carbon dioxide)

Table 6. *Classification of organisms as autotrophs based on carbon assimilation occurring via a $C_1 + C_5$ condensation*

Cell carbon via either 'ribulose bisphosphate pathway'
or 'ribulose monophosphate pathway'

	Energy source
Obligate	
Nitrosomonas winogradskyi	NH_4^+
Thiobacillus neapolitanus	$S_2O_3^{2-}$
Methanomonas methanica	CH_4
Facultative	
(a) Non-NAD$^+$ reducing oxidations:	
Nitrobacter winogradskyi	NO_2^-
Thiobacillus novellus	$S_2O_3^{2-}$
(b) NAD$^+$ reducing oxidations:	
Pseudomonas saccharophila	H_2
Pseudomonas oxalaticus	H.COOH

Table 7. *Autotrophy seen as embracing all organisms which are able to assimilate C_1 compounds as their sole carbon source*

Categories of assimilation	Carbon source	Example
'Ribulose bisphosphate pathway'	CO_2	*Nitrosomonas europaea* (growing on NH_4^+)
'Ribulose monophosphate pathway'[a] *and* a heterotrophic (possibly phospho-enolpyruvate carboxylase CO_2 fixing mechanism)	$HCHO + CO_2$	*Methylomonas methanica* (growing on CH_4 or CH_3OH)
'Serine pathway'[a] and one or more heterotrophic CO_2 fixing mechanisms as well as CO_2 fixing activity within serine pathway	$HCHO + CO_2$	*Methylosinus trichosporium* (growing on CH_4 or CH_3OH)
Unknown pathways in anaerobic CO_2 reducing microbes	CO_2	(a) *Methanobacterium* spp. (growing on H_2) (b) *Clostridium aceticum* (growing on H_2)

[a] There is more than one variant of the main pathway.

Table 7 is an expanded view of autotrophy which illustrates the proposition that all microbes primarily dependent on C_1 compounds as carbon sources should be brought under the umbrella of autotrophy – whether such organisms are obligate or facultative. This latter item we consider of minor importance in autotrophy as no unique feature can account for obligateness, whether it be in conventional autotrophs or methylotrophs (some use only methane, some only methanol or other C_1 compound as an energy source) and even if there were such a feature it would still not detract from the central feature of C_1 compound assimilation.

Below we construct a new concept of autotrophy in the light of the discussion that we have developed.

A NEW DEFINITION OF AUTOTROPHY

In our view this term should be retained in the vocabulary with, if anything, a *broadened* meaning. While bastardisation of language by casual misuse is to be decried, usage and usefulness over the decades is good cause for acceptance in common technical parlance (Gowers, 1954). In our use an autotroph should be able to obtain the bulk of its biosynthetic carbon from carbon dioxide or the metabolism of a one-carbon compound. Embraced within this definition should be the conceptual acceptance that while specialised enzyme systems are the basis of autotrophic carbon assimilation and that energy to drive these involves conventional energetic biochemistry, the substrates oxidised for energy generation are unusual in that they are usually available as such substrates only to organisms possessing the specialised C_1-assimilation enzyme systems.

Thus an autotroph is an entity using a particular energy-yielding substrate to drive specialised biosynthetic mechanisms dependent on C_1 substrates such as carbon dioxide or formaldehyde.

This definition lessens in no degree the usefulness of terms such as 'chemolithotrophic' which describes precisely a particular basis for autotrophic metabolism.

In our view Rittenberg attacked the 'wrong' target in setting out to demolish 'obligate autotrophy' in the sense of Winogradsky's 'Anorgoxydant'. The concept of organisms restricted in growth physiology to unique substrates (e.g. ammonia or iron) and to carbon dioxide as a main carbon substrate, or uniquely dependent on a C_1 compound like methane, is still valid, though the submarine-like exclusion of organic nutrients or universal organic compound toxicity has not been upheld by work in recent decades. The 'obligate autotroph' *sensu* Rittenberg could still exist but could well be very difficult to find, and if found would not substantially alter our understanding of autrotropy or the force of our arguments concerning the relative capacity for survival in natural environments of such organisms. What would then be open to discussion would be the nature of the selection pressures that produced an organism adapted to exclude absolutely exogenous organic nutrients from its metabolism

CONCLUSION

The definition of autotrophy over the years has veered between two standpoints:

(1) That *autotrophs are organisms which grow on inorganic nutrients*, no specification of pathways of C_1 compound assimilation being made. Following more detailed biochemical studies on the organisms in such a category, anomalies have presented themselves; e.g. hydrogen utilisers which fix carbon dioxide are included in the same category as such metabolically dissimilar microbes as anaerobic methane producers, but are separated from organisms of overall metabolic similarity such as the formate oxidising *Pseudomonas oxalaticus*.

(2) That *autotrophs are organisms which fix carbon dioxide as their prime carbon source, via the Benson–Calvin pathway and obtain energy from the oxidation of inorganic chemical compounds*. This more recent definition still suffers from its preciseness, excluding obviously similar organisms either on grounds of the nature of the energy source and/or different (but rather similar) pathways of carbon assimilation.

Our conclusion is that all micro-organisms which are able to assimilate C_1 compounds as their prime carbon source for cellular biosynthesis, irrespective of their energy source, should be included within a new definition of autotrophy. Such a definition covering both photo- and chemotrophic organisms could be: *autotrophs are micro-organisms which can synthesise all their cellular constituents from one or more C_1 compounds*. In this context we consider that there are at least three categories of autotrophic microbes: (1) those possessing the 'ribulose bisphosphate pathway'; (2) those possessing the 'ribulose monophosphate pathway'; (3) those possessing the 'serine pathway'. Leniency in the application of this definition to metabolic cripples (e.g. vitamin requirers) is obviously good common sense. Our summary of the carbon pathway interrelationships is given in Fig. 1.

REFERENCES

ALEEM, M. I. H. (1970). Oxidation of inorganic nitrogen compounds. *Annual Reviews of Plant Physiology*, **21**, 67–90.

ANDREESEN, J. R., SCHAUPP, A., NEURAUTER, C., BROWN, A. & LJUNGDAHL, L. G. (1973). Fermentation of glucose, fructose and xylose by *Clostridium thermoaceticum*: effects of metals on growth yield, enzymes and the synthesis of acetate from carbon dioxide. *Journal of Bacteriology*, **114**, 743–51.

ANTHONY, C. (1975). The biochemistry of methylotrophic micro-organisms. *Science Progress, Oxford*, **62**, 167–206.

COLBY, J. & ZATMAN, L. J. (1975). Enzymological aspects of the pathways for the trimethylamine oxidation and C_1 assimilation in obligate methylotrophs and restricted facultative methylotrophs. *Biochemical Journal*, **148**, 513–20.

COX, E. C. & GIBSON, T. C. (1974). Selection for high mutation rates in chemostates. *Genetics*, **77**, 169–84.

COX, R. B. & QUAYLE, J. R. (1975). The autotrophic growth of *Micrococcus denitrificans* on methanol. *Biochemical Journal*, **150**, 569–71.

DAVEY, J. F., WHITTENBURY, R. & WILKINSON, J. F. (1972). The distribution in the methylobacteria of some key enzymes concerned with intermediary metabolism. *Archiv für Mikrobiologie*, **87**, 359–66.

DAVIES, S. L. & WHITTENBURY, R. (1970). Fine structure of methane and other hydrocarbon-utilizing bacteria. *Journal of General Microbiology*, **61**, 227–32.

DAVIS, D. H., DOUDOROFF, M. & STANIER, R. Y. (1969). Proposal to reject the genus *Hydrogenomonas*: taxonomic implications. *International Journal of Systematic Bacteriology*, **19**, 375–90.

GOWERS, E. (1954). *The Complete Plain Words*. London: HMSO.

GREENFIELD, S. & CLAUS, G. W. (1972). Nonfunctional tricarboxylic acid cycle and the mechanism of glutamate biosynthesis in *Acetobacter suboxydans*. *Journal of Bacteriology*, **112**, 1295–301.

GUAY, R. & SILVER, M. (1975). *Thiobacillus acidophilus* sp. nov.; isolation and some physiological characteristics. *Canadian Journal of Microbiology*, **21**, 281–8.

HOLT, S. C., CONTI, S. F. & FULLER, R. C. (1966). Effect of light intensity on the formation of the photochemical apparatus in the green bacterium *Chloropseudomonas ethylicum*. *Journal of Bacteriology*, **91**, 344–55.

HORIUCHI, T., TOMIZAWA, T. & NOVICK, A. (1962). Isolation and properties of bacteria capable of high rates of β-galactosidase synthesis. *Biochimica et Biophysica Acta*, **55**, 152–63.

KELLY, D. P. (1967a). Problems of the autotrophic microorganisms. *Science Progress, Oxford*, **55**, 35–51.

KELLY, D. P. (1967b). The incorporation of acetate by the chemoautotroph *Thiobacillus neapolitanus* strain C. *Archiv für Mikrobiologie*, **58**, 99–116.

KELLY, D. P. (1971). Autotrophy: concepts of lithotrophic bacteria and their organic metabolism. *Annual Reviews of Microbiology*, **25**, 177–210.

KNOBLOCH, K., ISHAQUE, M. & ALEEM, M. I. H. (1971). Oxidative phosphorylation in *Micrococcus denitrificans* under autotrophic growth conditions. *Archiv für Mikrobiologie*, **76**, 114–25.

KUENEN, J. G. (1972). Een studie van kleurloze zvarfelbacterien uit het Groninger Wad. Doctoral Thesis, Groningen.

KUENEN, J. G. & VELDKAMP, H. (1973). Effects of organic compounds on growth of chemostat cultures of *Thiomicrospira pelophila*, *Thiobacillus thioparus* and *Thiobacillus neapolitanus*. *Archiv für Microbiologie*, **94**, 173–90.

LONDON, J. (1963). *Thiobacillus intermedius* nov. sp. a novel type of facultative autotroph. *Archiv für Mikrobiologie*, **46**, 329–37.

LONDON, J. & RITTENBERG, S. C. (1967). *Thiobacillus perometabolis* nov. sp., a non-autotrophic *Thiobacillus*. *Archiv für Mikrobiologie*, **59**, 218–25.

NEILSON, A. H., HOLM-HANSEN, O. & LEWIN, R. A. (1972). An obligately autotrophic mutant of *Chlamydomonas dysosmos*; a biochemical elucidation. *Journal of General Microbiology*, **71**, 141–8.

PAN, P. & UMBREIT, W. W. (1972). Growth of obligate autotrophic bacteria on glucose in a continuous flow-through apparatus. *Journal of Bacteriology*, **109**, 1149–55.

PECK, H. D. (1968). Energy-coupling mechanisms in chemolithotrophic bacteria. *Annual Reviews of Microbiology*, **22**, 489–518.

PEETERS, T. L., LIU, M. S. & ALEEM, M. I. H. (1970). The tricarboxylic acid cycle in *Thiobacillus denitrificans* and *Thiobacillus A2*. *Journal of General Microbiology*, **64**, 29–35.

PFEFFER, W. (1897). *Pflanzenphysiologie*, vol. I, 2 Aufl. Leipzig: Verlag W. Engelmann.

PRINGSHEIM, E. G. (1967). Die Mixotrophie von Beggiatoa. *Archiv für Mikrobiologie*, **59**, 247–54.

QUAYLE, J. R. (1961). Metabolism of C_1 compounds in autotrophic and heterotrophic microorganisms. *Annual Reviews of Microbiology*, **15**, 119–52.

QUAYLE, J. R. (1972). The metabolism of one-carbon compounds by microorganisms. *Advances in Microbial Physiology*, **7**, 119–203.

RITTENBERG, S. C. (1969). The roles of exogenous organic matter in the physiology of chemolithotrophic bacteria. *Advances in Microbial Physiology*, **3**, 151–96.

RITTENBERG, S. C. (1972). The obligate autotroph – the demise of a concept. *Antonie van Leeuwenhoek*, **38**, 457–78.

RITTENBERG, S. C. & GOODMAN, N. S. (1969). Mixotrophic growth of *Hydrogenomonas entropha*. *Journal of Bacteriology*, **98**, 617–22.

SCHLEGEL, H. G. (1966). Physiology and biochemistry of Knallgasbacteria. *Advances in Comparative Physiology and Biochemistry*, **2**, 185–236.

SCHLEGEL, H. G. (1975). Mechanisms of chemo-autotrophy. In *Marine Ecology*, ed. O. Kinne, vol. 2, part 1, pp. 9–60. New York: Wiley.

SMITH, A. J., LONDON, J. & STANIER, R. Y. (1967). Biochemical basis of obligate autotrophy in blue-green algae and thiobacilli. *Journal of Bacteriology*, **94**, 972–83.

STOKES, J. E. & HOARE, D. S. (1969). Reductive pentose cycle and formate assimilation in *Rhodopseudomonas palustris*. *Journal of Bacteriology*, **100**, 890–4.

STUKAS, P. E. & DECICCO, B. T. (1970). Autotrophic and heterotrophic metabolism of *Hydrogenomonas*: regulation of autotrophic growth by organic substrates. *Journal of Bacteriology*, **101**, 339–45.

SUZUKI, I. (1974). Mechanisms of inorganic oxidation and energy coupling. *Annual Reviews of Microbiology*, **28**, 85–101.

TABITA, R. & LUNDGREN, D. G. (1971). Utilization of glucose and the effect of organic compounds on the chemolithotroph *Thiobacillus ferrooxidans*. *Journal of Bacteriology*, **108**, 328–33.

TAYLOR, B. F. & HOARE, D. S. (1969). New facultative *Thiobacillus* and a re-evaluation of the heterotrophic potential of *Thiobacillus novellus*. *Journal of Bacteriology*, **100**, 487–97.

TAYLOR, B. F. & HOARE, D. S. (1971). *Thiobacillus denitrificans* as an obligate chemolithotroph. *Archiv für Mikrobiologie*, **80**, 262–76.

TAYLOR, G. T., KELLY, D. P. & PIRT, S. J. (1975). Intermediary metabolism in methanogenic bacteria. *Proceedings of Congress on the Microbial Metabolism of Gases*, Göttingen. (In press.)

TAYLOR, I. J. & ANTHONY, C. (1976). A biochemical basis for obligate methylotrophy; properties of a mutant of *Pseudomonas* AM1 lacking 2-oxoglutarate dehydrogenase. *Journal of General Microbiology*, **93**, 259–65.

WATSON, S. W., GRAHAM, L. B., REMSEN, C. C. & VALOIS, F. W. (1971). A lobular, ammonia-oxidizing bacterium, *Nitrosolobus multiformis* Nov. Gen. Nov. Sp. *Archiv für Mikrobiologie*, **76**, 183–203.

WHITTENBURY, R., DALTON, H., ECCLESTON, M. & REED, H. L. (1975). The different types of methane oxidizing bacteria and some of their more unusual properties. In *Microbial Growth on C_1 Compounds*, pp. 1–9. Tokyo: The Society of Fermentation Technology.

WHITTENBURY, R. & NICOLL, J. M. (1971). A new, mushroom-shaped budding bacterium. *Journal of General Microbiology*, **66**, 123–6.

WHITTENBURY, R., PHILLIPS, K. C. & WILKINSON, J. F. (1970). Enrichment, isolation and some properties of methane-utilizing bacteria. *Journal of General Microbiology*, **61**, 205–18.

WIERINGA, K. T. (1940). The formation of acetic acid from carbon dioxide and hydrogen by anaerobic spore forming bacteria. *Antonie van Leeuwenhoek*, **6**, 251–62.

WOLFE, R. S. (1972). Microbial formation of methane. *Advances in Microbial Physiology*, **6**, 107–46.

WOODS, D. D. & LASCELLES, J. (1954). The no-man's land between the autotrophic and heterotrophic ways of life. In Autotrophic Micro-organisms, ed. B. A. Fry & J. L. Peel, pp. 1–27. *Symposia of the Society for General Microbiology*, **4**. London: Cambridge University Press.

ELECTRON TRANSPORT AND ATP SYNTHESIS IN THE PHOTOSYNTHETIC BACTERIA

O. T. G. JONES

Department of Biochemistry, University of Bristol, Bristol BS8 1TD, UK

INTRODUCTION

A study of the mechanisms of photosynthetic energy conservation in photosynthetic bacteria has proved remarkably fruitful and has increased the understanding of such processes in higher plants although, superficially, the processes of photosynthesis in eukaryotes and in photosynthetic bacteria differ widely. The most obvious difference between eukaryotic and prokaryotic photosynthesis is that illumination of green plants leads to the evolution of oxygen, whereas no oxygen is evolved by the bacteria; indeed in many cases oxygen inhibits the growth of the bacteria or prevents the development of the photosynthetic pigments. A short account of photosynthetic electron flow and energy conservation in higher plants is given below, for the purpose of comparison with bacterial systems.

CHLOROPLAST PHOTOSYNTHETIC ELECTRON FLOW

Higher plants catalyse the splitting of water to yield oxygen+electrons+protons, thus:

$$2H_2O \rightarrow O_2 + 4e + 4H^+.$$

The electrons derived from water are subsequently used for the reduction of $NADP^+$, and the NADPH in turn is used by the enzymes of carbon dioxide fixation.

Two photosystems (Photosystems I and II) are involved in green plant photosynthesis (see Fig. 1). Photosystem I contains a pigment (a special form of chlorophyll *a*) absorbing light maximally at around 700 nm, called P700 (Kok, 1956); Photosystem II contains a form of chlorophyll *a* absorbing light maximally around 680 nm (Doring, Renger, Vater & Witt, 1969) called P680. Hill & Bendall (1960) suggested how these two reactions may be linked when they proposed their 'Z-scheme' which remains the basis for most formulations of photosynthetic electron flow in higher plants, such as given in Fig. 1. The light-driven reactions are

6

localised in the lamellar system, or inner membrane, of the chloroplast. Light energy is absorbed by an array of light-harvesting pigments, carotenoids and chlorophylls a and b, and migrates to the reaction centres where it is used with great efficiency in electron transfer reactions. The electron transfer reactions of higher plants and the associated mechanisms of energy conservation have been reviewed recently by several authors (e.g. Trebst, 1974; Avron, 1975; Jagendorf, 1975; Bearden & Malkin, 1975; Radmer & Kok, 1975; Bendall, 1976; Crofts & Wood, 1977) and a detailed analysis is inappropriate here, but some of the important characteristics of the processes will be noted below.

Light absorbed by P700 excites this pigment and causes an electron to be transferred from it to the primary acceptor, producing a strong reductant which has an electron paramagnetic resonance (epr) signal characteristic of an iron–sulphur protein and an oxidation–reduction potential near -530 mV (see Bearden & Malkin, 1975). At least two distinct membrane-bound low-potential iron centres can be detected in chloroplast particles prepared so that they are enriched in Photosystem I activity, with mid-point potentials determined by two groups of workers as $E^{0'} = -530$ mV and $E^{0'} = -580$ mV (Ke, Hansen & Beinert, 1973) or $E^{0'} = -553$ mV and $E^{0'} = -594$ mV (Evans, Reeves & Cammack, 1974). The precise physiological function of the very low-potential bound iron–sulphur centre, detected in dark dithionite titrations is not clearly established, and only the centre with a mid-point potential near -530 mV is found to be light-reduced. A rapid change in absorbance at 430 nm has been found on illuminating Photosystem I-enriched chloroplast fragments and the pigment, called P430 (Hiyama & Ke, 1971), has very similar kinetic properties to the bound iron–sulphur compound that is the primary acceptor of Photosystem I; thus the change at 430 nm is likely to be another spectral manifestation of the reduction of the iron–sulphur protein.

Between the primary acceptor of Photosystem I and NADP$^+$ ($E^{0'} = -320$ mV) lie two carriers (Fig. 1). Ferredoxin is a soluble low-potential iron–sulphur protein with an $E^{0'}$ equal to -420 mV (Arnon, 1965) and is easily leached out of disrupted chloroplasts. Ferredoxin-NADP$^+$ reductase is a soluble flavoprotein with an FAD prosthetic group.

The electron lost by P700 to the primary acceptor leads to the formation of P700$^+$, a weak oxidant ($E^{0'} = +500$ mV; Ruuge & Izawa, 1972) and this component is re-reduced by an electron supplied from Photosystem II (see Fig. 1). The primary reaction of Photosystem II is similar to that of Photosystem I, with the crucial difference that the redox

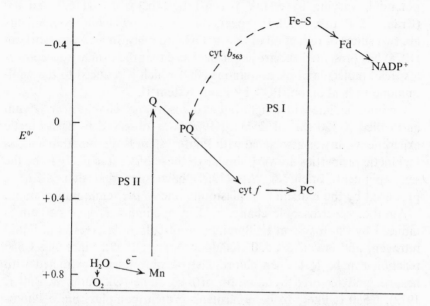

Fig. 1. A representation of the electron transport chain in green plant photosynthesis, giving approximate mid-point oxidation–reduction potentials (in volts) of the components. Mn, manganese complex involved in oxygen evolution; PS, Photosystem; Q, quencher of the fluorescence of PS II; PQ, plastoquinone; cyt. f, cytochrome f; PC, plastocyanin; Fe–S, an iron–sulphur protein; Fd, ferredoxin; cyt b_{563}, a cytochrome with an absorption maximum at 563 nm.

potential span covered is much higher (Bendall (1976) calculates that at pH 5.5 the mid-point potential of P680 is around $+910$ mV). The primary electron acceptor does not give the epr spectrum of an iron–sulphur compound. It was characterised as a compound that quenches the fluorescence of Photosystem II (Duysens & Sweers, 1963; Duysens, 1964) and designated Q. It was postulated that when Q was reduced (by Photosystem II activity) electrons were unable to leave P680 to Q and energy was instead lost from the excited P680 in the form of fluorescence. Q can be photoreduced at the temperature of liquid nitrogen, which is consistent with its role as the primary acceptor of Photosystem II (Kok, 1963; Kok, Malkin, Owens & Forbush, 1966). Kok et al. (1966) determined the mid-point potential of Q by measuring the fluorescence yield of Photosystem II following illumination at varied oxidation–reduction potentials. The value for $E^{\circ\prime}$, $+180$ mV, was higher than expected and later determinations under anaerobic conditions showed that Q was present as two components, $E^{\circ\prime} = -35$ mV and $E^{0\prime} = -270$ mV, each of which acted as a one-electron carrier with a pH-dependent mid-point potential, varying by 60 mV per pH unit between pH 6.0 and 8.0 (Cramer & Butler, 1969). It appears that the two quenching components are two states of reduction of Q, e.g. QH and QH_2. Indeed van Gorkom (1974) has presented spectroscopic evidence that the primary acceptor is a special molecule of plastoquinone (PQ) which is reduced to the semi-quinone radical anion, PQ^{-} by Photosystem II.

An intermediate with a light-induced absorbance change near 320 nm and called X-320 (Stiehl & Witt, 1969) was observed in flash kinetic experiments. It was associated with Photosystem II illumination and, as its kinetic properties are very similar to those of Q, it is likely to be the same pigment. Stiehl & Witt (1969) have suggested that X-320 is produced by the reduction of plastoquinone to the semiquinone anion.

Another spectroscopic change with a maximum around 550 nm is induced by Photosystem II illumination of chloroplasts, even in liquid nitrogen, and is called C550 (Knaff & Arnon, 1969). Since this C550 reaction can be lost when chloroplast membranes are extracted with hexane, and restored by addition of β-carotene (Okayama & Butler, 1972), C550 is likely to be a carotene–protein complex. Since Photosystem II activity persists in chloroplast fragments that lack C550, it is likely that C550 is not itself an obligatory electron transport carrier undergoing oxidation reactions, but, because of its close association with P680, it acts as an indicator, and reflects the oxidation–reduction state of the quencher Q (see Butler, 1973; Bendall, 1976).

Between Q and P700 is a chain of electron transport carriers (see Fig. 1). Direct measurements of absorbance at 255 nm suggest that a pool of 10 to 20 equivalents of plastoquinone accepts electrons from Q, and membrane preparations from chloroplasts catalyse the cytochrome-f-dependent reduction of plastocyanin by plastoquinol-1 (Wood & Bendall, 1976). Similarly the light-induced oxidation of cytochrome f by Photosystem I light requires the addition of plastocyanin (e.g. Gorman & Levine, 1966; Hind, 1968). Cytochrome f is an insoluble, membrane-bound c-type cytochrome, $E^{\circ\prime} = +365$ mV (Bendall, Davenport & Hill, 1971) shown to be oxidised by Photosystem II light (Duysens, 1964), firmly placing it between the two photosystems. Plastocyanin is a soluble copper protein, $E^{\circ\prime} = +370$ mV (see Katoh, 1971).

Re-reduction of oxidised P680 is achieved through the water-splitting reaction. The P680/$P680^{+}$ couple has a sufficiently high oxidation–reduction potential to be reduced by the oxygen–water couple ($E^{\circ\prime} = +800$ mV). The oxidation of water to oxygen, in a flashing light regime, shows a clear damped series of oscillations of the oxygen field, with a period of four flashes. Kok, Forbush & McGloin (1970) interpret this as suggesting that four positive charges are accumulated on a carrier Z from four successive oxidations of one P680. Oxygen appears only on every fourth flash. The nature of Z is uncertain, but there is some good evidence that manganese is involved in this process.

Two cytochromes of the b-type found in chloroplasts are closely associated with photosynthetic electron transport. Cytochrome $b_{559(HP)}$ ($E^{0'} = +350$ mV) is oxidised at the temperature of liquid nitrogen when Photosystem II is activated, but as discussed in the review by Cramer & Horton (1975) its precise function around Photosystem II remains to be elucidated. Cytochrome b_{563} is thought to be involved in a cyclic flow of electrons around Photosystem I. Instead of being used for the reduction of NADP$^+$, some electrons may cycle, via b_{563}, to re-reduce plastoquinone or cytochrome f (see Fig. 1).

The coupling of photosynthetic electron flow to phosphorylation in chloroplasts has been well reviewed (e.g. Trebst, 1974; Jagendorf, 1975; Bendall, 1976; Crofts & Wood, 1977). The most widely accepted mechanisms are based upon the chemiosmotic hypothesis developed by Mitchell (1968) which is discussed elsewhere in this volume. In very brief summary it requires that the reactions occur on a membrane which encloses an inner space. This membrane is poorly permeable to protons. During the flow of reducing equivalents along an electron transport chain an alternation of electron carriers (e.g. cytochrome f) and hydrogen carriers (e.g. plastoquinone) in the chain leads to the uptake or release of protons. The carriers are postulated as being placed on opposite sides of the membrane, so that electron flow has the effect of translocating protons from one side of the membrane to the other. This creates both a pH gradient across the membrane and also a membrane potential that may be partly collapsed by the movement of the other ions. The electrochemical activity gradient for protons was called the protonmotive force (pmf) which can be calculated from the equation:

$$\text{pmf} = \Delta\psi - 2.3 \frac{RT}{F} \Delta\text{pH},$$

where R is the gas constant, T is the absolute temperature, F is the Faraday constant, $\Delta\psi$ is the membrane potential, and ΔpH is the pH gradient.

The protonmotive force is the energy available for driving ATP synthesis, through a membrane-bound ATPase which removes vectorially the hydroxyl and proton components of water from ADP and P_i, to abolish the pmf. Evidence supporting the chemiosmotic view of energy conservation by chloroplasts is outlined below.

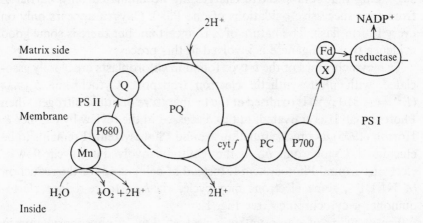

Fig. 2. The orientation of the chloroplast electron flow system within the membrane, showing how hydrogen ions are accumulated. (Modified from Trebst, 1974.) Abbreviations as in Fig. 1.

In chloroplasts it is well established that photosynthetic phosphorylation requires intact vesicles and that electron transport components are arranged on and in the thylakoid membrane so that vectorial electron flow takes place across the membrane. This localisation has been demonstrated: by the use of specific antibodies, which are unable to penetrate the intact thylakoid membrane; by the use of non-penetrant labels such as p-(diazonium)-benzene sulphonic acid; by using polar, non-penetrant, electron acceptors and donors; and by the accessibility of components to attack by trypsin. Such evidence has been conveniently summarised by Trebst (1974) who suggests that the vectorial electron flow may be represented as shown in Fig. 2, where electron flow along the zigzag pathway leads to proton accumulation in the vesicle. Witt and his collaborators (reviewed by Witt, 1971) produced direct evidence, using a spectroscopic shift at 518 nm as a measure of the membrane potential, that activation of either Photosystem I or Photosystem II produced a charge separation across the chloroplast membrane, with the positive charge inside. Each photosystem contributed equally to the electric field generated, which means that donor site of Photosystem I must be located on the opposite side of the membrane from the acceptor site of Photosystem II, as shown in Fig. 2.

A considerable uptake of protons can be measured, using a glass electrode, when chloroplasts are illuminated (Hind & Jagendorf, 1963; Neumann & Jagendorf, 1964) and Jagendorf & Uribe (1965) showed that chloroplasts catalysed ATP synthesis in the dark when a pH gradient was induced across the membranes by transfer of chloroplasts from acidic (pH 4) to basic (pH 8) conditions. Thus it is clear that chloroplasts are able to use light energy to form a hydrogen ion gradient and that a hydrogen ion gradient can be used to drive ATP synthesis.

Jagendorf (1975) discusses the relative contribution to the proton-motive force of the ΔpH and $\Delta \psi$ components. He reviews the evidence that proton movement, in steady illumination, is accompanied by chloride entry and cation extrusion (largely K^+ and Mg^{2+}) resulting in a relatively low contribution by the steady-state membrane potential in the light, although values for the pH gradient depend upon the external pH, being about 2.5–3.5 at pH 8.0 outside.

Consideration of the electron transport pathway illustrated in Fig. 2 suggests that $2H^+$ should be released for every electron flowing to $NADP^+$ or to an alternative Photosystem I acceptor and indeed a number of workers have recorded values close to this theoretical value (e.g. Izawa & Hind, 1967; Schwartz, 1968; Crofts, Wraight & Fleischmann, 1971; Dilley, 1971). There are reports of ratios considerably higher than 2.0 and these are discussed by Jagendorf (1975) who suggests that in addition to the chemiosmotic proton translocation some other proton-binding reactions may take place within the complex milieu of the chloroplast.

BACTERIAL PHOTOSYNTHETIC ELECTRON FLOW

The photosynthetic bacteria, the Chromatiaceae, the Rhodospirillaceae and the Chlorobiaceae, differ amongst themselves in the nature of their light-harvesting, or antenna pigments. Even the structures of the predominant chlorophylls in the photosynthetic membranes differ between the Chromatiaceae and Rhodospirillaceae, which contain only bacteriochlorophylls a or b, and the Chlorobiaceae where a homologous series of chlorobium chlorophylls is found. Chlorobium chlorophylls have spectroscopic properties closer to those of higher plants than those providing the bulk bacteriochlorophyll of the Chromatiaceae and the Rhodospirillaceae. In all cases, however, energy harvested by the antenna pigments is transferred to a reaction centre which contains bacteriochlorophyll a, or very rarely bacteriochlorophyll b, and where electron transfer reactions take place. A scheme outlining the basic pathways in bacterial photosynthesis is given in Fig. 3. It is apparent that only one light reaction is involved and that electrons lost from the cyclic electron flow system during the formation of NADH are replaced not by the water-splitting reaction found in chloroplasts, but by electron

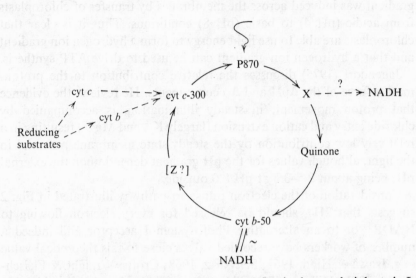

Fig. 3. A generalised scheme for photosynthetic electron flow in photosynthetic bacteria. In most cases it is likely that NADH is formed by a reversed electron flow process. X is likely to be an iron protein or an iron–quinone complex; Z is an intermediate suggested from kinetic evidence.

transport from reduced substrates provided in the growth medium. The general properties of the cyclic flow system are in many ways analogous to those of Photosystem I of higher plants.

In the sulphur bacteria (the Chlorobiaceae and Chromatiaceae) the reducing substrates required to replace electrons lost from the cyclic system are compounds such as hydrogen sulphide or thiosulphate; in the case of the non-sulphur bacteria (the Rhodospirillaceae) these substrates are organic molecules such as malate or succinate. Electrons flow via an electron transport system that probably involves cytochromes, flavo-proteins and iron–sulphur proteins, from the reduced substrates to top up the cyclic system. Many members of the Rhodospirillaceae are capable of aerobic growth in the dark, when energy seems to be obtained by electron-transport-dependent phosphorylation linked to the oxidation of organic substrates. Synthesis of the bacterial chlorophylls and caroten-oids is suppressed under these aerobic conditions (see Cohen-Bazire, Sistrom & Stanier, 1957).

The electron transport systems and light-harvesting pigments of photosynthetic bacteria are associated with a deeply invaginated intra-cytoplasmic membrane, which forms small closed vesicles, the chroma-tophores, when the cells are disrupted by passage through a French press or by sonication. Chromatophores retain the capacity for photosynthetic energy conversion, although the polarity of their membranes is reversed when compared with that of the membranes of the intact cell. (Oelze & Drews (1972) have conveniently reviewed the literature on the arrangement and composition of membranes of the photosynthetic bacteria.)

Light energy absorbed by light-harvesting pigments migrates to the bacterial reaction centre, a complex commonly referred to as P870, where P stands for pigment, almost certainly a form of bacteriochlorophyll, and 870 is the wavelength in nanometres at which a rapid bleaching is observed when the complex loses an electron following illumination (Duysens, 1954; Clayton, 1962). The amount of bacteriochlorophyll involved in such changes is about 2–5% of the total, the remaining bacteriochlorophyll probably exercising only a light-harvesting function. The wavelength band of the maximum bleaching may vary somewhat in different species of bacteria, but P870 is commonly used as a working abbreviation. This oxidation is very rapid, even at temperatures below 4 K, and proceeds with a quantum yield near 1.0 (Clayton, 1973; Wraight & Clayton, 1974; Parson & Cogdell, 1975). The rapid oxidation of a cytochrome of the c-type follows as the cytochrome donates an electron to P870$^+$ (Parson, 1968; Dutton, Petty, Bonner & Morse, 1975b). Because a fraction containing P870 can be purified from the membranes of a number of photosynthetic bacteria it has proved possible to characterise the general properties of the bacterial reaction centre in greater detail than has yet proved possible for higher plant reaction centres.

Reed & Clayton (1968) and Clayton & Wang (1971) described the purification of a reaction centre free from light-harvesting bacteriochlorophyll from a carotenoid-less mutant of *Rhodopseudomonas sphaeroides* (mutant R-26). The molecular weight of the preparation is near 70 000 and it contains three subunits. The particle is a chromoprotein containing four molecules of bacteriochlorophyll, two molecules of bacteriophaeophytin (bacteriochlorophyll in which Mg^{2+} is replaced by $2H^+$) together with one equivalent of ubiquinone and one equivalent of non-haem iron per mole of P870. No cytochromes are present in the most purified preparations. Parson & Cogdell (1975) have summarised the results of a large number of physical measurements made on the R-26 reaction centre. They conclude that P870 contains a complex of four bacteriochlorophyll molecules which interact strongly and two bacteriophaeophytin molecules which interact with the bacteriochlorophylls less intimately. The planes of the bacteriochlorophyll molecules are not parallel. When the complex loses an electron, giving the bleaching at 870 nm, one pair of bacteriochlorophylls share the unpaired

electron remaining (Norris, Druyan & Katz, 1973). The exciton interaction of the other bacteriochlorophyll pair is broken causing a small
shift to the blue in their absorption maxima at 800 nm. At the same time
the absorption band at 760 nm due to bacteriophaeophytin moves
slightly to the red. By using extremely short flashes of light it has proved
possible to make measurements of changes in the R-26 reaction centre
within picoseconds after illumination (Rockley, Windsor, Cogdell &
Parson, 1975; Dutton et al., 1975a) and to detect excited states of the
bacteriochlorophyll complex. The first intermediate form is detected in
less than 8 ps, but may not be a pure excited singlet state. It gives rise
within about 10 ps to an intermediate state with an absorption spectrum
that has been interpreted as arising from the formation of an anion-
cation diradical of two bacteriochlorophyll molecules (i.e. BChl$^+$—
BChl$^-$). This gives rise, over 120–250 ps to the formation of oxidised
bacteriochlorophyll and reduction of the primary acceptor. A simplified
version of the change is given below.

$$[BChl—BChl]X \xrightarrow{hv} [BChl\underline{*}BChl]X \xrightarrow{\sim 10\ ps}$$

$$[B^{+\cdot}Chl—B^{-\cdot}Chl]X \xrightarrow{120-250\ ps} [BChl\underline{\pm}BChl]X^-$$

In this representation P870$^+$ is synonymous with the oxidised bacteriochlorophyll dimer.

The oxidation–reduction potential of the P870 to P870$^+$ transition is
generally agreed to be within the region of $+450$ mV to $+490$ mV in
most bacteria (Kuntz, Loach & Calvin, 1964; Cusanovich & Kamen,
1968; Case & Parson, 1971; Dutton, 1971; Jackson, Cogdell & Crofts,
1973) although lower values have been found in the Chlorobiaceae
(Knaff, Buchanan & Malkin, 1973).

The primary acceptor which receives an electron from P870 has been
incompletely characterised, but the presence of non-haem iron and
ubiquinone in the purified reaction centre preparations encourages the
view that one of these, or an association of the two, is the reaction
partner of P870. In chromatophores there have been many determinations of the mid-point potential of the primary acceptor, based generally
upon the technique of poising the membranes at a range of redox potentials and observing the potentials at which the extent of the P870 photo-
oxidation becomes attenuated, presumably because the primary acceptor
is already reduced. The published values have been tabulated by Parson
& Cogdell (1975) and show considerable variation both within the same
species and between different species, ranging from -15 mV to
-160 mV. Since the mid-point potential is thought to be pH-dependent,

some discrepancies are explicable on the grounds of variation in the pH selected for the titration. In general it appears unlikely that the primary reductant is sufficiently reducing to drive the reduction of pyridine nucleotides, and in this respect bacteria differ from higher plants. An exception appears to be the primary acceptor of *Chlorobium limicola*, with a recently determined $E^{0'}$ more negative than -450 mV (Prince & Olson, 1976).

Electron paramagnetic resonance spectra have supported the view that the primary acceptor has a non-haem iron component (e.g. Dutton & Leigh, 1973; Dutton, Leigh & Reed, 1973). A new signal can be detected by epr spectroscopy following flash illumination of chromatophores at cryogenic temperatures and, in *Chromatium vinosum*, titration of the extent of this signal following illumination at different Eh showed that it had the same mid-point potential as that of the primary acceptor determined by photochemical activity (Evans, Lord & Reeves, 1974). This epr signal was produced after illumination of either chromatophores or isolated *Rhodopseudomonas sphaeroides* reaction centres (Dutton *et al.*, 1973), but in isolated reaction centres the mid-point potential was independent of pH whilst in membranes it had a dependency of -60 mV per pH unit. The reaction centre acceptor was also unaffected by the addition of the chelating agent σ-phenanthroline, whereas in the membrane σ-phenanthroline increased the $E^{0'}$ by approximately 30 mV. Apparently, significant changes in the organisation of the primary acceptor follow as a result of its extraction from the membrane during purification of the reaction centre and it may not be a completely reliable guide to mechanisms in the intact chromatophore.

Optical absorbance measurements made on *R. sphaeroides* reaction centres suggest that a semiquinone anion (and not a non-haem iron compound) is formed in the primary photochemical reaction, giving an absorption maximum near 450 nm in carefully constructed difference spectra (Slooten, 1972; Clayton & Straley, 1972; Romijn & Amesz, 1976). In iron-depleted reaction centres of *R. sphaeroides* photooxidation of P870 was associated with an epr signal with the characteristics of a semiquinone of ubiquinone (Loach & Hall, 1972; Feher, Okamura & McElroy, 1972). It has also proved possible to deplete reaction centres of ubiquinone by rigorous extraction with detergent in the presence of σ-phenanthroline, causing a loss of photochemical activity in the reaction centre, and to titrate back ubiquinone to restore activity. Ubiquinone was required at one equivalent per reaction centre (Okamura, Isaacson & Feher, 1975). The kinetics of the return of photosynthetic competence to P870, following a flash, were not the same as those for the decay of the 450 nm change and Wraight, Cogdell & Clayton (1975) explain this apparent inconsistency by suggesting that ubiquinone may be the primary acceptor, but that additional quinone molecules act as adventitious acceptors when the primary quinone is reduced.

In summary, it appears possible that in bacterial reaction centres both iron and ubiquinone are involved in the primary photo-act, possibly in the form of an iron–quinone complex.

The mid-point potential of the primary acceptor is pH-dependent and so the observation that its reduction does not involve a proton over the rapid time scale of normal photosynthetic electron flow, but only over the time scale of slow equilibrium redox titrations (Dutton & Wilson, 1974), becomes very important; an electron may have entered and left the acceptor before protons can overcome some kinetic barrier and interact with the acceptor. It has been possible to carry out photo-reductions at high pH ranges to determine the pK of the acceptor. At high pH, above the pK of the acceptor (pH 10 for *Rhodopseudomonas sphaeroides*, pH 9.0 for *Rhodospirillum rubrum*, and pH 8.0 for *Chromatium vinosum*) an estimate of the 'operational' mid-point potential was made (Prince & Dutton, 1976) for a change of the type:

$$X + e^- \rightleftharpoons X^-$$

and not

$$X + e^- + H^+ \rightleftharpoons XH.$$

The measured mid-point potentials at high pH were much lower than those recorded at near-neutral pH. Values for $E^{0'}$ for the 'operational' mid-point potential of the primary acceptor were: -180 mV for *Rhodopseudomonas sphaeroides*, -200 mV for *Rhodospirillum rubrum* and -160 mV for *Chromatium vinosum*.

Fortunately the nature of the group that donates electrons to re-reduce $P870^+$ is more unambiguously established than is the primary acceptor. Duysens (1954) first observed the photo-oxidation of a c-type cytochrome in *Rhodospirillum rubrum* and it appears that a soluble cytochrome of the c-type is always associated with P870, although the properties of the cytochrome c involved may vary widely in different species of bacteria; indeed in many cases two pools of c-cytochromes are closely associated with P870. One or both the cytochromes usually has a mid-point potential at around $+300$ mV and is often referred to as cytochrome c_2. This contrasts with the organisation of Photosystem I in higher plants where the donor to P700 is plastocyanin, a copper protein. It appears that algae are at an intermediate evolutionary stage since they contain a c-type cytochrome that can donate to P700 (Wood, 1976) although plastocyanin too can be present.

When *Chromatium vinosum* chromatophores are suspended at an ambient redox potential around $+300$ mV, cytochrome c_{555} is observed to become oxidised, following a rapid flash illumination, with the same kinetics as the re-reduction of P870$^+$ ($t_{\frac{1}{2}} = 2\mu s$; Parson, 1968), indicating that c_{555} acts as electron donor to P870. At low ambient redox potential (around 0 mV) a different cytochrome, c_{553}, becomes oxidised after flash illumination, even faster than c_{555} ($t_{\frac{1}{2}} \sim 1 \mu s$). If a second flash, is given in the 'dead-time' before c_{555} has been re-reduced by dark electron flow, then c_{555} becomes oxidised (Parson & Case, 1970). The mid-point potentials of the cytochromes are: c_{555}, $E^{0'} = +330$ mV; c_{553}, $E^{0'} = +10$ mV (see Cusanovich, Bartsch & Kamen, 1968). It is clear that in *C. vinosum*, P870$^+$ can accept electrons from either one of two cytochromes. The rate of re-reduction of the low-potential cytochrome (10 s) is much slower than that of the high-potential c_{555} (50 ms). It is possible that c_{553} normally accepts electrons only from an electron transport system connected to reduced substrates whereas c_{555} may be re-reduced through a cyclic pathway, as shown in Fig. 3 (Dutton & Jackson, 1972). Very similar arrangements of high- and low-potential cytochromes acting as alternative donors to P870 appear to operate in some members of the Rhodospirillaceae such as *Rhodopseudomonas viridis* (Case, Parson & Thornber, 1970) and *Rhodopseudomonas gelatinosa* (Dutton, 1971), with some uncertainty in the case of *Rhodopseudomonas capsulata* (Evans & Crofts, 1974). Alternative arrangements for reactions between cytochromes c and P870 are possible, as described below.

Repeated flash illumination of chromatophores of *Rhodopseudomonas sphaeroides* indicates that two cytochromes of the c-type, both with a mid-point potential of $+295$ mV, are oxidised by P870$^+$. After the first flash one molecule of cytochrome c is rapidly oxidised with biphasic kinetics, with a fast phase of $t_{\frac{1}{2}} = 20$–40 μs. A second brief flash causes the oxidation of a second cytochrome c, with similar oxidation kinetics to the first. Further flashes evoke only small responses in the cytochrome c region until sufficient time has elapsed for the c-type cytochromes to be re-reduced by cyclic electron flow (Dutton et al., 1975b). A very similar situation appears in reaction centre preparations of *Chlorobium limicola*, in which the reaction centre (mid-point potential $+250$ mV) is re-reduced by either of the two identical cytochromes c (mid-point potential $+165$ mV), the cytochrome oxidations showing biphasic kinetics (Prince & Olson, 1976).

In the cyclic photosynthetic electron flow system between the primary acceptor and cytochrome c (see Fig. 3) a considerable uncertainty exists about the nature of the intermediates. The primary acceptor, X, reduces a secondary acceptor Y, generally believed to be ubiquinone. Ubiquinone is present in amounts well in excess of P870 and has been implicated in photosynthetic electron flow following various types of experiment. Thus when *Chromatium vinosum* chromatophores are exposed to double pulses of light, closely spaced, it is possible to determine the rate of electron flow from X to Y. When the pulses are very close, the first electron has not had time to leave X to reduce Y before the second pulse excites the reaction centre, and so no P870 bleaching can occur (Parson & Case, 1970). By varying the time interval between pulses it is possible to determine $t_{\frac{1}{2}}$ of flow between X and Y ($t_{\frac{1}{2}} = 60 \ \mu s$). Removal of ubiquinone from chromatophores abolishes the reduction of Y by X and the addition of pure ubiquinone restores the reaction in *Chromatium vinosum* (Halsey & Parson, 1974). In *Rhodopseudomonas sphaeroides* it has been shown that Y is likely to be the component responsible for the rapid H^+ binding that follows flash illumination of chromatophores (Cogdell, Jackson & Crofts, 1972), a property expected of a hydrogen carrier like ubiquinone.

The involvement of cytochrome b in cyclic electron flow was suggested after the demonstration by Nishimura (1963) that the addition of antimycin A to illuminated chromatophores of *R. sphaeroides* or *Rhodospirillum rubrum* resulted in a steady-state increase in the level of reduction of a b cytochrome, together with increased oxidation of cytochrome c_2, and a similar involvement of cytochrome b in *Rhodopseudomonas capsulata* has been observed. Refinements of spectroscopic techniques have shown that a multiplicity of up to five cytochromes of the b-type may exist in one organism such as *R. sphaeroides* (Jones, 1969; Dutton & Jackson, 1972; Saunders & Jones, 1975), but by measuring cytochrome changes following flash illumination of chromatophores at varied fixed redox potentials Dutton & Jackson (1972) concluded that a cytochrome b with $E^{0'} = +50 \ \text{mV}$ is rapidly reduced ($t_{\frac{1}{2}} = 1\text{--}2 \ \text{ms}$) when antimycin A is present and that this cytochrome b is in the cyclic electron flow system. In uncoupled chromatophores the rate of re-oxidation of this cytochrome, in the absence of antimycin A, is equal to the rate of re-reduction of cytochrome c, as expected if it is indeed placed adjacent to cytochrome c in the cyclic system (Prince & Dutton, 1975). Using somewhat similar methods Evans & Crofts (1974) implicated a cytochrome b with a mid-point oxidation–reduction potential of $+60 \ \text{mV}$ in a cyclic photosynthetic flow system in *R. capsulata*. There is

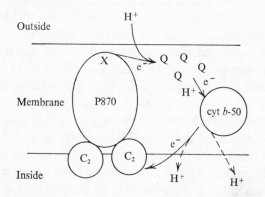

Outside

H⁺

Membrane

P870

Inside

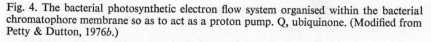

Fig. 4. The bacterial photosynthetic electron flow system organised within the bacterial chromatophore membrane so as to act as a proton pump. Q, ubiquinone. (Modified from Petty & Dutton, 1976*b*.)

now some evidence that cytochrome *b* may be present, and involved in photosynthetic electron flow, in organisms that were previously believed to lack cytochrome *b*; Knaff & Buchanan (1975) found that chromatophores prepared from either *Chromatium vinosum* or *Chlorobium thiosulphatophilum* contained one molecule of cytochrome *b* per reaction centre. This cytochrome was reduced on illumination. The photoreduction was enhanced by added antimycin A, under conditions where cyclic photophosphorylation was inhibited, which suggests that cytochrome *b* has a role in photosynthetic flow in Chromatiaceae and Chlorobiaceae as well as in Rhodospirillaceae.

Energy conservation in the photosynthetic bacteria

Coupled chromatophores prepared from photosynthetic bacteria fulfil many of the requirements for energy conservation by a chemiosmotic mechanism (see Crofts, 1974). It is very probable that the electron transport carriers are organised in a zigzag fashion across the chromatophore membrane, as shown in Fig. 4, which is based on schemes proposed by Crofts, Crowther & Tierney (1975) and Petty & Dutton (1976*b*). The evidence in favour of some such scheme is good. Illumination of chromatophores from *Rhodospirillum rubrum* was found to induce hydrogen ion uptake from the medium (von Stedingk & Baltscheffsky, 1966), a process that could be repeated with chromatophores prepared from a wide range of photosynthetic bacteria. This proton uptake was stimulated when valinomycin was added to chromatophores that had been suspended in high potassium chloride concentrations, and the addition of nigericin, which permits K^+/H^+ exchange, caused proton efflux from illuminated chromatophores suspended in high potassium chloride concentrations (Jackson, Crofts & van Stedingk, 1968). Photophosphorylation and hydrogen ion uptake were inhibited by the addition of uncoupling agents and by the detergent Triton X-114. Photophosphorylation was not inhibited by nigericin or valinomycin alone, but when both were present photophosphorylation was inhibited. Jackson *et al.* (1968) explained their results by suggesting that illumination gave rise to a transmembrane gradient of H^+ activity, linked closely to electron flow. In the absence of other ion movements this gave rise to a membrane potential. Valinomycin collapsed the membrane potential, when potassium chloride was present, by permitting the efflux of K^+ from the chromatophores, but the H^+ gradient was not dissipated. Nigericin addition then caused the collapse of this pH gradient by exchange of H^+ for K^+. The use of these ionophorous antibiotics indicated that the high energy state of chromatophores, induced by light, had two components; a membrane potential and a pH gradient. Since the H^+ movements in the absence of valinomycin and potassium chloride were relatively small, it was likely that in chromatophores, unlike chloroplasts, the membrane potential normally contributed a major capacity for energy storage and that chromatophore membranes were not readily permeable to chloride or other ions normally present in chromatophore preparations. Evidence of the formation of a membrane potential and a pH gradient by illuminated chromatophores has now been supplied following the use of a variety of methods for the measurement of these two parameters. The development of a transmembrane pH gradient has been shown (e.g. Gromet-Elhanan, 1972) by following the distribution of fluorescent amines across the membrane (Schuldiner, Rottenberg & Avron, 1972; Deamer, Prince & Crofts, 1972) and the development of a membrane potential has been determined by changes in the fluorescence of 8-anilinonaphthalene-1-sulphonate (ANS) (e.g. Vainio, Baltscheffsky, Baltscheffsky & Azzi, 1972) or in the distribution of phenyldicarbaundecaborane (PCB^-) (e.g. Isaev *et al.*, 1970). It has been confirmed that in chromatophores the membrane potential and pH gradient are interchangeable forms of conserved energy. It has also been shown that the addition of ATP or pyrophosphate to chromatophores in the dark causes the development of a membrane potential and a transmembrane pH gradient in a way that is sensitive to the action of uncouplers and oligomycin (see also Gromet-Elhanan & Leiser, 1973; Barsky *et al.*, 1975).

The wavelength maxima of the absorption bands of the carotenoid pigments of cells or chromatophores of many species of photosynthetic bacteria shift to longer wavelengths under conditions where the coupled system is in an energised state following illumination (e.g. Vredenberg & Amesz, 1966) or after the addition of ATP or pyrophosphate (Baltscheffsky, 1969). Similar spectral changes were induced in chromatophores by imposing diffusion potentials across the chromatophore membrane, using ionic gradients of K^+ operating through the ionophore valinomycin (Jackson & Crofts, 1969). These spectral shifts may arise from electrochromic effects that reflect the formation of electric fields across the membrane. The extent of the 'carotenoid shift' increased linearly with the applied potential and a calibration curve could be constructed and used to determine the size of the transmembrane potential ($\Delta\psi$) that was developed on the illumination of chromatophores. Values for $\Delta\psi$ of over 400 mV were found for *Rhodopseudomonas sphaeroides*, positive on the inside of the chromatophore, decaying on steady illumination possibly as a result of the compensatory movement of ions across the membrane. Very similar values have been obtained for *R. capsulata* (Casadio, Baccarini-Melandri & Melandri, 1974). Using the distribution of the fluorescent amine 9-aminoacridine to measure ΔpH across the membrane and the carotenoid shift to determine the change in membrane potential ($\Delta\psi$), Casadio, Baccarini-Melandri & Melandri (1974a) determined the total protonmotive force available to chromatophores of *R. capsulata*. They found that in low salt medium $\Delta\psi$ diminished over a period of 3 min illumination, from 306 mV to 156 mV, whilst 2.3 RT/F ΔpH rose from 'small' to 129 mV. The overall protonmotive force changed only from 306 to 285 mV. In the presence of potassium nitrate and valinomycin $\Delta\psi$ was 0 after 3 min and 2.3 RT/F ΔpH was 198 mV.

Casadio, Baccarini-Melandri, Zannoni & Melandri (1974b) illuminated *R. capsulata* chromatophores and measured $\Delta\psi$ and ΔpH in the steady state together with the steady-state concentrations of ATP, ADP and P_i. They determined the phosphorylation potential, using the relationships

$$\Delta G' = \Delta G^{0'} + RT\ln\frac{[\text{ATP}]}{[\text{ADP}][P_i]}$$

and $\Delta G' = -nF\Delta E$, where E is known as the phosphorylation potential. Assuming $H^+/\text{ATP} = 2$ (that is, that $n = 2$) the phosphorylation potential was calculated to be 325 mV, whilst the protonmotive force was measured as 419 mV, suggesting a fairly efficient degree of coupling between these energy forms. In experiments where the steady-state level of the protonmotive force was decreased by adding small amounts of uncouplers, the relationship between phosphate potential and proton-motive force remained reasonably linear (Casadio, Baccarini-Melandri & Melandri, 1975). There is little doubt that in chromatophores suffi-cient energy can be preserved in an electrochemical hydrogen ion gradient to drive the synthesis of ATP. It can be shown that the addition of ADP and P_i to chromatophores of *R. sphaeroides* that have been briefly illuminated increases the rate of decay of the electric potential across the membrane, in a way that is sensitive to oligomycin, and this stimulation is prevented if the coupling factor is removed from the chromatophores (Saphon, Jackson & Witt, 1975). This observation supports the view that the fall in the electric potential results from an H^+ current through a proton translocating ATPase, resulting in ATP synthesis.

The distribution of the electron transport carriers across the chroma-tophore membrane is not symmetrical in species where it has been examined and can explain how the hydrogen ion gradient is developed. Antibodies to cytochrome c_2 of *R. sphaeroides* and *R. capsulata* did not react with their respective antigens and prevent the photo-oxidation of cytochrome c_2 by chromatophores until the chromatophores had been lysed with cholate, showing that cytochrome c_2 was located inside the chromatophores (Prince *et al.*, 1975). Antiserum against highly purified reaction centres from *R. sphaeroides* did not bind to intact chromato-phore membranes but did react with membranes from which ATPase was removed by EDTA treatment (Reed, Raveed & Reporter, 1975), indicating that the reaction centre is located beneath the site of ATPase attachment to the membrane. Since cytochrome c_2 is on the inside of the chromatophore this must be where it interacts with the reaction centre. This immunological evidence suggests that reaction centre must span, or almost span, the chromatophore membrane, as depicted in Fig. 4. Since the secondary acceptor Y, which is probably ubiquinone, takes up protons from the medium this, also, is believed to be on the outside of the chromatophore. The reaction centre is arranged in the membrane so as to perform both redox and electrical work and to act as the electro-genic component of a proton pump, such as is required by Mitchell's chemiosmotic hypothesis. The uptake of a proton on reduction of ubiquinone forms the complementary H-carrying component. The properties of an in-vitro reconstituted system are in accordance with this proposed cross-membrane orientation of reaction centre. Purified reaction centres from *Rhodospirillium rubrum* after incorporation into phospholipid liposomes or planar bilayer membranes, generate a transmembrane electric potential difference upon illumination (Drachev, Kondrashin, Samuilov & Skulachev, 1975). The addition of the carriers UQ_6 and cytochrome c markedly increased the photo-effect, presumably through the return of reducing equivalents from the primary electron acceptor to oxidised P870.

There is evidence that a single flash illumination of *Rhodopseudomonas sphaeroides* leads to the binding of only one proton by the secondary acceptor (ubiquinone), with a stoichiometry of H^+/e^- of 1.0. Petty & Dutton (1976a) suggest that the product of reduction of ubiquinone is ubisemiquinone, which is then responsible for binding a proton. At ambient redox potentials below about $+240$ mV, where cytochrome c_2 is reduced, the $UQ^-/UQ^\cdot H$ couple has an apparent pK of 8.5. At higher pH the proton uptake diminishes and ubiquinone appears to act only as an electron carrier. The pK of this $UQ^-/UQ^\cdot H$ couple is affected by the physical state of the reaction centre through some charge inter-action, and this pK falls to 7.5 if P870 remains oxidised after a flash (e.g. if the ambient redox potential was around $+380$ mV). The cytochrome b that is involved in cyclic electron flow in *R. sphaeroides* with $E^{0\prime} = +50$ mV has a pH-dependent mid-point potential, with a pK at 7.4. That is, at a pH below 7.4 one proton is taken on with one electron reduction. Above pH 7.4 the reduction is independent of pH. By a suitable manipulation of pH conditions and ambient redox potentials Petty & Dutton (1976b) have obtained evidence that suggests that at pH greater than 7.4 a proton is released into the chromatophore when ubiquinone semiquinone reduces cytochrome b (an antimycin-A-insen-sitive reaction), or at a pH below 7.4, when cytochrome b reduces cyto-chrome c_2 (an antimycin-A-sensitive reaction). The two different pro-cesses may be represented:

$$UQ^\cdot H + \text{ferri } b \rightarrow UQ + \text{ferro } b^{(-)} + H^+ \quad \text{(above pH 7.4)}$$

$$UQ^\cdot H + \text{ferri } b \rightarrow UQ + \text{ferro } b^\cdot H \quad \text{(below pH 7.4)}$$

$$\text{ferro } b^\cdot H + \text{ferri } c_2 \rightarrow \text{ferri } b + \text{ferro } c_2^{(-)} + H^+$$

The experiments of Dutton and his collaborators lead them to suggest that cyclic electron flow is completed by the return of an electron from cytochrome b-50 to cytochrome c_2. However Crofts (1974) has reviewed evidence that suggests the presence in *R. capsulata* and *R. sphaeroides* of a second span of electron flow that acts as a proton pump. In brief, there is an antimycin-A-sensitive slow phase of the carotenoid change. In the simple loop, illustrated in Fig. 4, when electrons have flowed back to cytochrome b, on the inside of the membrane from the primary accep-tor on the outside, then the membrane potential should collapse and this collapse should be antimycin-A-insensitive. However, there is also an antimycin-A-sensitive slow phase of H^+ uptake in the presence of valinomycin. Uncouplers stimulate electron flow between cytochrome b and c. Crofts suggests an intermediate (Z), between cytochromes b and c, that can bind protons on the outside of the membrane. There is no lack of electron transport carriers in photosynthetic bacteria, the metabolic role of which has yet to be discovered, which may function as Z. Amongst these are iron–sulphur proteins, a carbon-monoxide-binding haem-protein (cytochrome c'), and a cytochrome b with a mid-point potential around $+240$ mV.

The formation of NAD(P)H

Since the $E^{0'}$ of the primary electron acceptor in bacterial photosynthesis (with the exception of the acceptor in the Chlorobiaceae) is well above that of the $NAD(P)^+/NAD(P)H$ couple it is probable that pyridine nucleotides are not reduced by a direct electron flow process such as occurs through Photosystem I in higher plants. It was shown by Keister & Yike (1967) that NAD^+ reduction by chromatophores of *Rhodospirillum rubrum* was dependent upon the formation of an energised state. It was inhibited by uncouplers, but not by oligomycin, in the light. In the dark the reaction was driven by ATP and was sensitive both to uncouplers and to oligomycin. Similar properties of the NADH formation reactions have been found in *Rhodopseudomonas sphaeroides*, *R. capsulata*, and *R. viridis* (Klemme, 1969; Keister & Minton, 1969; Jones & Whale, 1970; Jones & Saunders, 1972). The likely mechanism for this reduction appears to be by a reversal of electron flow. A high-energy state is generated by light, or by added ATP, and the energised state reverses the flow of electrons from some reduced electron transport intermediate to make NADH. Since the reaction is inhibited not only by uncouplers but also by a combination of valinomycin and nigericin, but not by either separately, it is likely that the energised state that is used is a combination of a pH gradient and a membrane potential.

In the case of *Chlorobium limicola* f. *thiosulfatophilum* direct photoreduction of NAD^+ appears possible, in a slow reaction which is dependent upon the addition of ferredoxin (Buchanan & Evans, 1969) and is not inhibited by uncouplers and therefore has properties like Photosystem I of chloroplasts. This pathway is possible because of the low mid-point potential of the primary acceptor in this organism, a potential much lower than found in the Rhodospirillaceae and Chromatiaceae.

Aerobic metabolism of the Rhodospirillaceae

Most members of the Rhodospirillaceae are able to grow aerobically in the dark, apparently by using parts of the photosynthetic electron transport system to catalyse oxidative phosphorylation. Oxygenation represses the formation of bacteriochlorophyll and carotenoids (Cohen-Bazire et al., 1957) and also the formation of the intracytoplasmic membranes (chromatophore membranes). Indeed the formation of bacteriochlorophyll appears to be coupled to the synthesis of these membranes. Aerobically grown cells have relatively little pigmentation yet retain approximately the same membrane concentration of b and c cytochromes that is found in the photosynthetic forms. Some changes in the spectroscopic properties of these cytochromes are, however, apparent, particularly in extracts of soluble cytochromes. Membranes from both types of cell have the capacity to catalyse ATP synthesis coupled to the oxidation of succinate or NADH with no reported increase in the P/O ratio in oxygen-adapted cells, although the fractional P/O ratios obtained in such experiments do not prove this point unambiguously.

The terminal oxidases in photosynthetically grown cells have not been clearly defined, but may be a b-type cytochrome that appears to give cytochrome o-type difference spectra when reduced chromatophore membranes are treated with carbon monoxide. Such carbon monoxide spectra have been noted in *Rhodospirillum rubrum* (Taniguchi & Kamen, 1964), *Rhodopseudomonas sphaeroides* (Kikuchi & Motokawa, 1968; Whale & Jones, 1970) and *R. viridis* (Saunders & Jones, 1973a). In *R. palustris*, cytochrome o appeared to be formed only in response to aeration (King & Drews, 1975), and a b-type cytochrome that does not bind carbon monoxide may act as one oxidase in photosynthetically grown *R. capsulata*, although an alternative oxidase is present that does bind carbon monoxide (Zannoni et al., 1974; LaMonica & Marrs, 1976). All these light-grown cells can oxidise cytochrome c. Even in photosynthetically grown cells more than one pigment may be present that can donate electrons to oxygen, as shown by the presence of respiration that is cyanide-sensitive and respiration that is relatively cyanide-insensitive within the same membrane. The presence of branched electron transport pathways is likely, although in the usual anaerobic photosynthetic growth conditions these oxidases almost certainly have some other function, and their capacity to act as oxidases may well be fortuitous. In *R. sphaeroides* only, aerobic growth leads to the development of a cytochrome with the same spectroscopic properties as cytochrome a/a_3 of mitochondria (Kikuchi & Motokawa, 1968). Potentiometric titration of membranes from aerobically grown *R. sphaeroides* at 607 nm showed that this cytochrome contained two components, $E^{0'} = +375$ mV and $E^{0'} = +200$ mV, in a ratio close to unity, as found in mammalian cytochrome oxidase (Saunders & Jones, 1973*b*). Aerobic growth of *R. sphaeroides* in a medium lacking added copper diminished the concentration of the *a*-type cytochromes, suggesting that this oxidase too contained copper. Electron paramagnetic resonance spectroscopy of aerobic membranes showed a copper signal with $E^{0'}$ near $+210$ mV, which corresponds reasonably with the value for the copper of cytochrome oxidase (O. T. G. Jones, W. J. Ingledew & R. C. Prince, unpublished observations). No such epr signal was found in membranes from photosynthetically grown cells, although the iron–sulphur signals were similar in both aerobically and photosynthetically grown cells. Growth of *R. sphaeroides* with low aeration reduced the concentration of this cytochrome oxidase in the membranes. No such oxidase has been observed in any other photosynthetic bacteria and no cytochrome *c* with redox properties corresponding to those of mammalian cytochrome *c* was detected in *R. sphaeroides* although a *c*-type $E^{0'} = +120$ mV was found.

Redox titration over a limited redox span of the cytochromes of membranes prepared from aerobically and photosynthetically grown cells of *R. sphaeroides* showed that the same three *b*-type cytochromes were present, with $E^{0'} = +185$ mV, $+41$ mV and -104 mV (Connelly, Jones, Saunders & Yates, 1973) and the same major cytochrome *c* was present, $E^{0'} = +295$ mV, in both samples, showing clearly that all the cytochromes that have been implicated in photosynthetic electron flow in *R. sphaeroides* (Dutton & Jackson, 1972) were present in the aerobic form. Later experiments (Saunders & Jones, 1975) show that in bacterio-chlorophyll-less cells two more *b*-type cytochromes of high potential can be detected, $E^{0'} = +390$ mV and $E^{0'} = +255$ mV. Neither component appeared to react with carbon monoxide and both were present in lower concentration in membranes from highly aerated cultures. The function of these high-potential cytochromes *b* is obscure, but their high potential makes them unlikely candidates for a photosynthetic flow system.

Antibody to cytochrome c_2 had little effect on cytochrome *c* oxidase activity of membranes from photosynthetically grown cells (G. Hauska & O. T. G. Jones, unpublished observations) which may indicate that cytochrome c_2 is not a donor to the oxidase(s) in this organism, a conclusion in accordance with its relatively slow kinetics of oxidation when anaerobic particles are mixed with oxygen (Connelly *et al.*, 1973).

In *R. capsulata* the relationship of aerobic and respiratory pathways

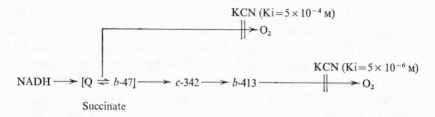

Fig. 5. A scheme for respiratory electron flow in *Rhodopseudomonas capsulata* (from Zannoni *et al.*, 1976), *c*-342 is a cytochrome *c* with $E^{0'} = +342$ mV, *b*-47 and *b*-413 are *b* cytochromes with $E^{0'} = +47$ mV and $+413$ mV respectively. Q, ubiquinone.

has been partly elucidated using mutants that are unable to grow photosynthetically (Marrs & Gest, 1973; Zannoni *et al.*, 1974; LaMonica & Marrs, 1976; Zannoni, Melandri & Baccarini-Melandri, 1976). A cytochrome *b*, $E^{0'} = +47$ mV and a cytochrome *c*, $E^{0'} = +342$ mV are present and involved in both respiratory and photosynthetic electron transport. A high-potential *b*, $E^{0'} = +413$ mV, found in aerobically grown cells of *R. capsulata* is absent from a mutant that grows normally photosynthetically but which is not capable of aerobic growth. Cytochrome *b*413 may act as an oxidase, receiving electrons from cytochrome *b*47. An alternative oxidase was present in wild-type cells. This oxidase accepted electrons at or near the ubiquinone level, and was relatively resistant to potassium cyanide, but sensitive to carbon monoxide inhibition (see Fig. 5). A cytochrome *b*, $E^{0'} = +260$ mV and a cytochrome *c*, $E^{0'} = +94$ mV were also present in these aerobic cells i.e. two cytochromes *b* of high potential and one cytochrome *c* of low potential are present in both *R. sphaeroides* and *R. capsulata* and in neither case do they appear to be involved in cyclic photosynthetic electron flow.

The components of the respiratory chain of *R. palustris* and *R. viridis* have not been so well characterised as those of *R. sphaeroides* and *R. capsulata*, but a pathway branching at the ubiquinone level has been found in *R. palustris* with two oxidases present that differ in sensitivity to potassium cyanide inhibition (King & Drews, 1975). Aerobic growth of *R. viridis* led to the development of a new *b*-type cytochrome and an oxidase with a greatly reduced K_m for oxygen (Saunders & Jones, 1973*a*).

Since bacteriochlorophyll-free membranes from aerobically grown cells of *R. sphaeroides* contain the same *b* and *c* cytochromes that are involved in photosynthetic cyclic electron flow, attempts have been made to reconstitute photosynthetic reactions in these membranes by adding back to them purified reaction centre (Jones & Plewis, 1974). A reaction centre–membrane complex was formed, illumination of which caused oxidation of cytochrome c and reduction of cytochrome b. Since both cytochrome changes were increased by the addition of antimycin A it appeared that a cyclic electron flow system had been reconstituted *in vitro*. The further addition of a purified light-harvesting pigment–protein complex made these two photoreactions more efficient at low light intensity, indicating energy transfer from the antenna-protein complex to the reaction centre (Hunter & Jones, 1976). A similar incorporation of pigment proteins into 'aerobic' membranes may proceed normally *in vivo* in the process of adapting aerobically growing cells into photosynthetic cells. A reconstitution of photosynthetic reactions has been achieved with membranes from aerobically grown *R. capsulata*, and in this preparation low rates of photophosphorylation and light-dependent NAD^+ reduction were found (Garcia, Drews & Kamen, 1974, 1975). Another component of photosynthetic energy conservation systems, coupling factor, can be purified from photosynthetically grown *R. capsulata* and then incorporated into aerobic membranes stripped of coupling factor by EDTA treatment, with reconstitution of oxidative phosphorylation activity, and the reverse reconstitution of 'aerobic' coupling factor into photosynthetic cells is also possible (Baccarini-Melandri & Melandri, 1972). The coupling factor from photosynthetically grown *R. sphaeroides* has also been shown to be interchangeable with the coupling factor from aerobically grown cells in reconstitution experiments (A. Baccarini-Melandri, B. A. Melandri & O. T. G. Jones, unpublished observations).

A picture arises of adaptation in the photosynthetic bacteria in which newly synthesised sections of electron transport chain and ATPase complexes may be incorporated into a pre-existing photosynthetic or respiratory electron transport chain. Appropriate modification occurs by the introduction of terminal oxidases or of light-trapping pigments. The electron transport carriers in both aerobically grown and photosynthetically grown cells of *R. sphaeroides* and *Rhodospirillum rubrum* appear to be orientated with the same polarity since respiration of cells of either growth-type is coupled to the outward translocation of protons, the same direction as proton translocation driven by light in whole photosynthetically grown cells (Scholes, Mitchell & Moyle, 1969). Many sections of the electron transport system are likely to be common to both aerobically and photosynthetically grown cells. An example is the NADH-dehydrogenase region which must incorporate an energy-conserving site even in photosynthetically grown cells, in order that NADH can be produced by reversed electron flow. It is an intriguing speculation that reaction centre bacteriochlorophyll and cytochrome oxidase may be alternative oxidases for the electron transport system of *Rhodopseudomonas sphaeroides*.

REFERENCES

ARNON, D. I. (1965). Role of ferredoxin in plant bacterial photosynthesis. In *Non-heme Iron Proteins: Role in Energy Conversion*, ed. A. San Pietro, pp. 137–73. Yellow Springs, Ohio: Antioch Press.

AVRON, M. (1975). The electron transport chain in chloroplasts. In *Bioenergetics of Photosynthesis*, ed. Govindjee, pp. 373–86. New York & London: Academic Press.

BACCARINI-MELANDRI, A. & MELANDRI, B. A. (1972). Energy transduction in photosynthetic bacteria. III. Coincidence of coupling factor of photosynthesis and respiration in *Rhodopseudomonas capsulata*. *FEBS Letters*, **21**, 131–4.

BALTSCHEFFSKY, M. (1969). Energy conversion-linked changes of carotenoid absorbance in *Rhodospirillum rubrum* chromatophores. *Archives of Biochemistry and Biophysics*, **130**, 646–52.

BARSKY, E. L., BONCH-OSMOLOVSKAYA, E. A., OSRTROUMOV, S. A., SAMUILOV, V. D. & SKULACHEV, V. P. (1975). A study on the membrane potential and pH gradient in chromatophores and intact cells of photosynthetic bacteria. *Biochimica et Biophysica Acta*, **387**, 388–95.

BEARDEN, A. J. & MALKIN, R. (1975). Primary photochemical reactions in chloroplast photosynthesis. *Quarterly Reviews of Biophysics*, **7**, 131–77.

BENDALL, D. S. (1976). Electron and proton transfer in chloroplasts. *MTP International Review of Science* (*Biochemistry*), Series 2, in press.

BENDALL, D. S., DAVENPORT, H. E. & HILL, R. (1971). Cytochrome components in chloroplasts of the higher plants. *Methods in Enzymology*, **23**, 327–44.

BUCHANAN, B. B. & EVANS, M. C. W. (1969). Photoreduction of ferredoxin and its use in $NAD(P)^+$ reduction by a subcellular preparation from the photosynthetic bacterium, *Chlorobium thiosulfatophilum*. *Biochimica et Biophysica Acta*, **180**, 123–9.

BUTLER, W. L. (1973). Primary photochemistry of Photosystem II of photosynthesis. *Accounts of Chemical Research*, **6**, 177–84.

CASADIO, R., BACCARINI-MELANDRI, A. & MELANDRI, B. A. (1974a). On the determination of the transmembrane pH difference in bacterial chromatophores using 9-aminoacridine. *European Journal of Biochemistry*, **47**, 121–8.

CASADIO, R., BACCARINI-MELANDRI, A. & MELANDRI, B. A. (1975). The degree of coupling of ATP synthetase in bacterial photophosphorylation. In *Electron Transfer Chains and Oxidative Phosphorylation*, eds. E. Quagliariello, S. Papa, F. Palmieri, E. C. Slater & N. Siliprandi, pp. 407–10. Amsterdam: North-Holland Publishing Co.

CASADIO, R., BACCARINI-MELANDRI, A., ZANNONI, D. & MELANDRI, B. A. (1974b). Electrochemical proton gradient and phosphate potential in bacterial chromatophores. *FEBS Letters*, **49**, 203–7.

CASE, G. D. & PARSON, W. W. (1971). Thermodynamics of the primary and secondary photochemical reactions in *Chromatium*. *Biochimica et Biophysica Acta*, **253**, 187–202.

CASE, G. D., PARSON, W. W. & THORNBER, J. P. (1970). Photo-oxidation of cytochromes in reaction centre preparations from *Chromatium* and *Rhodopseudomonas viridis*. *Biochimica et Biophysica Acta*, **223**, 122–8.

CLAYTON, R. K. (1962). Primary reactions in bacterial photosynthesis. I. The nature of light-induced absorbancy changes in chromatophores: evidence for a special bacteriochlorophyll component. *Photochemistry and Photobiology*, **1**, 201–10.

CLAYTON, R. K. (1973). Primary processes in bacterial photosynthesis. *Annual Reviews of Biophysics and Bioengineering*, **2**, 131–56.

CLAYTON, R. K. & STRALEY, S. C. (1972). Photochemical electron transport in photosynthetic reaction centres. IV. Observations related to the reduced photoproducts. *Biophysical Journal*, **12**, 1221–34.

CLAYTON, R. K. & WANG, R. T. (1971). Photochemical reaction centres from *Rhodopseudomonas sphaeroides*. *Methods in Enzymology*, **23**, 696–704.

COGDELL, R. J., JACKSON, J. B. & CROFTS, A. R. (1972). The effect of redox potential on the coupling between rapid hydrogen-ion binding and electron transport in chromatophores from *Rhodopseudomonas sphaeroides*. *Journal of Bioenergetics*, **4**, 413–29.

COHEN-BAZIRE, G., SISTROM, W. R. & STANIER, R. Y. (1957). Kinetic studies of pigment synthesis by non-sulphur purple bacteria. *Journal of Cellular and Comparative Physiology*, **49**, 25–68.

CONNELLY, J. L., JONES, O. T. G., SAUNDERS, V. A. & YATES, D. W. (1973). Kinetic and thermodynamic properties of membrane-bound cytochromes of aerobically and photosynthetically grown *Rhodopseudomonas sphaeroides*. *Biochimica et Biophysica Acta*, **292**, 644–53.

CRAMER, W. A. & BUTLER, W. L. (1969). Potentiometric titration of the fluorescence yield of spinach chloroplasts. *Biochimica et Biophysica Acta*, **172**, 503–10.

CRAMER, W. A. & HORTON, P. (1975). Recent studies of the chloroplast cytochrome b-559. *Photochemistry and Photobiology*, **22**, 304–8.

CROFTS, A. R. (1974). The electron transport system as a H^+ pump in photosynthetic bacteria. In *Perspectives in Membrane Biology*, ed. S. Estrada, O. & C. Gitler, pp. 373–412. New York & London: Academic Press.

CROFTS, A. R., CROWTHER, D. & TIERNEY, G. V. (1975). Electrogenic electron transport in photosynthetic bacteria. In *Electron Transfer Chains and Oxidative Phosphorylation*, ed. E. Quagliariello, S. Papa, F. Palmieri, E. C. Slater & N. Siliprandi, pp. 233–41. Amsterdam: North-Holland.

CROFTS, A. R. & WOOD, P. M. (1977). Comparative electron transport and photophosphorylation processes in plants and bacteria. *Current Topics in Bioenergetics*, **7**, in press.

CROFTS, A. R., WRAIGHT, C. A. & FLEISCHMANN, D. E. (1971). Energy conservation in photochemical reactions of photosynthesis and its relation to delayed fluorescence. *FEBS Letters*, **15**, 89–100.

CUSANOVICH, M. A., BARTSCH, R. G. & KAMEN, M. D. (1968). Light-induced electron transport in *Chromatium* strain D. II. Light-induced absorbance changes in *Chromatium* chromatophores. *Biochimica et Biophysica Acta*, **153**, 397–417.

CUSANOVICH, M. A. & KAMEN, M. D. (1968). Light-induced electron transport in *Chromatium* strain D. I. Isolation and characterisation of *Chromatium* chromatophores. *Biochimica et Biophysica Acta*, **153**, 376–96.

DEAMER, D. W., PRINCE, R. C. & CROFTS, A. R. (1972). The response of fluorescent amines to pH gradients across liposome membranes. *Biochimica et Biophysica Acta*, **274**, 323–35.

DILLEY, R. A. (1971). Coupling of ion and electron transport in chloroplasts. *Current Topics in Bioenergetics*, **4**, 237–71.

DORING, G., RENGER, G., VATER, J. & WITT, H. T. (1969). Properties of the photoactive chlorophyll a_{II} in photosynthesis. *Zeitschrift für Naturforschung*, **24B**, 1139–43.

DRACHEV, L. A., KONDRASHIN, A. A., SAMUILOV, V. D. & SKULACHEV, V. P. (1975). Generation of electric potential by reaction centre complexes from *Rhodospirillum rubrum*. *FEBS Letters*, **50**, 219–22.

DUTTON, P. L. (1971). Oxidation–reduction potential dependence of the interaction of cytochromes, bacteriochlorophyll and carotenoids at 77 °K in chromatophores of *Chromatium* D and *Rhodopseudomonas gelatinosa*. *Biochimica et Biophysica Acta*, **226**, 63–80.

DUTTON, P. L. & JACKSON, J. B. (1972). Thermodynamic and kinetic characterisation of electron transfer components *in situ* in *Rhodopseudomonas sphaeroides* and *Rhodospirillum rubrum*. *European Journal of Biochemistry*, **30**, 495–510.

DUTTON, P. L., KAUFMANN, K. J., CHANCE, B. & RENTZEPIS, P. M. (1975a). Picosecond kinetics of the 1250 nm band of the *Rhodopseudomonas sphaeroides* reaction centre: the nature of the primary photochemical intermediary state. *FEBS Letters*, **60**, 275–80.

DUTTON, P. L. & LEIGH, J. S. (1973). Electron spin resonance characterisation of *Chromatium* D hemes, non-heme irons and the components involved in primary photochemistry. *Biochimica et Biophysica Acta*, **314**, 178–90.

DUTTON, P. L., LEIGH, J. S. & REED, D. W. (1973). Primary events in the photosynthetic reaction centre from *Rhodopseudomonas sphaeroides* strain R-26: triplet and oxidised states of bacteriochlorophyll and the identification of the primary electron acceptor. *Biochimica et Biophysica Acta*, **292**, 654–64.

DUTTON, P. L., PETTY, K. M., BONNER, H. S. & MORSE, S. D. (1975b). Cytochrome c_2 and reaction centre of *Rhodopseudomonas sphaeroides* Ga membranes. Extinction coefficients, content, half-reduction potentials, kinetics and electric field alterations. *Biochimica et Biophysica Acta*, **387**, 536–56.

DUTTON, P. L. & WILSON, D. F. (1974). Redox potentiometry in mitochondrial and photosynthetic bioenergetics. *Biochimica et Biophysica Acta*, **346**, 165–212.

DUYSENS, L. N. M. (1954). Reversible photo-oxidation of a cytochrome pigment in photosynthesising *Rhodospirillum rubrum*. *Nature, London*, **173**, 692–3.

DUYSENS, L. N. M. (1964). Photosynthesis. *Progress in Biophysics and Molecular Biology*, **14**, 1–104.

DUYSENS, L. N. M., HUISKAMP, W. J., VOS, J. J. & VAN DER HART, J. M. (1956). Reversible changes in bacteriochlorophyll in purple bacteria upon illumination. *Biochimica et Biophysica Acta*, **19**, 188–90.

DUYSENS, L. N. M. & SWEERS, H. E. (1963). Mechanism of two photochemical reactions in algae as studied by means of fluorescence. In *Studies on Microalgae and Photosynthetic Bacteria*, ed. Japanese Society of Plant Physiology, pp. 353–72. Tokyo: University of Tokyo Press.

EVANS, E. H. & CROFTS, A. R. (1974). *In situ* characterisation of photosynthetic electron transport in *Rhodopseudomonas capsulata*. *Biochimica et Biophysica Acta*, **357**, 89–102.

EVANS, M. C. W., LORD, A. V. & REEVES, S. G. (1974). The detection and characterisation by electron paramagnetic resonance spectroscopy of iron–sulphur proteins and other electron transport components in chromatophores from the purple bacterium *Chromatium*. *Biochemical Journal*, **138**, 177–83.

EVANS, M. C. W., REEVES, S. G. & CAMMACK, R. (1974). Determination of the oxidation–reduction potential of the bound iron sulphur proteins of the primary electron acceptor complex of Photosystem I in spinach chloroplasts. *FEBS Letters*, **49**, 111–14.

FEHER, G., OKAMURA, M. Y. & MCELROY, J. D. (1972). Identification of an electron acceptor in reaction centres of *Rhodopseudomonas sphaeroides* by EPR spectroscopy. *Biochimica et Biophysica Acta*, **267**, 222–6.

GARCIA, A. F., DREWS, G. & KAMEN, M. D. (1974). On reconstitution of bacterial photophosphorylation *in vitro*. *Proceedings of the National Academy of Sciences, USA*, **71**, 4213–16.

GARCIA, A. F., DREWS, G. & KAMEN, M. D. (1975). Electron transport in an *in vitro* reconstituted bacterial photophosphorylating system. *Biochimica et Biophysica Acta*, **387**, 129–34.

GORMAN, D. S. & LEVINE, R. P. (1966). Photosynthetic electron transport chain of *Chlamydomonas reinhardi*. VI. Electron transport in mutant strains lacking either cytochrome 553 or plastocyanin. *Plant Physiology*, **41**, 1648–56.

GROMET-ELHANAN, Z. (1972). Changes in the fluorescence of atebrin and of anilino-naphthalene sulfonate reflecting two different light-induced processes in *Rhodospirillum rubrum* chromatophores. *European Journal of Biochemistry*, **25**, 84–8.

GROMET-ELHANAN, Z. & LEISER, M. (1973). Interchangeability of the membrane potential with the pH gradient in *Rhodospirillum rubrum* chromatophores. *Archives of Biochemistry and Biophysics*, **159**, 583–9.

HALSEY, Y. D. & PARSON, W. W. (1974). Identification of ubiquinone as the secondary electron acceptor in the photosynthetic apparatus of *Chromatium vinosum*. *Biochimica et Biophysica Acta*, **347**, 404–16.

HILL, R. & BENDALL, F. (1960). Function of two cytochrome components in chloroplasts: a working hypothesis. *Nature, London*, **186**, 136–7.

HIND, G. (1968). The site of action of plastocyanin in chloroplasts treated with detergent. *Biochemica et Biophysica Acta*, **153**, 235–40.

HIND, G. & JAGENDORF, A. T. (1963). Separation of light and dark stages in photophosphorylation. *Proceedings of the National Academy of Sciences, USA,* **49,** 715–22.

HIYAMA, T. & KE, B. (1971). A new photosynthetic pigment 'P430': its possible role as the primary electron acceptor of Photosystem I. *Proceedings of the National Academy of Sciences, USA,* **68,** 1010–13.

HUNTER, C. N. & JONES, O. T. G. (1976). Effect of added light-harvesting pigments on the reconstitution of photosynthetic reactions in membranes from bacteriochlorophyll-less mutants of *Rhodopseudomonas sphaeroides. Biochemical Society Transactions,* **4,** 669–70.

ISAEV, P. I., LIBERMAN, E. A., SAMUILOV, V. D., SKULACHEV, V. P. & TSOFINA, L. M. (1970). Conversion of biomembrane produced energy into electric form. III. Chromatophores from *Rhodospirillum rubrum. Biochimica et Biophysica Acta,* **216,** 22–9.

IZAWA, S. & HIND, G. (1967). The kinetics of the pH rise in illuminated chloroplast suspensions. *Biochimica et Biophysica Acta,* **143,** 377–90.

JACKSON, J. B., COGDELL, R. J. & CROFTS, A. R. (1973). Some effects of *o*-phenanthroline on electron transport in chromatophores from photosynthetic bacteria. *Biochimica et Biophysica Acta,* **292,** 218–25.

JACKSON, J. B. & CROFTS, A. R. (1969). The high energy state in chromatophores from *Rhodopseudomonas sphaeroides. FEBS Letters,* **4,** 185–9.

JACKSON, J. B., CROFTS, A. R. & VON STEDINGK, L.-V. (1968). Ion transport induced by light and antibiotics in chromatophores from *Rhodospirillum rubrum. European Journal of Biochemistry,* **6,** 41–54.

JAGENDORF, A. T. (1975). Mechanism of photophosphorylation. In *Bioenergetics of Photosynthesis,* ed. Govindjee, pp. 413–92. New York & London: Academic Press.

JAGENDORF, A. T. & URIBE, E. (1966). ATP formation caused by acid–base transition of spinach chloroplasts. *Proceedings of the National Academy of Sciences, USA,* **55,** 170–7.

JONES, O. T. G. (1969). Multiple light induced reactions of cytochromes b and c in *Rhodopseudomonas sphaeroides. Biochemical Journal,* **114,** 793–9.

JONES, O. T. G. & PLEWIS, K. M. (1974). Reconstitution of light-dependent electron transport in membranes from a bacteriochlorophyll-less mutant of *Rhodopseudomonas sphaeroides. Biochimica et Biophysica Acta,* **357,** 204–14.

JONES, O. T. G. & SAUNDERS, V. A. (1972). Energy linked electron transfer in *Rhodopseudomonas viridis. Biochimica et Biophysica Acta,* **275,** 427–36.

JONES, O. T. G. & WHALE, F. R. (1970). The oxidation and reduction of pyridine nucleotides by *Rhodopseudomonas sphaeroides* and *Chlorobium thiosulphatophilum. Archiv für Mikrobiologie,* **72,** 48–59.

KATOH, S. (1971). Plastocyanin. *Methods in Enzymology,* **23,** 408–13.

KE, B., HANSEN, R. E. & BEINERT, H. (1973). Oxidation–reduction potentials of bound iron–sulphur proteins of Photosystem I. *Proceedings of the National Academy of Sciences, USA,* **70,** 2941–5.

KEISTER, D. L. & MINTON, N. J. (1969). Energy linked reactions in photosynthetic bacteria. III. Further studies on energy linked nicotinamide-adenine dinucleotide reduction by *Rhodospirillum rubrum* chromatophores. *Biochemistry,* **8,** 167–73.

KEISTER, D. L. & YIKE, N. J. (1967). Energy linked reactions in photosynthetic bacteria. I. Succinate-linked ATP-driven NAD$^+$ reduction by *Rhodospirillum rubrum* chromatophores. *Archives of Biochemistry and Biophysics*, 121, 415–22.

KIKUCHI, G. & MOTOKAWA, Y. (1968). Cytochrome oxidase of *Rhodopseudomonas sphaeroides*. In *Structure and Function of Cytochromes*, ed. K. Okunuki, M. D. Kamen & I. Sezuku, pp. 174–81. Baltimore: University Park Press.

KING, M. T. & DREWS, G. (1975). The respiratory electron transport system of heterotrophically-grown *Rhodopseudomonas palustris*. *Archives of Microbiology*, 102, 219–31.

KLEMME, J.-H. (1969). Studies on the mechanism of NAD-photoreduction by chromatophores of the facultative phototroph, *Rhodopseudomonas capsulata*. *Zeitschrift für Naturforschung*, 24B, 67–76.

KNAFF, D. B. & ARNON, D. I. (1969). Spectral evidence for a new photoreactive component of the oxygen-evolving system in photosynthesis. *Proceedings of the National Academy of Sciences, USA*, 63, 963–9.

KNAFF, D. B. & BUCHANAN, B. B. (1975). Cytochrome *b* and photosynthetic sulphur bacteria. *Biochimica et Biophysica Acta*, 376, 549–60.

KNAFF, D. B., BUCHANAN, B. B. & MALKIN, R. (1973). Effect of oxidation–reduction potential on light-induced cytochrome and bacteriochlorophyll reactions in chromatophores from the photosynthetic bacteria *Chlorobium*. *Biochimica et Biophysica Acta*, 325, 94–101.

KOK, B. (1956). On the reversible absorption change at 705 μm in photosynthetic organisms. *Biochimica et Biophysica Acta*, 22, 399–401.

KOK, B. (1963). Fluorescence studies. In *Photosynthetic Mechanisms of Green Plants*, ed. B. Kok & A. T. Jagendorf, pp. 45–55. Washington, DC: National Academy of Sciences/National Research Council.

KOK, B., FORBUSH, B. & MCGLOIN, M. (1970). Cooperation of charges in photosynthetic O$_2$ evolution. I. A linear four step mechanism. *Photochemistry and Photobiology*, 11, 457–75.

KOK, B., MALKIN, S., OWENS, O. & FORBUSH, B. (1966). Observations on the reducing side of Photosystem II. *Brookhaven Symposia in Biology*, 19, 446–59.

KUNTZ, I. D., LOACH, P. A. & CALVIN, M. (1964). Absorption changes in bacterial chromatophores. *Biophysical Journal*, 4, 227–49.

LOACH, P. A. & HALL, R. L. (1972). The question of the primary electron acceptor in bacterial photosynthesis. *Proceedings of the National Academy of Sciences, USA*, 69, 786–90.

LAMONICA, R. F. & MARRS, B. L. (1976). The branched respiratory system of photosynthetically grown *Rhodopseudomonas capsulata*. *Biochimica et Biophysica Acta*, 423, 431–9.

MARRS, B. & GEST, H. (1973). Genetic mutations affecting the respiratory electron transport system of the photosynthetic bacterium *Rhodopseudomonas capsulata*. *Journal of Bacteriology*, 114, 1045–51.

MITCHELL, P. (1968). *Chemiosmotic Coupling in Oxidative and Photosynthetic Phosphorylation*. Bodmin: Glynn Research Ltd.

NEUMANN, J. & JAGENDORF, A. T. (1964). Light induced pH changes related to phosphorylation by chloroplasts. *Archives of Biochemistry and Biophysics*, **107**, 109–19.

NISHIMURA, M. (1963). Studies on the electron transport systems in photosynthetic bacteria. II. The effect of HQNO and antimycin A on the photosynthetic and respiratory electron transfer systems. *Biochimica et Biophysica Acta*, **66**, 17–21.

NORRIS, J. R., DRUYAN, M. E. & KATZ, J. J. (1973). Electron nuclear double resonance of bacteriochlorophyll free radical *in vitro* and *in vivo*. *Journal of the American Chemical Society*, **95**, 1680–2.

OELZE, J. & DREWS, G. (1972). Membranes of photosynthetic bacteria. *Biochimica et Biophysica Acta*, **265**, 209–39.

OKAMURA, M. Y., ISAACSON, R. A. & FEHER, G. (1975). Primary acceptor in bacterial photosynthesis: obligatory role of ubiquinone in photoactive reaction centres of *Rhodopseudomonas sphaeroides*. *Proceedings of the National Academy of Sciences, USA*, **72**, 3491–5.

OKAYAMA, S. & BUTLER, W. L. (1972). Extraction and reconstitution of Photosystem II. *Plant Physiology*, **49**, 769–74.

PARSON, W. W. (1968). The role of P870 in bacterial photosynthesis. *Biochimica et Biophysica Acta*, **153**, 248–59.

PARSON, W. W. & CASE, G. D. (1970). In *Chromatium*, a single photochemical reaction centre oxidises both cytochrome *c*-552 and cytochrome *c*-555. *Biochimica et Biophysica Acta*, **205**, 232–45.

PARSON, W. W. & COGDELL, R. J. (1975). The primary photochemical reaction of bacterial photosynthesis. *Biochimica et Biophysica Acta*, **416**, 105–49.

PETTY, K. M. & DUTTON, P. L. (1976*a*). Properties of the flash induced proton binding encountered in membranes of *Rhodopseudomonas sphaeroides*: a functional pK on the ubisemiquinone? *Archives of Biochemistry and Biophysics*, **172**, 335–45.

PETTY, K. M. & DUTTON, P. L. (1976*b*). Ubiquinone–cytochrome *b* electron and proton transfer: a functional pK on cytochrome b_{50} in *Rhodopseudomonas sphaeroides* membranes. *Archives of Biochemistry and Biophysics*, **172**, 346–53.

PRINCE, R. C., BACCARINI-MELANDRI, A., HAUSKA, G. A., MELANDRI, B. A. & CROFTS, A. R. (1975). Assymetry of an energy transducing membrane. The location of cytochrome c_2 in *Rhodopseudomonas sphaeroides* and *Rhodopseudomonas capsulata*. *Biochimica et Biophysica Acta*, **387**, 212–27.

PRINCE, R. C. & DUTTON, P. L. (1975). A kinetic completion of the cyclic photosynthetic electron pathway of *Rhodopseudomonas sphaeroides*: cytochrome *b* → cytochrome c_2 oxidation-reduction. *Biochimica et Biophysica Acta*, **387**, 609–13.

PRINCE, R. C. & DUTTON, P. L. (1976). The primary acceptor of bacterial photosynthesis: its operating midpoint potential? *Archives of Biochemistry and Biophysics*, **172**, 329–34.

PRINCE, R. C. & OLSON, J. M. (1976). Some thermodynamic and kinetic properties of the primary photochemical reactants in a complex from a green photosynthetic bacterium. *Biochimica et Biophysica Acta*, **423**, 357–62.

RADMER, R. & KOK, B. (1975). Energy capture in photosynthesis: Photosystem II. *Annual Reviews of Biochemistry*, **44**, 409–33.

REED, D. W. & CLAYTON, R. K. (1968). Isolation of a reaction centre fraction from *Rhodopseudomonas sphaeroides*. *Biochemical and Biophysical Research Communications*, **30**, 471–5.

REED, D. W., RAVEED, D. & REPORTER, M. (1975). Localisation of photosynthetic reaction centres by antibody binding to chromatophore membranes from *Rhodopseudomonas sphaeroides* strain R-26. *Biochimica et Biophysica Acta*, **387**, 368–78.

ROCKLEY, M. G., WINDSOR, M. W., COGDELL, R. J. & PARSON, W. W. (1975). Picosecond detection of an intermediate in the photochemical reaction of bacterial photosynthesis. *Proceedings of the National Academy of Sciences, USA*, **72**, 2251–5.

ROMIJN, J. C. & AMESZ, J. (1976). Photochemical activities of reaction centres from *Rhodopseudomonas sphaeroides* at low temperature and in the presence of chaotropic agents. *Biochimica et Biophysica Acta*, **423**, 164–73.

RUUGE, E. K. & IZAWA, S. (1972). The oxidation–reduction potential of P_{700} in unbroken chloroplasts. *Federation Proceedings of the American Societies for Experimental Biology*, **31**, 901.

SAPHON, S., JACKSON, J. B. & WITT, H. T. (1975). Electrical potential changes, H^+ translocation and phosphorylation induced by short flash excitation in *Rhodopseudomonas sphaeroides* chromatophores. *Biochimica et Biophysica Acta*, **408**, 67–82.

SAUNDERS, V. A. & JONES, O. T. G. (1973a). Oxidative phosphorylation and effects of aerobic conditions on *Rhodopseudomonas viridis*. *Biochimica et Biophysica Acta*, **305**, 581–9.

SAUNDERS, V. A. & JONES, O. T. G. (1973b). Properties of the cytochrome a-like material developed in the photosynthetic bacterium *Rhodopseudomonas sphaeroides* when grown aerobically. *Biochimica et Biophysica Acta*, **333**, 439–45.

SAUNDERS, V. A. & JONES, O. T. G. (1975). Detection of two further b-type cytochromes in *Rhodopseudomonas sphaeroides*. *Biochimica et Biophysica Acta*, **396**, 220–8.

SCHOLES, P., MITCHELL, P. & MOYLE, J. (1969). The polarity of proton translocation in some photosynthetic microorganisms. *European Journal of Biochemistry*, **8**, 450–4.

SCHULDINER, S., ROTTENBERG, H. & AVRON, M. (1972). Determination of ΔpH in chloroplasts. 2. Fluorescent amines as a probe for the determination of ΔpH in chloroplasts. *European Journal of Biochemistry*, **25**, 64–70.

SCHWARTZ, M. (1968). Light induced proton gradient links electron transport and phosphorylation. *Nature, London*, **219**, 915–19.

SLOOTEN, L. (1972). Electron acceptors in reaction centre preparations from photosynthetic bacteria. *Biochimica et Biophysica Acta*, **275**, 208–18.

STIEHL, H. H. & WITT, H. T. (1969). Quantitative treatment of the function of plastoquinone in photosynthesis. *Zeitschrift für Naturforschung*, **24B**, 1588–98.

TANIGUCHI, S. & KAMEN, M. D. (1964). The oxidase system of heterotrophic-ally-grown *Rhodospirillum rubrum*. *Biochimica et Biophysica Acta*, **96**, 395–428.

TREBST, A. (1974). Energy conservation in photosynthetic electron transport of chloroplasts. *Annual Reviews of Plant Physiology*, **25**, 423–58.

VAINIO, H., BALTSCHEFFSKY, M., BALTSCHEFFSKY, H. & AZZI, A. (1972). Energy-dependent changes in membranes of *Rhodospirillum rubrum* chromatophores as measured by 8-anilino-naphthalene-1-sulphonic acid. *European Journal of Biochemistry*, **30**, 301–6.

VAN GORKOM, H. J. (1974). Identification of the reduced primary electron acceptor of Photosystem II as a bound semiquinone anion. *Biochimica et Biophysica Acta*, **347**, 439–42.

VON STEDINGK, L.-V. & BALTSCHEFFSKY, H. (1966). The light-induced, reversible pH change in chromatophores from *Rhodospirillum rubrum*. *Archives of Biochemistry and Biophysics*, **117**, 400–4.

VREDENBERG, W. J. & AMESZ, J. (1966). Absorption bands of bacteriochloro-phyll types in purple bacteria and their response to illumination. *Biochimica et Biophysica Acta*, **126**, 244–53.

WHALE, F. R. & JONES, O. T. G. (1970). The cytochrome system of hetero-trophically-grown *Rhodopseudomonas sphaeroides*. *Biochimica et Bio-physica Acta*, **223**, 146–57.

WITT, H. T. (1971). Coupling of quanta, electrons, fields, ions and phos-phorylation in the functional membrane of photosynthesis. *Quarterly Reviews of Biophysics*, **4**, 365–477.

WOOD, P. M. (1976). Electron transport between plastoquinone and cyto-chrome c-552 in *Euglena* chloroplasts. *FEBS Letters*, **65**, 111–16.

WOOD, P. M. & BENDALL, D. S. (1976). The reduction of plastocyanin by plastoquinol-1 in the presence of chloroplasts. *European Journal of Bio-chemistry*, **61**, 337–44.

WRAIGHT, C. A. & CLAYTON, R. K. (1974). The absolute quantum efficiency of bacteriochlorophyll photo oxidation in reaction centres of *Rhodo-pseudomonas sphaeroides*. *Biochimica et Biophysica Acta*, **333**, 246–60.

WRAIGHT, C. A., COGDELL, R. J. & CLAYTON, R. K. (1975). Some experi-ments on the primary electron acceptor in reaction centres from *Rhodo-pseudomonas sphaeroides*. *Biochimica et Biophysica Acta*, **396**, 242–9.

ZANNONI, D., BACCARINI-MELANDRI, A., MELANDRI, B. A., EVANS, E. H., PRINCE, R. C. & CROFTS, A. R. (1974). Energy transduction in photo-synthetic bacteria. The nature of cytochrome c oxidase in the respiratory chain of *Rhodopseudomonas capsulata*. *FEBS Letters*, **48**, 152–5.

ZANNONI, D., MELANDRI, B. A. & BACCARINI-MELANDRI, A. (1976). Energy transduction in photosynthetic bacteria. X. Composition and function of the branched oxidase system in wild type and respiration deficient mutant of *Rhodopseudomonas capsulata*. *Biochimica et Biophysica Acta*, **423**, 413–30.

ENERGY COUPLING IN SUBSTRATE AND GROUP TRANSLOCATION

W. A. HAMILTON

*Department of Microbiology, University of Aberdeen,
Aberdeen AB9 1AS, UK*

INTRODUCTION

There are two problems unique to the passage of low molecular weight polar entities across biological membranes. First, there must exist a mechanism to increase specifically the permeability of the membrane to the transported solute. Secondly, where there is accumulation against a concentration gradient, the cell or organelle must couple the energy derived from metabolism to the translocation of the solute across the membrane. There exist two general mechanisms of transport which solve these problems in quite separate ways; they are classified as group translocation and substrate translocation.

In group translocation, the essential vectorial character of transport is achieved through the association of an enzyme, or group of enzymes, with the membrane such that the substrate approaches from the aqueous phase on one side of the membrane, and the product is released on the other. In substrate translocation, the diffusion of the chemically un-modified solute through the membrane is facilitated by the presence within the membrane of a specific carrier or porter.

The subject of microbial transport has been widely reviewed over recent years. In particular those articles by Harold (1972), Simoni & Postma (1975) and Hamilton (1975) have considered the problems associated with energy coupling. In view of this extensive background, the present article will largely confine itself to new findings published since the beginning of 1974.

GROUP TRANSLOCATION

The most extensively studied group translocation mechanism is the phosphotransferase system (PTS) which transports certain sugars into a range of anaerobic and facultative bacteria (Hays, Simoni & Roseman, 1973; Simoni, Hays, Nakazawa & Roseman, 1973; Simoni, Nakazawa, Hays & Roseman, 1973; Simoni & Roseman, 1973; Hamilton, 1975; Kornberg & Jones-Mortimer, this volume). Extracellular sugar or sugar alcohol is converted by the membrane-bound Enzyme II complex, usually to the 6-phosphate derivative which appears in the intracellular compartment. The phosphate group is transferred from phosphoenolpyruvate through the action of Enzyme I and the small molecular weight heat-stable protein, HPr. The major drop in free energy is at the final step, where simultaneous with the conversion to sugar-6-phosphate by Enzyme II, the translocation takes place across the membrane.

Initially it was believed that the PTS was quite separate structurally and functionally from any substrate translocation mechanisms, but this conclusion seems to be no longer tenable.

(a) Kornberg & Riordan (1976) and Postma (1976) have shown that free galactose can enter *Escherichia coli* and *Salmonella typhimurium* by facilitated diffusion on Enzyme II, without the involvement of Enzyme I or HPr.

(b) In *Arthrobacter pyridinolis* there are two pathways for the uptake of fructose; a PTS giving, intracellular fructose-1-phosphate, and a substrate translocation of free fructose (Wolfson, Sobel, Blanco & Krulwich, 1974). It was shown, however, that a mutant lacking this second system and so unable to transport free fructose, could not induce the fructose PTS and thus was unable to grow on that sugar (Wolfson & Krulwich, 1974).

(c) Kornberg (1973a,b; Kornberg & Jones-Mortimer, this volume) has discussed the inhibition exerted in *E. coli* on the transport of both PTS and non-PTS sugars by the rapid uptake of glucose on its specific PTS. Bag (1974) has recently extended this analysis to *Vibrio cholerae*.

(d) An interesting corollary to these studies is the demonstration of the inhibition of α-methyl glucoside transport on the glucose PTS in *E. coli* by the development, through either respiration or ATP hydrolysis, of a so-called energised membrane state (del Campo, Hernandez-Asensio & Ramirez, 1975; Singh & Bragg, 1976a).

In addition to its role in transport, Enzyme II has also been implicated in chemotaxis (Adler & Epstein, 1974; Koshland, this volume).

Another group translocation system is involved in the uptake of purines by enteric bacteria. Jackman & Hochstadt (1976) have extended their studies with *S. typhimurium* to experiments with membrane vesicles, in which they have shown the accumulation of guanine and hypoxanthine as the 5′-monophosphate derivatives through the action of the membrane-bound phosphoribosyltransferase. In a more extended study with *E. coli* whole cells, Yagil & Beacham (1975) looked at the transport of extracellular adenosine 5′-monophosphate (AMP). A

5'-nucleotidase in the periplasmic space first dephosphorylates to give adenosine. The combined action of membrane-bound purine nucleoside phosphorylase and phosphoribosyltransferase then produces free adeninine, which is translocated and appears intracellularly as AMP.

In contrast, purines and pyrimidines have been reported by other workers to be transported into *E. coli* and the yeast *Saccharomyces cerevisiae* by substrate translocation mechanisms that possibly involve the co-transport of protons (Roy-Burman & Visser, 1975; Chevallier, Jund & Lacroute, 1975; Reichert, Schmidt & Foret, 1975).

SUBSTRATE TRANSLOCATION

It is now clear that there are two quite distinct mechanisms of substrate translocation found among micro-organisms. A wide range of sugars, organic acids, amino acids and small peptides are transported in some Gram-negative bacteria by a mechanism involving the participation of soluble binding proteins. These are released from the periplasmic space by osmotic shock, or during the formation of spheroplasts or membrane vesicles so that such structures lack these binding protein transport reactions. From studies with uncouplers, and inhibitors and mutations affecting respiration and adenosine triphosphatase (ATPase) activity, it has been concluded that the energy source for these binding protein systems is either ATP itself or a closely related source of covalent bond energy (Berger & Heppel, 1974; Boos, 1974*a,b*; Harold, 1976; Oxender & Quay, 1976).

The other broad class of substrate translocating mechanisms requires only membrane-associated components. The coupling of metabolic energy operates through what has been described as the energised state of the membrane. The main body of this paper will be concerned with the identification of this energised state, and the discussion of the mechanisms of its production from metabolism, and of its coupling to transport.

Binding proteins and substrate translocation

In something of a classic paper, Berger (1973) was able to demonstrate that the transport of proline by *E. coli* was driven by a high-energy membrane state, derived from either respiration or the action of ATPase on glycolytically-produced ATP, and sensitive to the action of uncouplers. Glutamine transport, on the other hand, was driven directly by phosphate-bond energy, formed by either oxidative phosphorylation or glycolysis. It was already known that these two transport systems also differed in that glutamine uptake was sensitive to osmotic shock, depended upon a periplasmic binding protein, and was absent from membrane vesicle preparations. Clearly the long-standing argument as to whether or not ATP is directly involved in active transport is greatly illuminated by this identification of two quite independent systems for substrate translocation, one of which requires ATP itself, or a closely related compound, while the other is driven by an energised state of the membrane.

The further characterisation of transport reactions in *E. coli* dependent upon binding proteins and phosphate-bond energy has extended the list of substrates to include glutamine, α,ε-diaminopimelic acid, arginine, histidine, ornithine (Berger & Heppel, 1974; Gutowski & Rosenberg, 1976), methionine (Kadner & Winkler, 1975), leucine, isoleucine and valine (Oxender, 1972), glycylglycine (Cowell, 1974), and ribose (Curtiss, 1974). There are also claims that glucose transport in *Pseudomonas aeruginosa* may require a shock-releasable binding protein (Guymon & Eagon, 1974), and that in *Brucella abortus* the uptake of that hexose may involve ATP directly (Rest & Robertson, 1974).

There is some evidence, however, contradictory to this clear-cut description of the binding protein transport systems being energised solely by phosphate-bond energy. From their studies of the action of colicin K on the transport capabilities of *E. coli*, Plate, Suit, Jetten & Luria (1974) concluded that the uptake of glutamine was in fact powered by the energised membrane state. More recently Singh & Bragg (1976b) came to the same conclusion from their studies of transport in a cytochrome-deficient mutant. In partial support of these contrary views, Lieberman & Hong (1976) deduced that both ATP and the energised membrane state were required for the uptake of binding-protein-dependent substrates.

What remains unclear at the present time is the role of the periplasmic binding proteins in facilitating the transport of materials across the hydrophobic barrier of the cytoplasmic membrane. Do they have some recognition function, perhaps concentrating the substrate in the periplasmic space? In this regard it is interesting to note that the binding proteins have been shown to play an integral role in certain chemotactic responses (Ordal & Adler, 1974a,b; Koshland, this volume). Are membrane carriers also involved, as in other substrate translocation mechanisms? What is the nature of the interaction between these carriers and the binding proteins? What is the mechanism of energy coupling involving ATP?

A number of binding protein systems have been examined more fully, and although much of the data is somewhat preliminary, and often more than a little confusing, these studies do extend our understanding and focus our attention on the outstanding problems in this exciting area of research.

(1) Miner & Frank (1974) studied the transport of glutamate by three strains of *E. coli*. One strain showed a sodium dependence for the uptake of the amino acid by both whole cells and vesicles. However a glutamate binding protein was released from the whole cells, although the binding itself was found to be sodium-independent. Willis & Furlong (1975) also presented data on a shock-releasable glutamate binding protein and the possible involvement of sodium in transport. They also suggested that glutamate transport might in some cases involve a binding protein, and in other cases might not. Clearly there must exist more than one system for glutamate transport in *E. coli*, and these systems must vary in some as yet undetermined way with regard to their sodium dependence and requirement for a periplasmic binding protein. The most recent work from Halpern's laboratory (Kahane, Marcus, Metzer & Halpern, 1976*a*, *b*) has demonstrated that the glutamate binding protein and the membrane carrier are under separate control of induction and repression. Further, these authors suggest that the binding protein and membrane carrier are integral components of the same transport mechanism which, in vesicles and in whole cells under certain nutritional conditions, can also operate with the membrane carrier alone.

(2) Postma, Cools & van Dam (1973) concluded on the basis of some rather indirect evidence that the transport of Krebs cycle intermediates by *Azotobacter vinelandii* is possibly powered by ATP. More recently Lo & Sanwal (1975) have been looking at the dicarboxylic acid uptake in *E. coli*. They have identified three genetic regions responsible for the synthesis of components of the system. *dct*A and *dct*B code for membrane-bound proteins which are necessary for transport activity in both cells and vesicles. Two membrane proteins, SBP1 and SBP2, have now been separated and purified; SBP2 has been identified as the product of gene *dct*A. The third genetic region, *cbt*, codes for a periplasmic binding protein which is required for transport by whole cells, but whose absence does not affect the uptake of dicarboxylic acids by vesicle preparations.

(3) The transport of the sugar galactose by *E. coli* appears to be mediated by up to six possible systems (Boos, 1974*b*; Postma, 1976). The highest affinity system is that for β-methylgalactoside which incorporates a binding protein. A genetic analysis has characterised three genes involved in this system, *mgl*A, *mgl*B and *mgl*C. The gene products of

*mgl*A and *mgl*C have not yet been identified, but *mgl*B codes for the galactose binding protein; all three are required for galactose transport, but only the *mgl*B gene product for galactose taxis (Ordal & Adler 1974*a*,*b*).

(4) Maltose transport and taxis in *E. coli* have been subject to similar genetic and biochemical analyses (Hazelbauer, 1975*a*, *b*; Szmelcman & Hofnung, 1975; Szmelcman, Schwartz, Silkavy & Boos, 1976). These functions are controlled by the *mal*B region of the chromosome. This region contains the genes *mal*F, *mal*E, *mal*K and *lam*B. The products of *mal*F and *mal*K are unknown, but *mal*E codes for a periplasmic binding protein, and *lam*B for the binding site on the outer membrane for the phage λ. Although the maltose binding protein is required for both transport and taxis, there are some *mal*E mutants which are defective in transport while retaining normal taxis response. It would appear, therefore, that there are two separate sites on the binding protein, one specific for transport and the other for taxis.

Loss of the phage receptor, by mutation or by antibody titration, leads also to the loss of maltose transport and taxis, but only at very low hexose concentration. It is suggested therefore that the *lam*B gene product is necessary for the passage of maltose across the outer membrane from low concentrations in the medium. Transport across the inner cytoplasmic membrane then involves a second transfer and the action of the *mal*F and *mal*K gene products.

(5) Similar data on the importance of outer membrane sites in certain transport reactions come from the study of iron uptake by *E. coli*. Following an earlier report by Wang & Newton (1971), Frost & Rosenberg (1975) have shown that the *ton*B locus codes for the outer membrane receptor involved in the action of phages T1 and Φ80, and the colicins B, I and V. *ton*B negative mutants are, however, also impaired in the transport of iron along with a chelator such as enterochelin. Similarly, the outer membrane receptor for phage BF23 and colicin E has been implicated in the transport of vitamin B_{12} (Bradbeer, Woodrow & Khalifah, 1976).

These data with maltose, and with iron and vitamin B_{12}, lend weight to the idea that in Gram-negative bacteria the involvement of periplasmic binding proteins in transport and taxis may be a feature of the extra complexity of the two hydrophobic membrane barriers in these organisms. Such a suggestion, however, does not explain the mechanism of energy coupling with ATP.

In this last respect, it is relevant to point out at this time the findings of Harold & Spitz (1975) that the transport of phosphate and arsenate, and possibly of glutamate and aspartate, are apparently energised directly by ATP in the Gram-positive *Streptococcus faecalis*, although there is no evidence for the involvement of binding proteins.

Energised membrane state and substrate translocation

Since the studies of microbial membrane transport came of age with the classic work of the Pasteur Institute on the β-galactoside permease in *E. coli* (Cohen & Monod, 1957), the field has been dominated by the controversy regarding the merits and demerits of the permease, respiration-linked and chemiosmotic models that have been put forward to describe the mechanisms involved. Each of these models assumes some form of membrane-associated carrier which specifically facilitates the permeation of the translocated substrate across the hydrophobic barrier of the phospholipid membrane. It is in the proposed mechanism of energy coupling that the fundamental differences lie. The arguments for and against these models have already been conducted extensively, if not exhaustively, in the literature (Harold, 1972; Kaback & Hong, 1973; Kaback, 1974; Simoni & Postma, 1975; Hamilton, 1975) and need not be gone over again now.

The acceptance by Kaback of Mitchell's chemiosmotic hypothesis and the abandonment of his own respiration-linked model of membrane transport (Kaback, 1974; Patel, Schuldiner & Kaback, 1975; Schuldiner & Kaback, 1975; Ramos, Schuldiner & Kaback, 1976) mean that we are free to devote our energies, both in this article and, more importantly, in our laboratory studies, to examining the experimental evidence for the chemiosmotic model and the validity of its various predictions.

It is pertinent to make the point that although the chemiosmotic hypothesis is most often associated with oxidative phosphorylation, it relates to all membrane-associated cellular energy transductions, and in fact had its beginnings in the study of microbial transport reactions (Mitchell, 1966, 1970; Greville, 1969; Harold, 1972). In briefest outline, the hypothesis states that the oxidative and photosynthetic electron transport chains, the bacteriorhodopsin of *Halobacterium halobium*, and the proton translocating Mg^{2+}-activated ATPase are organised anisotropically within the cell, or organelle, membrane such that they can generate reversibly an electrochemical gradient of protons and of charge across the proton-impermeable membrane (see papers by Garland, C. W. Jones, Kroger, Haddock, Aleem, O. T. G. Jones and Oesterhelt, this volume). This electrochemical gradient is given the name protonmotive force. This is the nature of the so-called energised membrane state, so far rather loosely referred to in this article, and fairly widely in the scientific literature. The phenomena of oxidative and photosynthetic phosphorylation and of reversed electron transport can therefore be explained in terms of the mechanistic and energetic coupling proposed by the hypothesis. The remainder of this article will consider in some detail the relevance of the concepts of the chemiosmotic hypothesis to the problems of substrate translocation in microbial systems, and the experimental findings that have helped both to prove the general validity of the hypothesis and to extend our understanding of its implications to cellular function and control.

The so-called β-galactoside permease of E. coli is the archetype substrate translocation system. The membrane carrier (or permease, or M protein) is coded for by the y gene in the lac operon and, in the absence of metabolic energy, catalyses specifically the facilitated diffusion of galactosides to a transmembrane equilibrium with equal concentrations in the medium and within the cell. Metabolic energy may be coupled to the system such that 'active transport' takes place with the substrate being accumulated intracellularly to a concentration many times that in the medium. The energised membrane state required for active transport we can now identify with the protonmotive force. It is sensitive to the action of uncouplers and can be produced aerobically (or anaerobically in the presence of an alternative electron acceptor) by the redox reactions of electron transport, or anaerobically by the action of ATPase on glycolytically-produced ATP. There is no evidence for the involvement of a periplasmic binding protein, and transport activity can be demonstrated and studied in sub-cellular membrane vesicle preparations.

This capacity of membrane vesicles to demonstrate the characteristic transport reactions of whole cells is important in two respects. First, it clearly differentiates transport systems which are positive in this assay from those requiring the participation of a binding protein. Secondly, the membrane vesicle represents a significant attempt to produce a partially purified transport complex which none the less retains the essential character of membrane transport, i.e. translocation in space without chemical modification. For this second reason membrane vesicles have been very widely used in transport studies since their potentialities were first demonstrated by Kaback & Stadtman (1966). In fact the use of vesicles led directly to the experimental results on which Kaback built his respiration-linked model, and it was the eventual demonstration of a protonmotive force in these structures that led Kaback to abandon his own hypothesis in favour of Mitchell's chemiosmotic model. This rise and fall of the respiration-linked transport model has been one of the most significant features of the last few action-packed years in the field of microbial transport. It is important, therefore, that we have a clear understanding of the concepts involved and the nature of their dependence on the experimental findings with membrane vesicles.

Membrane vesicles as an experimental system

The respiration-linked model of transport proposed that the substrate translocation carriers existed in oxidised or reduced form, and that these differed in their affinity for the substrate and in their functional location within the membrane. The carriers were linked to specific segments of the respiratory chain, either directly or via a shunt pathway, such that respiration powered transport and accumulation. Under anaerobic conditions transport was still driven by the respiratory chain, using nitrate or fumarate as an alternative electron acceptor. Oxidative phosphorylation was not involved, nor could ATP drive transport other than, possibly, through reversed electron transport; substrate translocation was obligatorily linked to the terminal respiratory chain. This model was based on experimental data obtained with vesicle preparations, and on certain assumptions that were made regarding the behaviour and properties of these vesicles.

Although vesicles could oxidise a number of electron donors, the capacity to drive transport was very much more restricted. For example, sugar and amino acid uptake in *E. coli* seemed to be powered specifically by the oxidation of D-lactate. In other organisms different primary electron donors were linked to nutrient transport, e.g. D-α-glycerophosphate in *Staphylococcus aureus* and L-malate in *Azotobacter vinelandii*. ATP was unable to energise transport, even when attempts were made to trap it within vesicles during their preparation by spheroplast lysis. It was claimed that the vesicles retained the membrane orientation and general impermeability of the whole cell, with the total population being capable of transport at the same rate as the parent cells. The possible significance of the transport characteristics of the electron donor itself was largely disregarded. These findings and conclusions have been presented in a large number of papers by Kaback and his colleagues, and reviewed a number of times, most notably by Kaback & Hong (1973), Kaback (1974) and Konings (1976).

However, an equally large number of papers from laboratories around the world (and now from Kaback himself) have described contrary results, made other assumptions, and come to different conclusions. These differing viewpoints have already been discussed by, for example, Harold (1974), Simoni & Postma (1975) and Hamilton (1975), and only some of the more recent papers will be considered here.

(1) Claims that a mixed population of vesicles is obtained with respect to membrane orientation have been made for *E. coli* (Futai, 1974*a*; Hare, Olden & Kennedy, 1974; Futai & Tanaka, 1975; Mével-Ninio & Yamamoto, 1974; Rosen & McClees, 1974), *Micrococcus lysodeikticus* (Gorneva & Ryabova, 1974), *Mycobacterium phlei* (Hinds & Brodie, 1974) and *Paracoccus denitrificans* (Burnell, John & Whatley, 1975*a*). The factors controlling orientation have been partially identified in some, but by no means in all cases. For example, sonication of either spheroplasts or an actual vesicle preparation can lead to the production of a population of inverted vesicles. In at least one case (Burnell *et al.*, 1975*a*), the growth conditions determine the orientation of the vesicles obtained.

(2) To this problem must be added that associated with the permeability of the membrane to the electron donor itself. Hare *et al.* (1974) found that in their mixed populations of vesicles the great majority of the NADH oxidase activity was associated with the inverted fraction, whereas proline transport, which was not driven by NADH oxidase, was exclusively a property of the vesicles that retained the membrane orientation of the whole cell. Their explanation that the poor ability of NADH to drive transport results from the impermeability of the membrane and consequent very low NADH oxidase activity of intact right-side-out vesicles, is supported by the work of several other investigators. Futai (1974*b*) trapped NAD^+ and alcohol dehydrogenase during the preparation of the vesicles and initiated oxidation by adding ethanol to the suspension. He found that the oxidation of intravesicular NADH produced in this manner was capable of driving amino acid transport at 60–80% of the rate achieved with D-lactate. Hampton & Freese (1974) identified two K_m values for the oxidation of NADH by *Bacillus subtilis*. A high-affinity system was found with open cell envelopes and inverted vesicles. With whole cells, however, and right-side-out vesicles, the kinetics of NADH oxidation were characteristic of a low-affinity system with a high K_m. A similar high K_m was obtained for NADH-driven amino acid transport in vesicles. The equivalent study of oxidation and transport driven by α-glycerophosphate, the most efficient electron donor with this organism, gave a single K_m value corresponding to that for α-glycerophosphate dehydrogenase. These authors concluded that the low-affinity system represents the passage of NADH through the membrane, and that if allowance is made for this, then NADH is as

efficient at powering amino acid uptake as is α-glycerophosphate. In their studies of Ca^{2+} transport in inverted *E. coli* vesicles Rosen & McClees (1974) and Tsuchiya & Rosen (1975) demonstrated that NADH was the most efficient energy donor, while being relatively ineffective in the uptake of proline by right-side-out vesicles.

The efficiency of D-lactate as an energy donor in vesicle transport reactions is almost certainly the result of its own rapid translocation through the membrane by a specific carrier mechanism (Matin & Konings, 1973; Harold & Levin, 1974; Nichols & Hamilton, 1976).

The relative inefficiency of succinate as an energy donor for transport by vesicles raises some interesting points. Kaback & Milner (1970) have shown that *E. coli* vesicles convert succinate quantitatively to fumarate and are, therefore, in the chemiosmotic terms we shall consider below, capable of causing the efflux of one pair of protons per molecule of succinate oxidised. The uptake of succinate itself, though, is by an electroneutral mechanism involving the influx of two protons (Gutowski & Rosenberg, 1976). One can thus predict that the oxidation of succinate by vesicles is likely to generate a membrane potential, but not a significant pH gradient. Consequently succinate should be effective as an energy donor for the accumulation of cationic species such as lysine, but less so for neutral amino acids such as alanine or proline whose transport is driven by both the membrane potential and the pH gradient (see Hamilton, 1975). Kaback (1974) has in fact made the point that while succinate is nearly 70% as effective as D-lactate in stimulating lysine uptake, it is only 40% as effective in stimulating alanine uptake. Further, he has now demonstrated that under conditions where D-lactate and ascorbate-phenazine methosulphate generate a pH gradient of around 110 mV and a potential of around 70 mV, succinate gives a potential of 64 mV but zero pH gradient (Ramos *et al.*, 1976).

Proline is often used as a typical neutral amino acid in transport assays. In experiments with vesicles oxidising succinate, however, it is in fact quite atypical. As pointed out by Kaback (1974), succinate is only 4% as effective in stimulating proline uptake as is D-lactate, and thus the inadequacy of succinate as an electron donor is apparently particularly marked with proline as the translocated amino acid. Succinate is, however, invariably present as the sodium salt and, as noted by Harold (1976), sodium ions inhibit proline uptake by *E. coli* vesicles. Thus if one assays the effectiveness of succinate as an energy donor in *E. coli* by employing proline as the assay amino acid and ignoring consideration of the transport of succinate itself, one will very possibly come to the wrong conclusion.

(3) Similar permeability effects may be expected to produce negative results in experiments designed to explore the efficiency of ATP in powering active transport in right-side-out vesicles. Also Short, Kaback & Kohn (1975) have shown that 60–80% of the ATPase activity of the cell is lost during the preparation of membrane vesicles. Positive evidence that transport is not in fact obligatorily linked to the respiratory chain comes from the stimulation of serine uptake by ATP trapped within *E. coli* vesicles during their preparation by van Thienen & Postma (1973). In their study of Ca^{2+} transport by inverted *E. coli* vesicles Tsuchiya & Rosen (1975) also found direct stimulation from ATP hydrolysis.

(4) A particularly important demonstration that transport in vesicles can be energised by reactions in addition to the redox reactions of the respiratory chain comes from the work of MacDonald & Lanyi (1975) who reported that vesicles of *H. halobium* accumulated leucine against a concentration gradient in response to illumination. The significance of this finding in the present context is that it highlights for us the true nature of the energised membrane state (derived from respiration, ATP hydrolysis or the action of bacteriorhodopsin) capable of driving active transport. *H. halobium* is a unique bacterium whose purple membrane contains the pigment bacteriorhodopsin which, by a light-dependent proton efflux, generates transmembrane gradients of pH and potential, i.e. a protonmotive force (Oesterhelt, this volume).

Thus we can see that the respiration-linked transport model's concept of exclusive coupling through particular segments of the respiratory chain, is based on essentially negative evidence obtained with an experimental system that is itself something of an artefact. As techniques and our understanding of the system and its inevitable limitations have improved, however, so have results with vesicles come closer to the reality of our experience with whole cells. On a more positive note, one might ask if the pH gradient and membrane potential characteristic of the protonmotive force have now been demonstrated in vesicles; produced from respiration, ATP hydrolysis or photosynthesis depending on the organism and the conditions? The answer is an emphatic 'yes'. Reeves (1971) first demonstrated respiration-dependent proton efflux with vesicle preparations. This finding has been extended by a number of laboratories, noticeably Kaback's. Schuldiner & Kaback (1975) and Ramos *et al.* (1976) using the technique of flow dialysis have now shown that *E. coli* vesicles can generate a respiration-dependent membrane potential and pH gradient, and that these relate to the transport capacity of the suspensions. Hellingwerf, Michels, Dorpema & Konings

(1975) measured the membrane potential developed in response to light by vesicles from *Rhodopseudomonas spheroides* and related it to the uptake of amino acids. A membrane potential was also developed on ATP hydrolysis by inverted vesicles of *E. coli* (Griniuviene, Dzheja & Grinius, 1975a), and West & Mitchell (1974a) measured the actual flux of protons resulting from the action of the ATPase.

The real significance of these findings is that they now establish the membrane vesicle as a truly useful experimental system, now capable of producing data of genuine relevance to the problems of cellular energetics.

CHEMIOSMOTIC MODEL FOR SUBSTRATE TRANSLOCATION

The chemiosmotic hypothesis states that cellular energy transducing reactions are coupled through the reversible formation of the proton-motive force, which is defined by the equation:

$$\Delta p = \Delta\psi - Z\Delta pH,$$

where Δp is the protonmotive force in mV, $\Delta\psi$ is the membrane potential in mV, ΔpH is the transmembrane pH gradient, and Z is the factor converting pH into mV, which at 25°C equals 59 (Mitchell, 1966; Greville, 1969). The specific application of this general theory to microbial substrate translocation, states that cations should be transported across membranes on monofunctional carriers, or uniports, in response to the membrane potential, inside negative; anions should enter on bifunctional carriers, or proton symports, in response to the pH gradient, inside alkaline; the transport of neutral entities also on proton symports should be driven by the combined force of both the potential and the pH gradient (Mitchell, 1970; Harold, 1972; Hamilton, 1975). A fundamental difference between this and other models is that the link between the exergonic reactions of scalar metabolism and the ender-gonic function of vectorial transport is indirect. This fact lends itself to an experimental approach which gives unequivocal data in favour of the chemiosmotic model.

Using first a qualitative, and more recently a quantitative approach, a number of laboratories have been able to show that cells and vesicles do in fact generate a protonmotive force, of the correct magnitude and sign, in response to metabolic activity. Further, transport activity has been shown to be driven by such a force, even when it is produced in non-metabolising suspensions by manipulation of the ionic environment in the presence of ionophores and uncouplers.

The protonmotive force in metabolising cells and vesicles

Proton extrusion associated with bacterial respiration has been well documented for a number of organisms (C. W. Jones, this volume; Haddock, this volume). Energy-linked proton efflux has also been noted in *Neurospora crassa* (Slayman & Slayman, 1975), *Saccharomyces cerevisiae* (Riemersma & Alsbach, 1975) and *Rhodotorula gracilis* (Misra & Höfer, 1975). Quantitative measurements of the membrane potential developed by respiration have come from several laboratories. Griniuviene *et al.* (1975*b*) reported a value of 140 mV in *E. coli* whole cells. With *E. coli* vesicles, Altendorff, Hirata & Harold (1975) found a potential with a probable magnitude in the region of 125 mV, while Schuldiner & Kaback (1975) and Ramos *et al.* (1976) have reported potentials of the order of 70 mV with a number of electron donors.

The action of ATPase on ATP produced by glycolysis in *Streptococcus faecalis* also generates a potential of 140–190 mV (Harold & Papineau 1972; Laris & Pershadsingh, 1974), and a pH gradient of 0.81 – 1.0 (Harold, Pavlasova & Baarda, 1970). Riebeling, Thauer and Jungermann (1975) have measured a similar pH gradient of 0.4 – 0.8 in *Clostridium pasteurianum*.

Estimates of the total protonmotive force have been obtained by Collins & Hamilton (1976) for *E. coli* and *Staphlyococcus aureus*. *E. coli* developed a protonmotive force of 230 mV, which was made up from a potential of 132 mV and a pH gradient of 1.65. In *S. aureus* the corresponding figures were 211 mV, 134 mV and 1.3. The adequacy of these driving forces to power ATP synthesis and active transport was discussed by the authors.

These data were obtained at a medium pH of 6.5. In their quantitation of the protonmotive force in *E. coli*, Padan, Zilberstein & Rottenberg (1976) recorded values of 122–129 mV at pH 7 over a range of potassium concentrations. When pH was varied at a constant potassium concentration, however, it was found that although the potential remained relatively constant at around 80 mV, the pH gradient decreased from 2.0 at pH 6 to zero at pH 7.65, and −0.51 (i.e. inside acid) at pH 9. A similar effect of medium pH was noted by Ramos *et al.* (1976) in their study of the protonmotive force developed by respiration on ascorbate-phenazine methosulphate in their *E. coli* vesicles. Over the pH range 5 to 8, the potential was approximately constant at around 70 mV; the pH gradient however decreased from 1.95 at pH 5.5 to zero at pH 7.5.

Of these studies, several linked the potential and pH gradients developed to the capacity to drive transport reactions (Slayman & Slayman, 1975; Misra & Höfer, 1975; Schuldiner & Kaback, 1975). In particular, Ramos *et al.* (1976) showed that the established inability of NADH oxidation to power transport was mirrored by its inability to develop a protonmotive force. On the addition of ubiquinone, however, NADH oxidation generated a protonmotive force of 131 mV and became an efficient energy source for transport reactions. Kashket & Wilson (1974) were able to show a direct quantitative dependence of thio-methyl-β-D-galactoside (TMG) uptake by *Streptococcus lactis* on the protonmotive force developed during fermentation. Their figures were in fact considerably lower than those claimed by other workers using closely related species. Kashket & Wilson (1974) measured a proton-motive force of 79 mV with glucose as the fermentable energy source, and 39 mV with arginine. In both cases the potential was around 37 mV, with a pH gradient of 0.75 with glucose and zero with arginine.

Proton fluxes associated with substrate translocation

There are essentially two experimental procedures that have been developed to test the validity of the chemiosmotic model for substrate translocation; a flux of protons may be assayed with those transport reactions proposed to operate through a proton symport, or the response to artificially produced gradients of potential and pH may be noted in non-metabolising suspensions.

The first demonstration of a proton flux associated with the uptake of neutral β-galactosides in resting *E. coli* was made by West (1970) and West & Mitchell (1972, 1973). In accordance with the predictions of the chemiosmotic model there was a strict 1:1 stoichiometry of protons to sugar, and the transport was electrogenic in that the proton/sugar/ carrier complex facilitated the passage of a positive charge across the membrane. Under metabolic conditions the uncharged carrier would therefore be expected to bind a proton and one molecule of sugar, whose translocation would then be driven by both the pH gradient and the membrane potential, i.e. by the total protonmotive force. In a study of a mutant which had lost the ability to accumulate β-galactosides against a concentration gradient although still having the capacity for facilitated diffusion, West & Wilson (1973) showed that the ability to co-transport protons was also lost.

These penetrating studies have stimulated a large number of laboratories to test for proton symport carrier mechanisms using the same basic experimental approach. Henderson & Skinner (1974) showed that proton uptake was associated with the transport of galactose, fucose and arabinose in appropriately induced strains of *E. coli*. The anionic glucose-6-phosphate is taken up by *E. coli* with a minimum proton stoichiometry of 1:1 (Essenberg & Kornberg, 1975). The transport of the dicarboxylic acids succinate, fumarate and malate in *E. coli* is electroneutral as a result of the observed 2:1 proton stoichiometry (Gutowski & Rosenberg, 1976). In the same cells there is evidence for an inducible low-affinity system which transports the monocarboxylic aspartate in an electrogenic manner with the same 2:1 proton stoichiometry; the constitutive high-affinity system does not appear to involve protons at all. The electroneutral exchange on a proton/sodium antiport by a mechanism directly equivalent to those above, appears to be the mechanism of sodium extrusion in *E. coli* (West & Mitchell, 1974*b*).

Equivalent findings have resulted from studies with other organisms. Harold & Spitz (1975) found that arsenate and phosphate entered *Streptococcus faecalis* through the action of a proton symport (or its thermodynamic equivalent, an hydroxyl antiport), but these authors concluded that this mechanism only functioned to preserve electroneutrality across the membrane. They invoked an energy coupling mechanism utilising ATP. Proton fluxes associated with hexose uptake in *Chlorella vulgaris* (Komor, 1973), *Neurospora crassa* (Slayman & Slayman, 1975) and *R. gracilis* (Höfer, Misra & Dahle, 1974) have also been recorded. Cuppoletti & Segel (1975) studied the kinetics of sulphate transport by *Penicillium notatum*. They obtained evidence for a 1:1 proton stoichiometry but concluded that the active form of the carrier was the ternary complex, carrier/SO_4^{2-}/Ca^{2+}/H^+. A similar complex has already been suggested by Willecke, Gries & Oehr (1973) for the transport of citrate by *Bacillus subtilis* as carrier/citrate^{3-}/Mg^{2+}/H^+. The work of Eddy's group on the co-transport of protons with a number of sugars and amino acids by various strains of *Saccharomyces* will be discussed below.

Substrate translocation driven by artificial gradients of potential and pH

The demonstrations of nutrient accumulation in response to imposed ion gradients in resting suspensions of *S. faecalis* and vesicles of *E. coli* by Harold and his colleagues, of *Streptococcus lactis* by Wilson and co-workers, and of *Staphylococcus aureus* in my own laboratory have already been reviewed (Hamilton, 1975). More recent findings are those given here. In whole-cell studies with *Clostridium pasteurianum*, Booth & Morris (1975) have demonstrated the accumulation of galactose by an electrogenic mechanism driven by a membrane potential and/or a transmembrane pH gradient. Gluconate on the other hand is taken up electroneutrally in response to a pH gradient only. Lactate flux across the *S. faecalis* cell membrane is also electroneutral and is driven by the pH gradient (Harold & Levin, 1974). It is interesting to note that in this organism, which obtains its energy by fermentation, the direction of lactate flux will normally be towards exit. Yet when a pH gradient, inside alkaline, is induced, lactate is accumulated in the non-metabolising cells. From this we can draw the important conclusion that the carrier is itself isotropic and responds passively to the transmembrane gradients of chemical and electrical potential, both with regard to the direction of flux and the final equilibrium attained.

In vesicle preparations from *E. coli* potassium diffusion potentials, inside negative and created by valinomycin treatment of potassium-loaded vesicles in low-potassium medium, can drive the uptake of proline, glycine, TMG, lactose, tyrosine and glutamate (Hirata, Altendorf & Harold, 1974; Schuldiner & Kaback, 1975). A potassium diffusion potential also drives leucine uptake in vesicles of *Halobacterium halobium* (MacDonald & Lanyi, 1975). An induced pH gradient, inside acid, drives the accumulation of Ca^{2+} by inverted *E. coli* vesicles (Tsuchiya & Rosen, 1976).

The development of a pH gradient, inside alkaline, across the membrane of *Paracoccus denitrificans* vesicles can drive the uptake of either phosphate or sulphate (Burnell, John & Whatley, 1975a, b). This work is of particular importance due to the authors' ability to prepare separate populations of right-side-out and of inverted vesicles. Whereas respiration is only able to drive anion uptake into right-side-out vesicles, artificially induced pH gradients cause accumulation in both populations. Thus again we have convincing evidence that the direction of flux is controlled by the driving force and does not result from any inherent anisotropy in the carrier itself.

In completing this section on the validity of the chemiosmotic hypothesis to the problems of cellular energetics in general and of membrane transport in particular, reference must be made to a large body of work on oxidative phosphorylation and the structure and function of the bacterial ATPase. Space restrictions dictate that this must be omitted from the present discussion, but interested readers should consult the articles of Simoni & Postma (1975), Rosen & Adler (1975) and Wilson, Alderete, Maloney & Wilson (1976).

UNSOLVED PROBLEMS AND THE DIRECTION
OF FUTURE RESEARCH

While there now seems no justifiable reason for omitting the chemiosmotic hypothesis from any analysis of cellular energetics, it must be appreciated that its primary function is to rationalise the thermodynamic properties of the system. This is especially true of experimental studies concerned with the steady-state transport equilibria achieved in resting suspensions of whole cells or subcellular membrane vesicles. We are still required, however, to identify and characterise the molecular mechanisms giving rise to membrane translocation and energy coupling. Further, the integrated functioning of the cell's transport and metabolic activities during growth will introduce aspects of kinetic steady states, and genotypic and phenotypic control that we in turn must integrate into our overall understanding of the energetic phenomena.

The nature of the carrier

Despite the comparatively long time that has elapsed since Fox & Kennedy (1965) first identified the β-galactoside permease with their M protein, progress has been remarkably slow in this area. Using the same techniques Fournier & Pardee (1974) found three proteins from the membranes of induced *Bacillus subtilis* which they identified as possible sub-units of the malate carrier. The solubilisation and partial purification of the SBP1 and SBP2 proteins from the dicarboxylic acid transport system in *E. coli*, have already been referred to (Lo & Sanwal, 1975).

At best, however, these proteins have only been assayed by their ability to bind their putative substrates. What is really required is the reconstitution of an active transport system, very much along the lines of the elegant work that has been performed with oxidative phosphorylation (Racker, 1975; Yoshida, Sone, Hirata & Kagawa, 1975). Such a transport system has now been achieved by this same Japanese group (Hirata, Sone, Yoshida & Kagawa, 1976). They have solubilised and partially purified the alanine carrier from a thermophilic bacterium and incorporated it into vesicles made from phospholipids extracted from the same organism. The development of a potassium diffusion potential was then capable of driving an uncoupler-sensitive alanine uptake.

As to the location of the carrier in the membrane and the nature of the translocation mechanism, increasingly our understanding of membrane structure suggests that models proposing protein diffusion across, or rotation within the membrane are unrealistic (see, for example, Kaprelyanz, Binyukov, Ostrovskii & Grigoryan, 1974). The exciting new discoveries with bacteriorhodopsin from *H. halobium* (Henderson & Unwin, 1975; Oesterhelt, this volume) and with the glucose transport carrier from erythrocytes (Kahlenberg, 1976), suggest that membrane transport is likely to be facilitated by a protein, spanning the membrane and enclosing a water-filled channel within a structure composed of an aggregate of subunits.

The mechanism of energy coupling

Earlier kinetic analyses of transport came to various conclusions as to whether energy coupling affected the entry or exit process. Wilson & Kusch (1972) for example, claimed that energy coupling reduces the exit rate of galactosides from *E. coli*. On the other hand, Lancaster, Hill & Struve (1975), studying the partial de-energisation of the same transport system, concluded that both entry and exit were affected. In their extensive studies of hexose transport by the alga *Chlorella vulgaris*, Komor, Haas, Komor & Tanner (1973) demonstrated that accumulation of the non-metabolised 6-deoxyglucose could only be explained kinetically on the basis of energy coupling affecting not either, but both, the affinity of the carrier for the sugar, and the velocity constants for the apparent flux of complexed and free carrier.

These same systems, however, have now been re-examined with quite different experimental rationales. West & Wilson (1973) showed that the nature of the lesion in the energy-uncoupled mutant studied by Wilson & Kusch (1972) was the loss of the ability to co-transport protons during sugar uptake. Schuldiner, Kerwar, Kaback & Weil (1975) and Schuldiner, Weil & Kaback (1976) have studied the specific binding of fluorescent dansylgalactoside derivatives to the galactoside carrier of *E. coli*. They have demonstrated the apparent energy dependence of this binding and suggested that this is indicative of the carrier bearing a net negative charge such that the binding site is only accessible to the medium after the generation of a membrane potential, inside negative.

Komor & Tanner (1974) have subjected their hexose uptake system in *C. vulgaris* to a penetrating analysis. From a study of the effects of pH on the maximal rate of transport of 6-deoxyglucose, these authors concluded that the carrier can exist in one of two forms; a protonated high-affinity form with a K_m for sugar of 0.3 mM, and an unprotonated low-affinity form with a K_m of 50 mM. The K_m for protons has been estimated at 0.14 μM; that is to say, at pH 6.85 half the carrier molecules will be protonated and half unprotonated. This has the effect of varying the stoichiometry such that at pH 6 it approaches one proton per molecule of sugar, while at pH 8 it has decreased to one proton per four molecules of sugar. This decrease in proton-to-sugar stoichiometry is reflected in the greatly decreased capacity for accumulation against the concentration gradient at the higher pH value.

In this connection it is perhaps worth drawing attention to a similar finding reported by Wilson & Kusch (1972). They noted that the parent *E. coli*, presumably with a 1:1 proton-to-sugar stoichiometry, showed decreased accumulation capacity above pH 7, and that at pH 8 it could only accumulate to the same extent as the energy-uncoupled mutant.

It is from studies of this nature that we shall arrive at a real understanding of the molecular mechanisms of energy coupling in substrate translocations.

The integration of transport with total cellular function

Inevitably the studies that have formed the substance of this review have been carried out on transport systems that, in one way or another, have been isolated from other cell functions. In the growing organism clearly the situation is very different, and it is salutary to consider the various factors that might be expected to modify transport patterns, either quantitatively or even qualitatively.

Many of the experimental designs we have considered have involved thermodynamic steady states in non-metabolising cells or vesicles. While these have been ideal in allowing a study of the chemiosmotic energy coupling mechanism involved, they will not necessarily lead to a true estimate of the likely intracellular concentration of translocated solute under conditions of active metabolism. Here the determining factors are more likely to be kinetic, with the amount accumulated being dependent upon the rates of entry, of exit, and of conversion to other metabolites. For example, Lee, Robins & Oxender (1975) showed that in merodiploid strains of *E. coli* with double the complement of the *dag*A gene, the transport of glycine, D-alanine and D-serine was increased, both in rate and, as a consequence, in the amount accumulated.

As mentioned briefly at the beginning of this article, and dealt with more fully by Kornberg and Jones-Mortimer (this volume), there are control mechanisms affecting both the synthesis and activity of transport reactions. Although repression appears to be more common in bacteria (Whiting, Midgley & Dawes, 1976; Templeton & Savageau, 1974), transinhibition effects have been noted (Kadner, 1975); they are more common in eukaryotic cells (Morrison & Lichstein, 1976). Clearly such effects must be superimposed on any 'simple' energetic analysis.

Another problem relates to the adequacy of the proton-motive force to account for some of the intracellular to extracellular concentration ratios that have actually been observed (see Cockburn, Earnshaw & Eddy, 1975). The chemiosmotic transport model states that anions such as phosphate and glutamate will experience electroneutral translocation by a proton symport mechanism. The driving force would therefore be the pH gradient which at 1.65 (Collins & Hamilton, 1976) could only power a concentration ratio of around 40. In accordance with the restriction imposed by these calculations, and as a result of their studies with ionophores and the ATPase inhibitor N,N'-dicyclohexylcarbodiimide (DCCD), Harold & Spitz (1975) concluded that the uptake of arsenate and phosphate, and possibly of glutamate and aspartate, by *Streptococcus faecalis* is driven by ATP and that the proton symport functions simply to preserve electroneutrality. This is the only recorded case of such a direct involvement of phosphate-bond energy in transport in a Gram-positive organism, but it of course parallels the findings with the binding protein systems identified in Gram-negative bacteria. In common with these systems Harold & Spitz (1975) have found that the uptake of arsenate by *S. faecalis* is essentially unidirectional with no evidence for exchange diffusion or counterflow.

Cockburn *et al.* (1975) and Collins, Jarvis, Lindsay & Hamilton (1976) have considered further the potential energy of accumulation inherent in the protonmotive force. The energy available for transport can be given by the expression: $(m+n)\Delta\psi-nZ\Delta pH$, where $\Delta\psi$ is the membrane potential, inside negative; $-Z\Delta pH$ is the pH gradient, inside alkaline, expressed in mV; m is the charge on the translocated solute, and n is the number of protons co-translocated. Where an anion enters on an electroneutral proton symport with a stoichiometry of 1:1, the maximum driving force is therefore the pH gradient. In *Saccharomyces carlsbergensis* this is of the order of 2 units and consequently the maximum possible phosphate concentration ratio by this mechanism would be 10^2. Cockburn *et al.* (1975) have however shown that the proton-to-phosphate stoichiometry can be as high as 3:1, giving energetic feasibility to concentration ratios up to 10^{10}. Seaston, Carr & Eddy (1976) have extended these and earlier studies of stoichiometries greater than 1:1 for the transport of neutral amino acids and sugars by *Saccharomyces* species. They have shown that the electrogenic transport of glycine is powered by a symport with a 2:1 proton to glycine stoichiometry. These authors have further suggested that energy coupling in the yeast amino acid permease serves mainly to increase the V_{max} value of the entry system. They point out that this is not compatible with the suggestion (Schuldiner *et al.*, 1975, 1976) that the free carrier in *E. coli* bears a net negative charge. Seaston *et al.* (1976) further discuss how the sign and the number of charges associated with the carrier are an important aspect of transport kinetics.

In our own studies with *E. coli* and amino acid transport (Collins *et al.*, 1976) we have argued that an increased ability to accumulate nutrient would be particularly advantageous during growth at low concentrations in the medium. Under alanine limitation in the chemostat we have isolated mutants with proton-to-alanine stoichiometries of 2:1 and 4:1; the parent organism used as an inoculum gave a 1:1 stoichiometry. The mutants arose spontaneously after several days in the chemostat and simply outgrew the parent due to their ability to grow at faster rates at the very low concentrations of alanine (10^{-7}–10^{-8} M) in the pot. The mutations were shown to affect only the transport mechanism specific for alanine, glycine and serine.

These findings are of great significance in that they extend the scope of the chemiosmotic transport model beyond its simple statement in terms of equilibrium thermodynamics. Clearly the existence of proton stoichiometries greater than one, removes the quantitative restriction that would otherwise have to be imposed. Further, the capacity to respond to environmental pressures by a mechanism under genetic control is the final accolade of respectability in a world dominated by molecular biology and ecology.

CONCLUSIONS

It is now 20 years since Cohen & Monod (1957) published their review 'Bacterial Permeases'. During that time there has been much activity in the field of microbial transport, no small degree of controversy, and often a considerable amount of confusion. In the last six or seven years, the activity and controversy have sometimes risen to fever pitch, but now we appear to have been compensated by at least some decrease in the confusion.

Even within one organism, inevitably *E. coli*, it is possible to identify two quite separate mechanisms of group translocation, and two of substrate translocation. The phosphotransferase system has yielded many of its inner secrets to extended biochemical and genetic studies in a number of laboratories throughout the world. The true nature of the involvement of the periplasmic binding proteins in the transport of some substrates through the outer and inner membrane barriers of Gram-negative bacteria remains something of a challenging enigma at the present time. The central role of the chemiosmotic hypothesis in controlling shock-resistant substrate translocations across the cytoplasmic membrane is now established beyond doubt.

Whereas these systems have been studied as unique functional entities, it is clear that their functioning and control are in fact interrelated, and that in certain cases they may even share common structural elements.

As to the future, the full range of biochemical and genetic technology can now be brought to bear on the outstanding problems concerning the molecular mechanisms which determine these cellular phenomena. It is with some confidence that this reviewer concludes that the researches of these last 20 years have given us a firm conceptual grasp of the true nature of vectorial transport on which to build the studies of the next two decades.

I should like to thank Dr I. R. Booth for his critical reading of this manuscript, and Drs F. M. Harold, H. R. Kaback and W. N. Konings for sending me copies of their papers prior to publication.

REFERENCES

ADLER, J. & EPSTEIN, W. (1974). Phosphotransferase-system enzymes as chemoreceptors for certain sugars in *Escherichia coli* chemotaxis. *Proceedings of the National Academy of Sciences, USA*, **71**, 2895–9.

ALTENDORF, K., HIRATA, H. & HAROLD, F. M. (1975). Accumulation of lipid-soluble ions and of rubidium as indicators of the electrical potential in membrane vesicles of *Escherichia coli. Journal of Biological Chemistry*, **250**, 1405–12.

BAG, J. (1974). Glucose inhibition of the transport and phosphoenolpyruvate-dependent phosphorylation of galactose and fructose in *Vibrio cholerae. Journal of Bacteriology*, **118**, 764–7.

BERGER, E. A. (1973). Different mechanisms of energy coupling for the active transport of proline and glutamine in *Escherichia coli. Proceedings of the National Academy of Sciences, USA*, **70**, 1514–18.

BERGER, E. A. & HEPPEL, L. A. (1974). Different mechanisms of energy coupling for the shock-sensitive and shock-resistant amino acid permeases of *Escherichia coli. Journal of Biological Chemistry*, **249**, 7747–55.

Boos, W. (1974*a*). Bacterial transport. *Annual Review of Biochemistry*, **43**, 123–46.

Boos, W. (1974*b*). Pro and contra carrier proteins; sugar transport via the periplasmic galactose-binding protein. In *Current Topics in Membranes and Transport*, ed. F. Bronner & A. Kleinzeller, vol. 5, pp. 51–136. New York & London: Academic Press.

BOOTH, I. R. & MORRIS, J. G. (1975). Proton-motive force in the obligately anaerobic bacterium *Clostridium pasteurianum*: a role in galactose and gluconate uptake. *FEBS Letters*, **59**, 153–7.

BRADBEER, C., WOODROW, M. L. & KHALIFAH, L. I. (1976). Transport of vitamin B_{12} in *Escherichia coli*: common receptor system for vit. B_{12} and bacteriophage BF 23 on the outer membrane of the cell envelope. *Journal of Bacteriology*, **125**, 1032–9.

BURNELL, J. N., JOHN, P. & WHATLEY, F. R. (1975a). The reversibility of active sulphate transport in membrane vesicles of *Paracoccus denitrificans*. *Biochemical Journal*, **150**, 527–36.

BURNELL, J. N., JOHN, P. & WHATLEY, F. R. (1975b). Phosphate transport in membrane vesicles of *Paracoccus denitrificans*. *FEBS Letters*, **58**, 215–18.

CHEVALLIER, M. R., JUND, R. & LACROUTE, F. (1975). Characterization of cytosine permeations in *Saccharomyces cerevisiae*. *Journal of Bacteriology*, **122**, 629–41.

COCKBURN, M., EARNSHAW, P. & EDDY, A. A. (1975). The stoicheiometry of the absorption of protons with phosphate and L-glutamate by yeasts of the genus *Saccharomyces*. *Biochemical Journal*, **146**, 705–12.

COHEN, G. N. & MONOD, J. (1957). Bacterial permeases. *Bacteriological Reviews*, **21**, 169–94.

COLLINS, S. H. & HAMILTON, W. A. (1976). Magnitude of the protonmotive force in respiring *Staphylococcus aureus* and *Escherichia coli*. *Journal of Bacteriology*, **126**, 1224–31.

COLLINS, S. H., JARVIS, A. W., LINDSAY, R. J. & HAMILTON, W. A. (1976). Proton movements coupled to lactate and alanine transport in *Escherichia coli*: isolation of mutants with altered stoichiometry in alanine transport. *Journal of Bacteriology*, **126**, 1232–44.

COWELL, J. L. (1974). Energetics of glycylglycine transport in *Escherichia coli*. *Journal of Bacteriology*, **120**, 139–46.

CUPPOLETTI, J. & SEGEL, I. H. (1975). Kinetics of sulfate transport by *Penicillium notatum*. Interactions of sulfate, protons, and calcium. *Biochemistry*, **14**, 4712–18.

CURTISS, S. J. (1974). Mechanism of energy coupling for transport of D-ribose in *Escherichia coli*. *Journal of Bacteriology*, **120**, 295–303.

DEL CAMPO, F. F., HERNANDEZ-ASENSIO, M. & RAMIREZ, J. M. (1975). Transport of α-methyl glucoside in mutants of *Escherichia coli* K12 deficient in Ca^{2+}. Mg^{2+}-activated ATPase. *Biochemical and Biophysical Research Communications*, **63**, 1099–105.

ESSENBERG, R. C. & KORNBERG, H. L. (1975). Energy coupling in the uptake of hexose phosphate by *Escherichia coli*. *Journal of Biological Chemistry*, **250**, 939–45.

FOURNIER, R. E. & PARDEE, A. B. (1974). Evidence for inducible, L-malate binding proteins in the membrane of *Bacillus subtilis*. Identification of presumptive components of C_4-dicarboxylate transport systems. *Journal of Biological Chemistry*, **249**, 5948–54.

FOX, C. F. & KENNEDY, E. P. (1965). Specific labelling and partial purification of the M protein, a component of the β-galactoside transport system of *Escherichia coli*. *Proceedings of the National Academy of Sciences, USA*, **54**, 891–9.

FROST, G. E. & ROSENBERG, H. (1975). Relationship between the *ton*B locus and iron transport in *Escherichia coli*. *Journal of Bacteriology*, **124**, 704–12.

FUTAI, M. (1974*a*). Orientation of membrane vesicles from *Escherichia coli* prepared by different procedures. *Journal of Membrane Biology*, **15**, 15–28.

FUTAI, M. (1974*b*). Stimulation of transport into *Escherichia coli* membrane vesicles by internally generated reduced nicotinamide adenine dinucleotide. *Journal of Bacteriology*, **120**, 861–5.

FUTAI, M. & TANAKA, Y. (1975). Localization of D-lactate dehydrogenase in membrane vesicles prepared by using a French press or ethylenediamine-tetraacetate-lysozyme from *Escherichia coli*. *Journal of Bacteriology*, **124**, 470–5.

GORNEVA, G. E. & RYABOVA, I. D. (1974). Membrane orientation in vesicles from *Micrococcus lysodeikticus*. *FEBS Letters*, **42**, 271–4.

GREVILLE, G. D. (1969). A scrutiny of Mitchell's chemiosmotic hypothesis of respiratory chain and photosynthetic phosphorylation. In *Current Topics in Bioenergetics*, ed. D. R. Sanadi, vol. 3, pp. 1–78. New York & London: Academic Press.

GRINIUVIENE, B., CHMIELIAUSKAITE, V., MELVYDAS, V., DZHEJA, P. & GRINIUS, L. (1975*b*). Conversion of *Escherichia coli* cell-produced metabolic energy into electric form. *Journal of Bioenergetics*, **7**, 17–38.

GRINIUVIENE, B., DZHEJA, P. & GRINIUS, L. (1975*a*). Anilinonaphthalene-sulfonate as a fluorescent probe of the energized membrane state in *Escherichia coli* cells and sonicated membrane particles. *Biochemical and Biophysical Research Communications*, **64**, 790–6.

GUTOWSKI, S. J. & ROSENBERG, H. (1975). Succinate uptake and related proton movements in *Escherichia coli* K12. *Biochemical Journal*, **152**, 647–54.

GUTOWSKI, S. J. & ROSENBERG, H. (1976). Energy coupling to active transport in anaerobically grown mutants of *Escherichia coli* K12. *Biochemical Journal*, **154**, 731–4.

GUYMON, L. F. & EAGON, R. G. (1974). Transport of glucose, gluconate, and methyl α-D-glucoside by *Pseudomonas aeruginosa*. *Journal of Bacteriology*, **117**, 1261–9.

HAMILTON, W. A. (1975). Energy coupling in microbial transport. *Advances in Microbial Physiology*, **12**, 1–53.

HAMPTON, M. L. & FREESE, E. (1974). Explanation for the apparent inefficiency of reduced nicotinamide adenine dinucleotide in energizing amino acid transport in membrane vesicles. *Journal of Bacteriology*, **118**, 497–504.

HARE, J. F., OLDEN, K. & KENNEDY, E. P. (1974). Heterogeneity of membrane vesicles from *Escherichia coli* and their subfractionation with antibody to ATPase. *Proceedings of the National Academy of Sciences, USA*, **71**, 4843–6.

HAROLD, F. M. (1972). Conservation and transformation of energy by bacterial membranes. *Bacteriological Reviews*, **36**, 172–230.

HAROLD, F. M. (1974). Chemiosmotic interpretation of active transport in bacteria. *Annals of the New York Academy of Sciences*, **227**, 297–311.

HAROLD, F. M. (1976). Membranes and energy transduction in bacteria. In *Current Topics in Bioenergetics*, vol. 6, ed. D. R. Sanadi, New York & London: Academic Press, in press.

HAROLD, F. M. & LEVIN, E. (1974). Lactic acid translocation: terminal step in glycolysis by *Streptococcus faecalis*. *Journal of Bacteriology*, **117**, 1141–8.

HAROLD, F. M. & PAPINEAU, D. (1972). Cation transport and electrogenesis by *Streptococcus faecalis*. The membrane potential. *Journal of Membrane Biology*, **8**, 27–44.

HAROLD, F. M., PAVLASOVA, E. & BAARDA, J. R. (1970). A transmembrane pH gradient in *Streptococcus faecalis*: origin and dissipation by proton conductors and N,N'-dicyclohexylcarbodiimide. *Biochimica et Biophysica Acta*, **196**, 235–44.

HAROLD, F. M. & SPITZ, E. (1975). Accumulation of arsenate, phosphate and aspartate by *Streptococcus faecalis*. *Journal of Bacteriology*, **122**, 266–77.

HAYS, J. B., SIMONI, R. D. & ROSEMAN, S. (1973). Sugar transport. V. A trimeric lactose-specific phospho-carrier protein of the *Staphylococcus aureus* phosphotransferase system. *Journal of Biological Chemistry*, **248**, 941–56.

HAZELBAUER, G. L. (1975a). Maltose chemoreceptor of *Escherichia coli*. *Journal of Bacteriology*, **122**, 206–14.

HAZELBAUER, G. L. (1975b). Role of the receptor for bacteriophage lambda in the functioning of the maltose chemoreceptor of *Escherichia coli*. *Journal of Bacteriology*, **124**, 119–26.

HELLINGWERF, K. J., MICHELS, P. A. M., DORPEMA, J. W. & KONINGS, W. N. (1975). Transport of amino acids in membrane vesicles of *Rhodopseudomonas spheroides* energized by respiratory and cyclic electron flow. *European Journal of Biochemistry*, **55**, 397–406.

HENDERSON, P. J. F. & SKINNER, A. (1974). Association of proton movements with the galactose and arabinose transport systems of *Escherichia coli*. *Biochemical Society Transactions*, **2**, 543–5.

HENDERSON, R. & UNWIN, P. N. T. (1975). Three-dimensional model of purple membrane obtained by electron microscopy. *Nature New Biology*, **257**, 28–32.

HINDS, T. R. & BRODIE, A. F. (1974). Relationship of a proton gradient to the active transport of proline with membrane vesicles from *Mycobacterium phlei*. *Proceedings of the National Academy of Sciences, USA*, **71**, 1202–6.

HIRATA, H., ALTENDORF, K. & HAROLD, F. M. (1974). Energy coupling in membrane vesicles of *Escherichia coli*. 1. Accumulation of metabolites in response to an electrical potential. *Journal of Biological Chemistry*, **249**, 2939–45.

HIRATA, H., SONE, N., YOSHIDA, M. & KAGAWA, Y. (1976). Solubilization and partial purification of alanine carrier from membranes of a thermophilic bacterium and its reconstitution into functional vesicles. *Biochemical and Biophysical Research Communications*, **69**, 665–71.

HÖFER, M., MISRA, P. C. & DAHLE, P. (1974). On the uniformity of the energy-dependent (non-phosphorylating) monosaccharide transport across the cell membrane of eukaryotes. *Proceedings of the Fourth International Symposium on Yeasts, Vienna 1974*, pp. 287–8.

JACKMAN, L. E. & HOCHSTADT, J. (1976). Regulation of purine utilization in bacteria. VI. Characterization of hypoxanthine and guanine uptake into isolated membrane vesicles from *Salmonella typhimurium. Journal of Bacteriology*, **126**, 312–26.

KABACK, H. R. (1974). Transport studies in bacterial membrane vesicles, *Science*, **186**, 882–92.

KABACK, H. R. & HONG, J.-S. (1973). Membranes and transport. *Critical Reviews of Microbiology*, **2**, 333–76.

KABACK, H. R. & MILNER, L. S. (1970). Relationship of a membrane-bound D-lactic dehydrogenase to amino acid transport in isolated bacterial membrane preparations. *Proceedings of the National Academy of Sciences, USA*, **66**, 1008–15.

KABACK, H. R. & STADTMAN, E. R. (1966). Proline uptake by an isolated cytoplasmic membrane preparation of *Escherichia coli. Proceedings of the National Academy of Sciences, USA*, **55**, 920–7.

KADNER, R. J. (1975). Regulation of methionine transport activity in *Escherichia coli. Journal of Bacteriology*, **122**, 110–19.

KADNER, R. J. & WINKLER, H. M. (1975). Energy coupling for methionine transport in *Escherichia coli. Journal of Bacteriology*, **123**, 985–91.

KAHANE, S., MARCUS, M., METZER, E. & HALPERN, Y. S. (1976*a*). Effect of growth conditions on glutamate-utilizing mutants of *Escherichia coli. Journal of Bacteriology*, **125**, 762–9.

KAHANE, S., MARCUS, M., METZER, E. & HALPERN, Y. S. (1976*b*). Glutamate transport in membrane vesicles of the wild-type strain and glutamate-utilizing mutants of *Escherichia coli. Journal of Bacteriology*, **125**, 770–5.

KAHLENBERG, A. (1976). Partial purification of a membrane protein from human erythrocytes involved in glucose transport. *Journal of Biological Chemistry*, **251**, 1582–90.

KAPRELYANZ, A. S., BINYUKOV, V. I., OSTROVSKII, D. N. & GRIGORYAN, G. L. (1974). Mobility of protein molecules in *Micrococcus lysodeikticus* membranes. *FEBS Letters*, **40**, 33–6.

KASHKET, E. R. & WILSON, T. H. (1974). Protonmotive force in fermenting *Streptococcus lactis* 7962 in relation to sugar accumulation. *Biochemical and Biophysical Research Communications*, **59**, 879–86.

KOMOR, E. (1973). Proton-coupled hexose transport in *Chlorella vulgaris. FEBS Letters*, **38**, 16–18.

KOMOR, E., HAAS, D., KOMOR, B. & TANNER, W. (1973). The active hexose-uptake system of *Chlorella vulgaris.* K_m values for 6-deoxyglucose influx and efflux and their contribution to sugar accumulation. *European Journal of Biochemistry*, **39**, 193–200.

KOMOR, E. & TANNER, W. (1974). The hexose-proton co-transport system of *Chlorella.* pH-dependent change in K_m values and translocation constants of the uptake system. *Journal of General Physiology*, **64**, 568–81.

KONINGS, W. N. (1976). Active solute transport in bacterial membrane vesicles. *Advances in Microbial Physiology*, in press.

KORNBERG, H. L. (1973*a*). Fine control of sugar uptake by *Escherichia coli.*

In *Rate Control in Biological Processes*, ed. D. D. Davies, *Society for Experimental Biology Symposium XXVII*, pp. 175–93. London: Cambridge University Press.

KORNBERG, H. L. (1973*b*). Carbohydrate transport by micro-organisms. *Proceedings of the Royal Society of London, Series B*, **183**, 105–23.

KORNBERG, H. L. & RIORDAN, C. (1976). Uptake of galactose into *Escherichia coli* by facilitated diffusion. *Journal of General Microbiology*, **94**, 75–89.

LANCASTER, J. R., HILL, R. J. & STRUVE, N. G. (1975). The characterization of energized and partially de-energized (respiration-independent) β-galactoside transport into *Escherichia coli*. *Biochimica et Biophysica Acta*, **401**, 285–98.

LARIS, P. C. & PERSHADSINGH, H. A. (1974). Estimations of membrane potentials in *Streptococcus faecalis* by means of a fluorescent probe. *Biochemical and Biophysical Research Communications*, **57**, 620–6.

LEE, M., ROBINS, J. C. & OXENDER, D. L. (1975). Transport properties of merodiploids covering the *dag*A locus in *Escherichia coli* K12. *Journal of Bacteriology*, **122**, 1001–5.

LIEBERMAN, M. A. & HONG, J.-S. (1976). Energization of osmotic shock-sensitive transport systems in *Escherichia coli* requires more than ATP. *Archives of Biochemistry and Biophysics*, **172**, 312–15.

LO, T. C. Y. & SANWAL, B. D. (1975). Membrane bound substrate recognition components of the dicarboxylate transport system in *Escherichia coli*. *Biochemical and Biophysical Research Communications*, **63**, 278–85.

MACDONALD, R. E. & LANYI, J. K. (1975). Light-induced leucine transport in *Halobacterium halobium* envelope vesicles: a chemiosmotic system. *Biochemistry*, **14**, 2882–9.

MATIN, A. & KONINGS, W. N. (1973). Transport of lactate and succinate by membrane vesicles of *Escherichia coli*, *Bacillus subtilis*, and a *Pseudomonas* species. *European Journal of Biochemistry*, **34**, 58–67.

MÉVEL-NINIO, M. & YAMAMOTO, T. (1974). Conversion of active transport vesicles of *Escherichia coli* into oxidative phosphorylation vesicles. *Biochimica et Biophysica Acta*, **357**, 63–6.

MINER, K. M. & FRANK, L. (1974). Sodium-stimulated glutamate transport in osmotically shocked cells and membrane vesicles of *Escherichia coli*. *Journal of Bacteriology*, **117**, 1093–8.

MISRA, P. C. & HÖFER, M. (1975). An energy-linked proton-extrusion across the cell membrane of *Rhodotorula gracilis*. *FEBS Letters*, **52**, 95–9.

MITCHELL, P. (1966). Chemiosmotic coupling in oxidative and photosynthetic phosphorylation. *Biological Reviews*, **41**, 445–502.

MITCHELL, P. (1970). Membranes of cells and organelles: morphology, transport and metabolism. In *Organization and Control in Prokaryotic and Eukaryotic Cells*, ed. H. P. Charles & B. C. J. G. Knight, *Society for General Microbiology Symposia*, **20**, pp. 121–66. London: Cambridge University Press.

MORRISON, C. E. & LICHSTEIN, H. C. (1976). Regulation of lysine transport by feedback inhibition in *Saccharomyces cerevisiae*. *Journal of Bacteriology*, **125**, 864–71.

NICHOLS, W. W. & HAMILTON, W. A. (1976). The transport of D-lactate by membrane vesicles of *Paracoccus denitrificans. FEBS Letters*, **65**, 107–10.

ORDAL, G. W. & ADLER, J. (1974*a*). Isolation and complementation of mutants in galactose taxis and transport. *Journal of Bacteriology*, **117**, 509–16.

ORDAL, G. W. & ADLER, J. (1974*b*). Properties of mutants in galactose taxis and transport. *Journal of Bacteriology*, **117**, 517–26.

OXENDER, D. L. (1972). Membrane transport. *Annual Reviews of Biochemistry*, **41**, 777–814.

OXENDER, D. L. & QUAY, S. C. (1976). Isolation and characterization of membrane binding proteins. In *Methods in Membrane Biology*, ed. E. D. Korn. New York & London: Plenum Press, in press.

PADAN, E., ZILBERSTEIN, D. & ROTTENBERG, H. (1976). The proton electrochemical gradient in *Escherichia coli* cells. *European Journal of Biochemistry*, **63**, 533–41.

PATEL, L., SCHULDINER, S. & KABACK, H. R. (1975). Reversible effects of chaotropic agents on the proton permeability of *Escherichia coli* membrane vesicles. *Proceedings of the National Academy of Sciences, USA*, **72**, 3387–91.

PLATE, C. A., SUIT, J. L., JETTEN, A. M. & LURIA, S. E. (1974). Effects of colicin K on a mutant of *Escherichia coli* deficient in Ca^{2+}, Mg^{2+}-activated adenosine triphosphatase. *Journal of Biological Chemistry*, **249**, 6138–43.

POSTMA, P. W. (1976). Involvement of the phosphotransferase system in galactose transport in *Salmonella typhimurium. FEBS Letters*, **61**, 49–53.

POSTMA, P. W., COOLS, A. & VAN DAM, K. (1973). The transport of Krebs-cycle intermediates in *Azotobacter vinelandii* under various metabolic conditions. *Biochimica et Biophysica Acta*, **318**, 91–104.

RACKER, E. (1975). Reconstitution, mechanism of action and control of ion pumps. *Biochemical Society Transactions*, **3**, 785–802.

RAMOS, S., SCHULDINER, S. & KABACK, H. R. (1976). The electrochemical gradient of protons and its relationship to active transport in *Escherichia coli* membrane vesicles. *Proceedings of the National Academy of Sciences, USA*, **73**, 1892–6.

REEVES, J. P. (1971). Transient pH changes during D-lactate oxidation by membrane vesicles. *Biochemical and Biophysical Research Communications*, **45**, 931–6.

REICHERT, U., SCHMIDT, R. & FORET, M. (1975). A possible mechanism of energy coupling in purine transport of *Saccharomyces cerevisiae. FEBS Letters*, **52**, 100–6.

REST, R. F. & ROBERTSON, D. C. (1974). Glucose transport in *Brucella abortus. Journal of Bacteriology*, **118**, 250–8.

RIEBELING, V., THAUER, R. K. & JUNGERMANN, K. (1975). The internal-alkaline pH gradient, sensitive to uncoupler and ATPase inhibitor in growing *Clostridium pasteurianum. European Journal of Biochemistry*, **55**, 445–53.

RIEMERSMA, J. C. & ALSBACH, E. J. J. (1975). Proton translocation during anaerobic energy production in *Saccharomyces cerevisiae*. *Biochimica et Biophysica Acta*, **339**, 274–84.

ROSEN, B. P. & ADLER, L. W. (1975). The maintenance of the energized membrane state and its relation to active transport in *Escherichia coli*. *Biochimica et Biophysica Acta*, **387**, 23–36.

ROSEN, B. P. & MCCLEES, J. S. (1974). Active transport of calcium in inverted membrane vesicles of *Escherichia coli*. *Proceedings of the National Academy of Sciences, USA*, **71**, 5042–6.

ROY-BURMAN, S. & VISSER, D. W. (1975). Transport of purines and deoxyadenosine in *Escherichia coli*. *Journal of Biological Chemistry*, **250**, 9270–5.

SCHULDINER, S. & KABACK, H. R. (1975). Membrane potential and active transport in membrane vesicles from *Escherichia coli*. *Biochemistry*, **14**, 5451–61.

SCHULDINER, S., KERWAR, G. K., KABACK, H. R. & WEIL, R. (1975). Energy-dependent binding of dansylgalactosides to the β-galactoside carrier protein. *Journal of Biological Chemistry*, **250**, 1361–70.

SCHULDINER, S., WEIL, R. & KABACK, H. R. (1976). Energy-dependent binding of dansylgalactosides to the *lac* carrier protein: direct binding measurements. *Proceedings of the National Academy of Sciences, USA*, **73**, 109–12.

SEASTON, A., CARR, G. & EDDY, A. A. (1976). The concentration of glycine by preparations of the yeast *Saccharomyces carlsbergensis* depleted of adenosine triphosphate. Effects of proton gradients and uncoupling agents. *Biochemical Journal*, **154**, 669–76.

SHORT, S. A., KABACK, H. R. & KOHN, L. D. (1975). Localization of D-lactate dehydrogenase in native and reconstituted *Escherichia coli* membrane vesicles. *Journal of Biological Chemistry*, **250**, 4291–6.

SIMONI, R. D., HAYS, J. B., NAKAZAWA, T. & ROSEMAN, S. (1973). Sugar transport. VI. Phosphoryl transfer in the lactose phosphotransferase system of *Staphylococcus aureus*. *Journal of Biological Chemistry*, **248**, 957–65.

SIMONI, R. D., NAKAZAWA, T., HAYS, J. B. & ROSEMAN, S. (1973). Sugar transport. IV. Isolation and characterization of the lactose phosphotransferase system in *Staphylococcus aureus*. *Journal of Biological Chemistry*, **248**, 932–40.

SIMONI, R. D. & POSTMA, P. W. (1975). The energetics of bacterial active transport. *Annual Reviews of Biochemistry*, **43**, 523–54.

SIMONI, R. D. & ROSEMAN, S. (1973). Sugar transport. VII. Lactose transport in *Staphylococcus aureus*. *Journal of Biological Chemistry*, **248**, 966–76.

SINGH, A. P. & BRAGG, P. D. (1976a). Transport of α-methyl glucoside in a cytochrome-deficient mutant of *Escherichia coli* K12. *FEBS Letters*, **64**, 169–72.

SINGH, A. P. & BRAGG, P. D. (1976b). Anaerobic transport of amino acids coupled to the glycerol-3-phosphate–fumarate oxidoreductase system in a cytochrome-deficient mutant of *Escherichia coli*. *Biochimica et Biophysica Acta*, **432**, 450–61.

SLAYMAN, C. W. & SLAYMAN, C. L. (1975). Energy coupling in the plasma membrane of *Neurospora*: ATP-dependent proton transport and proton-dependent sugar cotransport. In *Molecular Aspects of Membrane Phenomena*, ed. H. R. Kaback, H. Neurath, G. K. Radda, R. Schwyzer & W. R. Wiley, pp. 233–48. Berlin: Springer-Verlag.

SZMELCMAN, S. & HOFNUNG, M. (1975). Maltose transport in *Escherichia coli* K12: involvement of the bacteriophage lambda receptor. *Journal of Bacteriology*, **124**, 112–18.

SZMELCMAN, S., SCHWARTZ, M., SILHAVY, T. J. & BOOS, W. (1976). Maltose transport in *Escherichia coli*. *European Journal of Biochemistry*, **65**, 13–19.

TEMPLETON, B. A. & SAVAGEAU, M. A. (1974). Transport of biosynthetic intermediates: regulation of homoserine and threonine uptake in *Escherichia coli*. *Journal of Bacteriology*, **120**, 114–20.

TSUCHIYA, T. & ROSEN, B. P. (1975). Characterization of an active transport system for calcium in inverted membrane vesicle of *Escherichia coli*. *Journal of Biological Chemistry*, **250**, 7687–92.

TSUCHIYA, T. & ROSEN, B. P. (1976). Calcium transport driven by a proton gradient in inverted membrane vesicles of *Escherichia coli*. *Journal of Biological Chemistry*, **251**, 962–7.

VAN THIENEN, G. & POSTMA, P. W. (1973). Coupling between energy conservation and active transport of serine in *Escherichia coli*. *Biochimica et Biophysica Acta*, **323**, 429–40.

WANG, C. C. & NEWTON, A. (1971). An additional step in the transport of iron defined by the *ton*B locus of *Escherichia coli*. *Journal of Biological Chemistry*, **246**, 2147–51.

WEST, I. C. (1970). Lactose transport coupled to proton movements in *Escherichia coli*. *Biochemical and Biophysical Research Communications*, **41**, 655–61.

WEST, I. C. & MITCHELL, P. (1972). Proton-coupled β-galactoside translocation in non-metabolizing *Escherichia coli*. *Journal of Bioenergetics*, **3**, 445–62.

WEST, I. C. & MITCHELL, P. (1973). Stoicheiometry of lactose-H^+ symport across the plasma membrane of *Escherichia coli*. *Biochemical Journal*, **132**, 587–92.

WEST, I. C. & MITCHELL, P. (1974*a*). The proton-translocating ATPase of *Escherichia coli*. *FEBS Letters*, **40**, 1–4.

WEST, I. C. & MITCHELL, P. (1974*b*). Proton/sodium ion antiport in *Escherichia coli*. *Biochemical Journal*, **144**, 87–90.

WEST, I. C. & WILSON, T. H. (1973). Galactoside transport dissociated from proton movement in mutants of *Escherichia coli*. *Biochemical and Biophysical Research Communications*, **50**, 551–8.

WHITING, P. H., MIDGLEY, M. & DAWES, E. A. (1976). The regulation of transport of glucose, gluconate and 2-oxogluconate and of glucose catabolism in *Pseudomonas aeruginosa*. *Biochemical Journal*, **154**, 659–68.

WILLECKE, K., GRIES, E. M. & OEHR, P. (1973). Coupled transport of citrate and magnesium in *Bacillus subtilis*. *Journal of Biological Chemistry*, **248**, 807–14.

WILLIS, R. C. & FURLONG, C. F. (1975). Interactions of a glutamate–aspartate

8

binding protein with the glutamate transport system of *Escherichia coli*. *Journal of Biological Chemistry*, **250**, 2581–6.

WILSON, D. M., ALDERETE, J. F., MALONEY, P. C. & WILSON, T. H. (1976). Protonmotive force as the source of energy for adenosine 5′-triphosphate synthesis in *Escherichia coli*. *Journal of Bacteriology*, **126**, 327–37.

WILSON, T. H. & KUSCH, M. (1972). A mutant of *Escherichia coli* K12 energy-uncoupled for lactose transport. *Biochimica et Biophysica Acta*, **255**, 786–97.

WOLFSON, E. B. & KRULWICH, T. A. (1974). Requirement for a functional respiration-coupled D-fructose transport system for induction of phosphoenolphyruvate: D-fructose phosphotransferase activity. *Proceedings of The National Academy of Sciences, USA*, **71**, 1739–42.

WOLFSON, E. B., SOBEL, M. E., BLANCO, R. and KRULWICH, T. A. (1974). Pathways of D-fructose transport in *Arthrobacter pyridinolis*. *Archives of Biochemistry and Biophysics*, **160**, 440–4.

YAGIL, E. & BEACHAM, I. R. (1975). Uptake of adenosine-5′-monophosphate by *Escherichia coli*. *Journal of Bacteriology*, **121**, 401–5.

YOSHIDA, M., SONE, N., HIRATA, H. & KAGAWA, Y. (1975). ATP synthesis catalyzed by purified DCCD-sensitive ATPase incorporated into reconstituted purple membrane vesicles. *Biochemical and Biophysical Research Communications*, **67**, 1295–1300.

THE PHOSPHOTRANSFERASE SYSTEM AS A SITE OF CELLULAR CONTROL

H. L. KORNBERG AND M. C. JONES-MORTIMER

*Department of Biochemistry, University of Cambridge,
Tennis Court Road, Cambridge CB2 1QW, UK*

INTRODUCTION

The most direct experimental procedure for defining the relationship between the rate at which a carbon source is utilised by some micro-organism and the rate at which that organism makes its cell components from that carbon source, is to study these parameters in cells grown in continuous culture, with the supply of the carbon source imposing the rate limitation for growth. In this technique, the concentration of the organic nutrient, though finite and varying with growth rate, is very small: if it were higher, the organisms would grow faster. It would thus be expected that, in organisms thus grown, any systems responsible for the recognition, uptake and retention of the nutrient by the cells would operate with maximal efficiency (see also Hamilton, this volume).

When this procedure was applied to two strains of *Escherichia coli* growing upon glucose as the limiting carbon source, a striking correlation was noted between the rates at which glucose had to be utilised, in order to maintain the cultures at their observed cell density at the various growth rates, and the rates at which cells harvested from the chemostats took up [^{14}C]glucose (Herbert & Kornberg, 1976). In particular, cells sampled from cultures growing at greater than 0.2 doubling/h took up the labelled sugar at rates virtually identical to those at which glucose was calculated to be used to account for the growth of the organisms; for example, cells harvested from a culture growing at 0.5 doubling/h took up [^{14}C]glucose nearly three times as rapidly as did cells harvested at 0.2 doubling/h. On the other hand, this equivalence did not hold for cultures harvested at growth rates less than 0.2 doubling/h: clearly, there was a lower limit below which the activity of the uptake system could not fall.

This simple experiment indicates that it is the rate at which the uptake system functions in the growing cultures that sets the pace of overall glucose utilisation and that thus limits the rate at which cell components are synthesised. Since the activity of this uptake system varies with

8-2

growth rate, it further follows that the organisms must 'know' how rapidly this uptake system is required to operate: some controls must affect the synthesis of its components, or their activity, or both.

A further phenomenon was revealed when the two cultures used for this experiment were grown in shake flasks, in which the supply of glucose is not limiting. Samples from cultures of one of these strains, designated B11 (Broda, 1967), took up [^{14}C]glucose or its analogue methyl-α-D-[^{14}C]glucoside much more slowly than did samples of the other, designated K2.1t (Brice & Kornberg, 1967); moreover, whereas strain K2.1t took up these labelled materials rapidly irrespective of the carbon source on which the cells had grown, strain B11 transported these substances well only when the organism had been grown on glucose (Kornberg & Reeves, 1972a, b). There must, therefore, be yet another type of regulation, that affects differently glucose transport by these two strains in batch but not in continuous culture.

In addition to these controls, that govern the rates at which a single carbon source is taken up and utilised by E. coli, there are also controls that determine whether one carbohydrate is taken up, and to what extent, in the presence of another. It has, for example, long been known that glucose represses the synthesis of enzymes of lactose utilisation (Monod, 1942): this has been held to account for the 'diauxie' observed when cultures grow upon small quantities of glucose mixed with excess lactose. Studies on the utilisation of glucose preferentially to many other sugars (McGinnis & Paigen, 1969, 1973; Kornberg, 1972, 1973) have further shown that glucose inhibits the continued utilisation of another sugar even by cultures previously exposed only to that other sugar; this inhibition is exerted on the functioning of the uptake system for that other sugar (Lengeler, 1966; Amaral & Kornberg, 1975).

It is the purpose of this article to review these various controls as they affect and are affected by the phosphotransferase (PT)-system that mediates the uptake of many, but not all, carbohydrates by E. coli. We shall concentrate largely on describing our work on the utilisation of glucose and fructose, since these two sugars are taken up via the PT-system, and since a combination of biochemical and genetical procedures has provided information on the minimum number of proteins that must be involved in the uptake of these sugars, the sugar specificity of these proteins, and the factors that affect their synthesis and activity.

There is evidence that the various mechanisms whereby carbohydrates traverse the E. coli plasma membrane may share certain features. In this context we discuss the belief that carbohydrate uptake via the PT-system is necessarily associated with its phosphorylation. An interrelationship between the uptake of sugars via the PT-system and that of sugars not transported via this system is also revealed by the behaviour of mutants impaired in pleiotropic components of the PT-system.

THE NATURE OF THE PT-SYSTEM

The PT-system was discovered in 1964 by S. Roseman and his colleagues (Kundig, Ghosh & Roseman, 1964) and has been extensively investigated since. Much valuable information on the components of this system has been obtained, largely in Professor Roseman's laboratory, through isolation of such components from extracts of *E. coli* and *Salmonella typhimurium* and their reconstitution with consequent restoration of the overall phosphorylating activity of the system. This information has been summarised in several review articles (Roseman, 1969, 1972, 1975; Kundig, 1974) and it is unnecessary for us to rehearse the details of this information. However, it is necessary to describe the basic features of the model that emerges from these studies and that has been advanced to account for the simultaneous uptake and phosphorylation of sugars by the PT-system.

There are several components needed to catalyse the phosphorylation of sugars, such as glucose or its non-catabolisable analogues, to the 6-phosphate esters, and of fructose to fructose-1-phosphate. A cytoplasmic enzyme, designated Enzyme I, catalyses the transfer of phosphate from phosphoenolpyruvate (PEP) to the N-1 position of a histidine moiety on a small protein HPr (reaction (1) below). The phospho-HPr thus formed serves as a general donor of phosphate to the various sugars that are substrates of the PT-system, but this ultimate phosphorylation of the sugars requires the participation of proteins that exhibit some specificity for the sugar substrates. Moreover, biochemical fractionation of these sugar-specific proteins by Kundig & Roseman (1971*a*, *b*) showed that more than one of these proteins is required for phosphate transfer from phospho-HPr (HPr$\sim$P) to glucose (G) or to methyl-α-D-glucoside (αMG) and that more than one combination of such proteins can effect the formation of the 6-phosphate esters. One system, with relatively low affinity for glucose (K_m approx. 0.1 mM), appears to consist of two membrane-bound components, designated II-A and II-B (reaction (2) below); a second system, with higher affinity for glucose (K_m approx. 5–10 μM), comprises both a cytoplasmic factor III and a membrane-bound Enzyme II-B' (reaction (3) below). These systems are further distinguished by their inducibility: the II-A/II-B system is constitutive in *E. coli* whereas the III/II-B′ combination can be either constitutive or inducible.

$$PEP + HPr \xrightarrow{\text{Enzyme I}} HPr \sim P + pyruvate \quad (1)$$

$$HPr \sim P + G \text{ or } \alpha MG \xrightarrow{\text{II-A/II-B}} HPr + G\text{-}6\text{-}P \text{ or } \alpha MG\text{-}6\text{-}P \quad (2)$$

$$HPr \sim P + G \text{ or } \alpha MG \xrightarrow{\text{III/II-B}'} HPr + G\text{-}6\text{-}P \text{ or } \alpha MG\text{-}6\text{-}P \quad (3)$$

$$\text{Sum: } PEP + G \text{ or } \alpha MG \longrightarrow Pyruvate + G\text{-}6\text{-}P \text{ or } \alpha MG\text{-}6\text{-}P$$

As intermediates in this overall transfer of phosphate from PEP to glucose or to methyl-α-D-glucoside, the sugar-specific proteins II-A and III are themselves phosphorylated and dephosphorylated. Kundig (1974) has suggested that the phosphate groups are bound to these proteins as acyl-phosphates; in the soluble protein III this is probably an aspartyl-phosphate. Phosphorylation of methyl-α-D-glucoside by these phosphorylated proteins was found to require not only the appropriate membrane protein II-B or II-B′ but also phospholipids, of which phosphatidylglycerol proved to be most effective.

Genetical investigations of the PT-system are based on the premise that information on a multi-protein assembly of this type can usefully be obtained by studies of mutants impaired in components of that assembly. Such information can be misleading. For example, mutants devoid of Enzyme I activity (*ptsI*) not only fail to grow on all sugars taken up via the PT-system (as predicted), but also grow poorly or not at all on succinate, lactose, glycerol, maltose and other substances that are known not to be substrates of the PT-system; the nature of this effect is discussed later. However, in those mutants, characterised as being specifically unable to take up e.g. methyl-α-D-glucoside, which are impaired in their growth on glucose but are unimpaired in growth on PT sugars such as fructose, sorbitol or mannitol, and where this lesion (now designated *ptsG*; Bachmann, Low & Taylor, 1976) has been shown to be the consequence of a change in a single gene that is located at min 24 on the *E. coli* linkage map and that can be brought into wild-type recipients by phage-mediated transduction to produce the phenotype of the donor mutant, it is safe to conclude that a component specific to the uptake of methyl-α-D-glucoside, and involved to a major extent in the uptake of glucose, has been altered (Kornberg & Smith, 1972). Moreover, such mutants, rendered permeable by treatment with toluene, also fail to catalyse the PEP-dependent phosphorylation of methyl-α-D-glucoside, and are greatly impaired in the phosphorylation of glucose: these are processes readily effected by their parent organisms (Kornberg

& Reeves, 1972*b*). Similar results have been obtained with mutants specifically impaired in the uptake of fructose, when that sugar is present in low concentration in the medium: this phenotype has been found to be associated also with an inability of toluene-treated cells to catalyse the PEP-dependent phosphorylation of fructose (Ferenci & Kornberg, 1971*a*) and to be specified by a gene *ptsF* located at min 46 on the revised linkage map (Bachmann *et al.*, 1976). It may thus be further concluded that the proteins affected in these mutants are sugar-specific components of the PT-system; this provides further evidence on the role of that system in the uptake of at least these carbohydrates by *E. coli*.

Although *ptsG* mutants are virtually completely unable to take up and phosphorylate methyl-α-D-glucoside, they are only partially affected in the uptake and phosphorylation of glucose. Such mutants still grow slowly on glucose and still take up [^{14}C]glucose, albeit at a rate considerably less than do wild-type cells. This uptake of glucose might be due to the lack of specificity shown by the systems that effect the active transport of galactose if such systems are present and induced; however, the *ptsG* strains studied by us lack the genes specifying the active transport systems for galactose (*galP mgl*; Kornberg & Riordan, 1976). Some other system must function in such strains to take up glucose. By the criteria used for characterising *ptsG* and *ptsF*, this system catalyses not only glucose uptake but also its PEP-dependent phosphorylation; it was first described by Curtis & Epstein (1970) who designated it *gptB*, and was characterised further by these workers (Curtis & Epstein, 1975) as an uptake system for mannose and glucosamine as well as for glucose. Independently, we showed that a system known to be capable of catalysing the PEP-dependent phosphorylation of fructose to fructose-6-phosphate, albeit with low affinity for this sugar (Ferenci & Kornberg, 1971*b*, 1974), was able much more readily to catalyse the PEP-dependent phosphorylation of mannose, glucosamine and glucose; the gene *ptsX* specifying this activity (Jones-Mortimer & Kornberg, 1974, 1976) was located at min 40 on the revised linkage map and is identical with *gptB*. In accordance with the suggestion of Bachmann *et al.* (1976), it will in future be designated *ptsM*.

Whereas *ptsG* mutants still grow on glucose, and incorporate [^{14}C]-glucose if this is added to cultures growing on e.g. glycerol, double mutants carrying *ptsG ptsM* also grow readily on glycerol but incorporate isotopic carbon from added [^{14}C]glucose to less than 5% of the extent to which it is incorporated by wild-type cells; moreover, such double mutants grow on glucose with a doubling time of the order of

10 h (Kornberg & Jones-Mortimer, 1975). It thus appears that about 75 out of every 100 glucose molecules taken up by our strains of wild-type *E. coli*, growing in batch culture, enter the cells via the PtsG-system, and about 20 of the remaining molecules enter via PtsM. These sugar-specific components of the PT-system are specified by genes *ptsG* and *ptsM* 16 min apart on the linkage map, and differ most obviously in their ability to react with methyl-α-D-glucoside. It is not easy to reconcile these findings with the model derived from fractionation of cell-free *E. coli* (Kundig, 1974; Roseman, 1975).

There are two possible interpretations of this model in terms of the genetical data. If, as Kundig (1974) suggests, 'the physiological role of the low affinity PTS (II-A/II-B system) seems primarily to be one of translocating sugars across the cell membrane', this role, and its demonstrated ability to transfer phosphate from HPr~P to methyl-α-D-glucoside, indicate that it is, or is part of, the system specified by *ptsG*. However, the III/II-B' system is clearly not that specified by *ptsM*, and indeed Kundig (1974) reports that, in *S. typhimurium*, protein III is specified by a gene *crr* co-transducible with *ptsI*, the function of which is discussed later.

On the other hand, it could be argued that the bulk of glucose and methyl-α-D-glucoside transport and phosphorylation is effected not via II-A/II-B but via the III/II-B' system. Since, as stated above, the component III is apparently specified by a gene (*crr*) located at min 51.5 on the revised *E. coli* linkage map (Bachmann *et al.*, 1976), the gene *ptsG* at min 24 must specify the membrane component II-B'. Dysfunction of *ptsG* (and hence of II-B') is known to cause severe impairment of glucose utilisation and abolition of methyl-α-D-glucoside uptake. It would be expected that dysfunction of the other partner, component III, would have identical effects. However, this appears not to be so: mutants of *E. coli* that lack the Crr-system but retain Enzyme I activity are virtually unaffected in both these functions, and similar results have been obtained with *S. typhimurium* (P. W. Postma, personal communication).

There may also be difficulties in equating the II-A/II-B system with that specified by *ptsM*. Kundig & Roseman (1971*b*) reported the isolation of three activities from the solubilised membrane component II-A, which were able to catalyse the transfer of phosphate from PEP (in the presence of HPr and Enzyme I) to D-glucose, D-mannose and D-fructose respectively. This accords with the specificity of the PtsM-system (Curtis & Epstein, 1975). However, methyl-α-D-glucoside was used as acceptor to measure this phosphate transfer to glucose: the PtsM-system does not readily recognise this glucose analogue as substrate for phosphate transfer. It may be that the concentration of methyl-α-D-glucoside employed in this assay (5 mM) was sufficiently high to overcome this reluctance. Similarly, 5 mM-fructose was used to measure its phosphorylation, but the product was not characterised to determine whether the 1- or 6-phosphate ester had been formed. Clearly, further work is required to reconcile the model of the PT-system derived from reconstitution studies, with the picture assembled from genetical evidence and from physiological studies of the growth of, and incorporation of isotopic materials by, specific strains of *E. coli* and mutants derived from them.

A further question on the completeness of the Roseman/Kundig model arises from studies of galactose utilisation by *galP mgl* strains of *E. coli* (Kornberg & Riordan, 1976). Such organisms still grow on galactose but at rates that are a function of the galactose concentration in the medium: half-maximal growth rates required more than 2 mM-galactose to be present. Introduction of *ptsG* into such organisms simultaneously led to the (expected) loss of the ability to take up methyl-α-D-glucoside, to slower growth on glucose, and (unexpectedly) also to much slower growth on galactose. This suggested that, in these organisms devoid of systems that normally actively transport galactose, galactose acts as an analogue of glucose and is taken up by a protein specified by (or very closely associated with) *ptsG*. However, in contrast to the uptake of methyl-α-D-glucoside, galactose transport through this system did not require phosphotransferase activity and, in mutants also devoid of galactokinase activity, led only to equilibration of external with internal galactose. It thus appears that a component associated with the PtsG-system can catalyse the diffusion of galactose. Consistent with this interpretation, mutants mapping in, or very close to, *ptsG* were isolated which grew normally on galactose although (being *ptsG*) they were greatly impaired in their growth on glucose. However, addition of glucose to such organisms growing on galactose powerfully inhibits their growth, and reduces the rate of doubling to that observed on glucose as sole carbon source. Apparently, glucose has a much higher affinity for the facilitated diffusion components of this system than has galactose. This implies that glucose, like galactose, might cross the plasma membrane without the simultaneous phosphorylation that is an essential feature of the Roseman model. With entirely different mutants, Postma (1976) has independently arrived at similar conclusions from his studies with *S. typhimurium*.

The suggestion that these are two components of the PtsG-system,

Table 1. *Rates of uptake of* [^{14}C]*glucose and of methyl-α-D-*[^{14}C]*glucoside* (α*MG*) *by suspensions of* E. coli, *and phosphotransferase activity of the cells, toluene-treated*

		Rate of uptake of[a]		Phosphotransferase activity with[a]	
Strain	Grown on	Glucose	αMG	Glucose	αMG
B11	Glycerol	6	3	6	3
	Fructose	10	3	12	3
	Glucose	48	15	52	17
K2.1t	Glycerol	25	19	26	23
	Fructose	86	44	93	53
	Glucose	70	38	68	49
K2.1t.X1 (PtsG⁻)	Fructose	4	2	1	1
K2.1t.X2 (Tgl⁻?)	Fructose	12	2	34	34

[a] All rates are nmoles of substrate taken up or phosphorylated/min/mg dry mass.

one of which represents a protein catalysing facilitated diffusion of hexoses and the other the PEP-dependent phosphorylating system, is strengthened by the work of Bourd, Erlagaeva, Bolshakova & Gershanovitch (1975) and is supported by the data of Table 1. Bourd *et al.* (1975) isolated mutants that grew on lactose or mannitol in the presence of the usually toxic methyl-α-D-glucoside. Among them were some that grew at normal rates on glucose, but that were impaired in the uptake of methyl-α-D-[^{14}C]glucoside; however, extracts of such mutants evinced enhanced abilities to catalyse the PEP-dependent phosphorylation of methyl-α-D-glucoside. The gene specifying this phenotype, designated *tgl*, was (like *ptsG*) co-transducible with *purB*. The behaviour of our mutant (Table 1) is very similar. Suspensions of the cells take up glucose at a much reduced rate, and barely take up methyl-α-D-glucoside; however, cells rendered permeable with toluene (Kornberg & Reeves, 1972*a*) appear to be unimpaired in their ability to phosphorylate both glucose and methyl-α-D-glucoside with concomitant dephosphorylation of PEP. It is tempting to conclude that this mutant is defective in the component specified by *tgl* that permits external glucose and its analogue to enter the cells, but possesses normal (or enhanced) levels of the phosphorylating components of the PtsG-system (i.e. *tgl ptsG*⁺); in contrast, the *galP mgl* double mutant further impaired in this region, described by Kornberg & Riordan (1976), would have the constitution *tgl*⁺ *ptsG*.

REGULATION OF PT-ACTIVITY

The change with growth rate in the rapidity with which cells, grown on limiting glucose in a chemostat, take up glucose (Herbert & Kornberg, 1976) shows that *E. coli* can adjust the rate at which glucose enters the cells to the rate at which glucose metabolism must occur to supply both biosynthetic precursors and metabolic energy. Two regulatory phenomena may be involved in this correlation of uptake with metabolism: 'coarse' controls that alter rates of uptake by the induction (preferential synthesis) or repression (inhibition of synthesis) of protein components of the PT-system; and 'fine' controls, that regulate the activity of such proteins already extant in the cells.

Coarse controls of PT-component synthesis

The fact that strains of *E. coli* such as K2.1t (Table 1) readily take up glucose and its non-catabolisable analogues even though the cells have not previously encountered these substances may have concealed the inducibility of the sugar-specific components of the PT-system. However, it has been long established that other strains of *E. coli* form the uptake system for methyl-α-D-glucoside inducibly (Kepes, 1960; Kessler & Rickenberg, 1963). Indeed, Winkler & Wilson (1967) showed that glucose-grown *E. coli* strain ML 308 contained a five-fold greater capacity to transport methyl-α-D-glucoside than did cells grown on other carbon sources. Similarly, the work of Fraenkel (1968), Hanson & Anderson (1968), Berkowitz (1971) and Lengeler (1975a, b) showed that the uptake systems for fructose and hexitols are inducible.

Studies with cells rendered permeable to small molecules (Kornberg & Reeves, 1972b) showed that it was the phosphorylating components of the uptake systems for glucose and fructose that were inducible. The procedure used involved direct measurements of either the formation of pyruvate, as the phosphate group was transferred from added PEP to an appropriate added sugar by the corresponding cellular PT-system, or the formation of the phosphorylated sugar by this system; it was thus not transport but PT-activity that was measured. Glycerol-grown strain B11, placed in medium containing fructose as the sole carbon source, formed the fructose-specific phosphotransferase (now recognized as the PtsF-system) before growth recommenced and maintained its synthesis thereafter, but did not synthesise glucose phosphotransferase; on the other hand, glycerol-grown cells placed in medium containing glucose as the sole carbon source formed the phosphotransferase for glucose and methyl-α-D-glucoside specified by *ptsG* but not the fructose phosphotransferase. It was a puzzle at the time that, under both growth conditions (though more so under the latter), a PT-activity catalysing the phosphorylation of mannose was also induced; it is now clear that this is a manifestation of the PtsM-system that is thus shown to be induced by glucose, and hence probably to play a role in the uptake of that hexose by wild-type cells.

In contrast to the inducibility of the sugar-specific components of the PT-system, Enzyme I and HPr are reported to be constitutively synthesised by *E. coli* (Kundig, 1974) although *E. coli* grown on hexoses or hexitols were found to contain about three times as much of these proteins as did cells grown on lactate (Saier, Simoni & Roseman, 1970). A promoter gene controlling the rate of expression of the *ptsI* and *ptsH* genes specifying these proteins has been reported to be present in *S. typhimurium* (Cordaro *et al.*, 1974).

Fine controls of PT-activity

A simple way of regulating the activity of a multi-component system is to modulate the availability of a reactant, such as PEP in the present context. This might be sensible from a physiological standpoint since PEP stands at a metabolic cross-roads and itself acts as allosteric effector of several enzymes of carbohydrate metabolism. Thus, it has been shown (Ashworth & Kornberg, 1966; Cánovas & Kornberg, 1966) that the anaplerotic fixation of carbon dioxide which is required to maintain the levels of intermediates of the tricarboxylic acid cycle in growing *E. coli*, occurs only through PEP-carboxylase and that even in cells growing on lactate or pyruvate, a special enzyme (PEP-synthetase) is needed to transform pyruvate to PEP so that it may then be transformed, with the aid of PEP-carboxylase, to the C_4-compound oxaloacetate (Cooper & Kornberg, 1967). At the same time, the two pyruvate kinases of *E. coli* (Malcovati & Kornberg, 1969; Kornberg & Malcovati, 1973) effect the catabolism of PEP in its conversion to pyruvate and acetyl-coenzyme A. And, thirdly, PEP is required for the biosynthesis of cell components such as aromatic amino acids. The supply of PEP from carbohydrate catabolism must therefore meet these various metabolic demands; in addition, it must also suffice to ensure that the external carbohydrate can be taken up by the PT-system.

That the availability of PEP can impose a limitation on growth was first demonstrated with mutants devoid of phosphofructokinase activity (Kornberg & Smith, 1970). In wild-type cells, the initial expenditure of 1 mol of PEP to take up 1 mol of glucose yields 1 mol of glucose-6-

phosphate that, via the predominant reactions of glycolysis, yields 2 mol of PEP: there is thus no obvious shortage of PEP. In mutants that lack phosphofructokinase (*pfk*), however, this is not so. The glucose-6-phosphate made in the initial reaction of the PT-system cannot be converted to fructose-1,6-bisphosphate by the usual route, but must traverse the oxidative pentose phosphate pathway, in which glyceraldehyde-3-phosphate is formed by a route by-passing phosphofructokinase, but which yields only 1 mol of PEP per mol of glucose taken up. Shortage of PEP is therefore likely to occur and, indeed, such mutants do not grow well on glucose although they grow readily on fructose (which enters the metabolic paths below the block and thus yields 2 mol of PEP per mol of hexose) and on glucose-6-phosphate (which is not taken up via the PT-system and hence causes no drain on PEP supplies). Moreover, suspensions of *pfk* mutants do not take up [^{14}C]glucose unless they are simultaneously supplied with (unlabelled) precursors of PEP. Perhaps most conclusively, the addition of unlabelled pyruvate to such suspensions of *pfk* mutants was found not to promote the uptake of [^{14}C]glucose if the mutants were devoid of PEP-synthetase activity (*pps*), but to stimulate glucose uptake better than any other carbon source tried after the mutants were transduced to *pps*$^{+}$. Since *pps* strains are not impaired in their ability to oxidise pyruvate and hence to form ATP through that oxidation, the differences observed between the behaviour of *pps* and *pps*$^{+}$ cells point directly to the availability of PEP, and not of some indirect product of pyruvate catabolism, as the determinant of [^{14}C]glucose uptake by *pfk* mutants. This interpretation has received independent support from the work of Vinopal & Fraenkel (1974) with similar mutants and also from studies with mutants devoid of key enzymes of gluconate catabolism (Kornberg & Soutar, 1973).

A similar explanation may underlie also the observation that pyruvate inhibits the growth of *pps* mutants of *E. coli* on substances taken up through the PT-system, whereas no such inhibition is observed during growth on substances taken up by active transport (Morgan & Kornberg, 1967). Since the inhibition is relieved by transduction of the *pps* mutants to *pps*$^{+}$, or by the provision of C$_4$-dicarboxylic acids (which are the products of the anaplerotic carboxylation of PEP in *E. coli*), it is likely that in the presence of external pyruvate, the rate of growth is restricted by the availability of PEP. The only source of PEP, in organisms growing on a sugar taken up through the PT-system, is that sugar phosphorylated: it is therefore further likely that exogenous pyruvate, or some catabolite of pyruvate, inhibits the PT-system. The inhibitory agent has not yet been unambiguously identified, but several pieces of evidence suggest that it is not pyruvate itself but possibly acetyl-coenzyme A. For example, *pps* mutants that also lack pyruvate dehydrogenase activity are not inhibited by pyruvate whereas *pps* mutants that also lack citrate synthase activity are unusually sensitive to that C_3-acid; moreover, the uptake of metabolisable and non-metabolisable substrates of the PT-system by acetate-grown *pps* cells is strongly inhibited not only by pyruvate but also by acetate (Morgan & Kornberg, 1969). Although the mechanism of this inhibition also remains to be understood, the effect of pyruvate is, at the physiological level, akin to the 'feedback inhibitions' long known to regulate the activity of biosynthetic routes (Umbarger, 1961): pyruvate or acetyl-coenzyme A are the end-products of the glycolytic pathway initiated by the PT-system.

Preferential uptake of glucose over other sugars

The addition of glucose to cultures of *E. coli* growing on a wide variety of other carbon sources rapidly inhibits the utilisation of those substrates. This phenomenon, which has been termed 'catabolite inhibition' by McGinnis & Paigen (1969, 1973) is due to interference by glucose with the continued uptake of the other substrates. Non-catabolisable analogues of glucose, such as methyl-α-D-glucoside, and 1- and 2-deoxy-D-glucose, were found to exert a similar inhibitory effect on carbohydrate transport, but only if the organisms had been previously induced for the uptake of such analogues (Kepes, 1960; Kessler & Rickenberg, 1963; Koch, 1964; Boniface & Koch, 1967; Winkler & Wilson, 1967). This suggests that it is either the actual process of translocation of glucose and its analogues into the cells, or elevation of the intracellular concentration of the 6-phosphate esters formed from these materials, that inhibits the uptake of other substrates. In accordance with either explanation, it has been found that, in mutants impaired in glucose-specific components of the PT-system, glucose and its non-catabolisable analogues do not inhibit the uptake of substances such as galactose (Asensio, Avigad & Horecker, 1963; Kornberg, 1973) and fructose (Kornberg, 1972, 1973).

Although these observations establish that it is not glucose *per se* that interferes with the uptake of other sugars, but that glucose has to be phosphorylated to be effective as a competitor for uptake, they do not locate the site at which the inhibitor acts. Amaral & Kornberg (1975) described a mutant of *E. coli* which, unusually, continued to use fructose in the presence of glucose; however, glucose still virtually excluded other sugars such as mannitol, sorbitol, lactose and galactose. Moreover, the growth of this mutant on glucose was normal, and the organism was not impaired in any glucose-specific component of the PT-system. The gene that specified this relief from catabolite inhibition specifically on fructose (*cif*) was found to be highly co-transducible with the gene *ptsF* that specifies the high-affinity uptake system for fructose. Moreover, during growth on fructose, *cif* mutants had lost their sensitivity not only to glucose but also to glucose-6-phosphate; since this latter compound is not a substrate of the PT-system, it is possible that the utilisation of glucose preferentially to other sugars is mediated through changes in the intracellular levels of glucose-6-phosphate. This would, of course, also explain why *ptsG* mutants (which cannot readily form glucose-6-phosphate from glucose) continue to use fructose, as well as other sugars, in the presence of glucose. A similar locus, specifying a glucose (or glucose-6-phosphate) sensitive control point in sorbitol utilisation, has been recently found to lie within the gene specifying the uptake system for that hexitol (J. Lengeler, personal communication).

Inducer exclusion

Apart from the inhibition that glucose exerts over the continued utilisation of other sugars, glucose is known to inhibit the induction of enzymes required for the catabolism of those other sugars. The mechanism of this so-called 'glucose effect' is considered further in the next section of this article. It is, however, relevant at this point to mention that the change from *cif*$^+$ to *cif* overcomes the 'glucose effect' specifically for the induction of enzymes of fructose catabolism (Amaral & Kornberg, 1975): both the PtsF-system and fructose-1-phosphate kinase were induced when *cif* mutants grew on a mixture of fructose and glucose, although these activities were induced to only a negligible extent in *cif*$^+$ organisms under these conditions. This strengthens the view (Adhya & Echols, 1966; Lengeler, 1966) that the failure to induce certain catabolic enzymes if glucose is also present is due, at least in part, to the inability of the sugars, necessary to induce those enzymes, sufficiently to enter the cells under these conditions.

The glucose effect

The effects of glucose on the utilisation of sugars which do not enter by a phosphotransferase mechanism are complex. They have been fully discussed by Magasanik (1970) for the case of lactose transport and will not be considered in detail here, except insofar as they may explain the phenotype of mutants deficient in Enzyme I. Three main effects have been observed: catabolite repression, transient repression and inducer exclusion.

The term *catabolite repression* is used to describe the effect of the carbon source used for growth on the expression of an inducible enzyme. The effect was first described by Epps & Gale (1942) who demonstrated that glucose inhibited the synthesis of amino acid decarboxylases in *E. coli*. That the effect is more general was shown by Magasanik (1961) and by McFall & Mandelstam (1963): under 'conditions of gratuity' (Monod, 1956), where they are not needed for the utilisation of the nutrients in the medium, the rate of synthesis of inducible enzymes is lower on carbon sources which permit faster growth. There is, therefore, a correlation between the intensity of carbon flux through catabolic pathways and the rate at which the genes specifying these enzymes are expressed.

If glucose (Paigen, 1966) or some other good carbon source (Tyler, Loomis & Magasanik, 1967) is added to a culture in which an inducible enzyme is being synthesised gratuitously during growth on a poor carbon source, the synthesis of that inducible enzyme is halted for a variable period (up to 30 min) and when synthesis begins again, it takes place at the differential rate typical of the rate of expression of the enzyme during growth on the new carbon source. This effect is known as *transient repression*. The effect cannot be due to the exclusion or expulsion by glucose of the inducer, since the same effect is observed during constitutive synthesis of the enzyme.

Reversal of catabolite and transient repression by cyclic AMP

Makman & Sutherland (1965) discovered the presence of cyclic 3',5'-AMP (cAMP) in *E. coli* cells, and observed that there was more of it in cells grown with a poor carbon source than in cells grown with a rich carbon source. This observation suggested that cAMP might play a role in the control of induced enzyme synthesis. That this might be so was established by Perlman & Pastan (1968) and by Ullmann & Monod (1968), who demonstrated that exogenous cAMP abolishes not only transient repression but also catabolite repression. Mutants which are unable to synthesise cAMP have lost the ability to grow on carbon sources the catabolic pathways for which are inducible (Perlman & Pastan, 1969), and cAMP is required for the in-vitro synthesis of inducible enzymes, such as β-galactosidase (Chambers & Zubay, 1969). This synthesis is inhibited by cyclic GMP.

Conversely, the extent of catabolite repression of an enzyme may be considered as an indirect measure of the steady-state concentration of cAMP; for example, Lis & Schleif (1973) showed that the concentration of cAMP required for the induction of the lactose operon is less than that required for the induction of the arabinose operon. Thus, if the effect of cAMP on transient repression is a true reversal of that phenomenon, and if cAMP has only a catalytic role in enzyme synthesis, then glucose must have some more drastic effect than merely halting cAMP synthesis: cAMP must also be removed from its site of action.

The control of cAMP levels in E. coli cells

That cAMP levels in the cell are controlled both at the levels of synthesis and excretion was first suggested by Makman & Sutherland (1965). Since then, relatively little attention has been given to the means available to cells for decreasing their cAMP content. This is mainly because about 98% of the cAMP present in a suspension of *E. coli* is found to be in the medium rather than in the cells (Peterkofsky & Gazdar, 1974). However, Saier, Feucht & McCaman (1975) have confirmed Makman & Sutherland's suggestion that there is control of cAMP excretion as well as of cAMP formation.

Glucose has been shown to inhibit cAMP synthesis in whole cells (Peterkofsky & Gazdar, 1974) and in toluene-treated preparations (Harwood & Peterkofsky, 1975) but it does not inhibit the adenyl cyclase present in cell-free extracts of *E. coli*. One possible explanation of this is that, in whole cells, adenyl cyclase is loosely bound to the membrane; this is supported by the findings of Tao & Lipmann (1969), who showed that 50% of the activity sedimented with the cell debris. It might then be that it is the interaction between glucose and some component of the PT-system, rather than glucose or glucose-6-phosphate *per se*, that affects adenyl cyclase activity. Although there are data that support the occurrence of such a mechanism, there also remain some inconsistencies to be resolved. For example, Peterkofsky & Gazdar (1974, 1975) showed that glucose inhibits cAMP synthesis in wild-type *E. coli* and overcomes the activation of adenyl cyclase by PEP both in these cells and in a *ptsI* mutant. Moreover, Saier, Feucht & Hofstadter (1976) and Harwood et al. (1976) showed that the appropriate active Enzyme II was required for a sugar to inhibit adenyl cyclase. This suggests (Peterkofsky & Gazdar, 1975) that the phosphorylated form of Enzyme I (phospho-EI) was required to activate an otherwise latent adenyl cyclase, and that dephosphorylation of the phospho-EI reversed this effect. However, it remains to be explained why glucose also inhibits the adenyl cyclase activity of *ptsH* mutants, in which (presumably) transfer of phosphate from phospho-EI to sugar would not occur.

Despite the fact that cAMP is required for the expression of genes specifying inducible enzymes (Perlman & Pastan, 1969) and co-ordinately controlled non-PT transport systems, and despite the evident cure it provides for transient and catabolite repression, one cannot unambiguously conclude that the intracellular cAMP level *per se* determines the differential rate of gene expression. Some Enzyme I-negative mutants grow on arabinose but not on lactose (Epstein & Curtis, 1972) which appears incompatible with the finding of Lis & Schleif (1973). Ullmann (1974) has demonstrated catabolite repression in *Bacillus megaterium* and its relief by added cAMP; this is particularly interesting since this organism is apparently incapable of synthesising cAMP (Setlow, 1973). Moreover, Haggerty & Schleif (1975) confirmed that glucose, but not glycerol, exerted catabolite repression on the synthesis of *lac* messenger RNA; however, glycerol was as effective as glucose in lowering cAMP levels. The observation by Bernlohr, Haddox & Goldberg (1974) of an inverse correlation between the cAMP and cGMP levels in *E. coli* cells suggests that a more balanced view of the effect of cyclic nucleotides may be required.

In the experiments demonstrating transient and catabolite repression, glucose is added to a culture already expressing the enzyme in question. If the experiment is carried out differently, i.e. if glucose and inducer are added simultaneously, induction is not observed (Cohn & Horibata, 1959), though it is observed if the transport system has previously been induced. Glucose can prevent the entry of the inducer, but only when the level of the transport system is low. This phenomenon, inducer exclusion, has already been considered in relation to substrates of the PT-system.

A further possible regulatory effect of glucose was described by Koch (1971), who suggested that glucose not only inhibits the uptake of galactosides but also stimulates their exit from the cell. It is difficult to reconcile this type of control with any simple type of allosteric control of the function of a transport protein, but it is what would be predicted if glucose, or its uptake by the PT-system, were to discharge the membrane potential. Attractive though this proposal may sound, it is unlikely since the export of cAMP, which is promoted by glucose, requires the energised state of the membrane (Saier *et al.*, 1975). The possibilities remain that the effect is due to inhibition by intracellular hexose phosphates, or as proposed in the model discussed on p. 234.

The pleiotropic phenotype of Enzyme I mutants

Mutants defective in Enzyme I of the PT-system may be divided into two classes. One class shows the expected phenotype and fails to grow on all carbon compounds that are taken up by a PT-mechanism, but grows on all other carbon sources tested. The second class not only fails to grow on PT sugars but also on some compounds, notably lactose, maltose and glycerol (Wang, Morse & Morse 1969, 1970) which are known from other evidence not to be transported by a phosphotransferase mechanism (for a review see Hamilton 1975, and this volume); however, mutants of this latter class can grow on other compounds not taken up by PT-systems, such as glucose-6-phosphate, gluconate, lactate and pyruvate. The first class comprises only about 5% of newly isolated *pts* mutants (unpublished observations). What are the differences between these two kinds of mutant, and how does the PT-system act to prevent the function of proteins involved in the utilisation of some carbohydrates?

The addition of cAMP to a culture of one of the apparently anomalous mutants (of the second class) may permit growth on lactose (Pastan & Perlman, 1969) or on glycerol (Berman, Zwaig & Lin, 1970). But there are mutants, apparently of this type, in which cAMP is not effective, though it is not clear whether the cells in question were more than usually impermeable to exogenous cAMP. In other instances the defect may be overcome by the addition of a suitable inducer (Pastan & Perlman, 1969) or by mutation to the constitutive synthesis of the enzyme system in question (Berman *et al.*, 1970; Wang & Morse, 1968; Wang *et al.*, 1969, 1970). The first of these observations suggests that the effect might be related to those other phenomena, catabolite and transient repression, that are known to be overcome by increasing the concentration of cAMP in the medium and hence presumably also in the cell. The second observation suggests that the effect is also related to the other means, namely inducer exclusion, by which PT sugars such as glucose regulate the utilisation of non-PT sugars.

The anomalous mutants of the PT-system have been shown (Saier & Roseman, 1972; Saier *et al.*, 1970, 1976) to possess small residual amounts of phosphotransferase Enzyme I activity; the synthesis of inducible enzymes in such mutants is abnormally sensitive to the inhibitory effects of PT sugars. In contrast, mutants in which the *pts* gene is deleted do not respond to cAMP and are not sensitive to any PT sugar; mutants which have lost the Enzyme II for a particular sugar do not show the effect with that sugar (Saier *et al.*, 1976), though they do with others.

Besides these means of overcoming the effect of a deficiency in Enzyme I on the assimilation of non-PT sugars, two kinds of mutation can restore growth on such sugars. The first type of mutation is sugar-specific: the lesions probably lie within a structural gene for a transport system for that sugar (M. H. Saier, personal communication) and are probably of the same general type as the mutant described by Amaral & Kornberg (1975). The second type of mutation is not sugar-specific, but allows the organism to assimilate all carbon compounds which do not require to be phosphorylated by the PT-system (Saier & Roseman, 1972; Cordaro & Roseman, 1972; Jones-Mortimer & Kornberg, 1974; Saier *et al.*, 1976). Though the gene affected in these mutants, *crr*, maps near the *ptsI* gene, it appears not to form part of the *ptsHI* operon (Cordaro & Roseman, 1972; Cordaro *et al.*, 1974). We suppose that the relatively rare first class of mutant which retains the ability to grow on non-PT sugars, is one in which the two genes *ptsI* and *crr* have been deleted. According to Kundig (1974), these *crr* mutants are defective in component III of the III/II-B′ system for glucose uptake, and also lack a hexose-6-phosphatase activity. Though it appears likely that these two functions are properties of a single protein species, it is not clear from the published evidence whether the polypeptide in question is coded for by the gene *crr* or whether the loss of these activities is a secondary consequence of the mutation.

These observations have led Saier & Stiles (1975) to propose a model for the effects of the PT-system on the entry of non-PT sugars into the cell. In this model, a central regulatory protein RPr is assumed to be able to interact allosterically with proteins specific for the transport of non-PT sugars, thereby inhibiting their activity. However, RPr can also be phosphorylated (by PEP, Enzyme I and HPr) and can thus be converted to RPr~P: this phosphorylated protein is believed to be no longer able to inhibit the uptake of non-PT sugars. If the RPr is the product of the *crr* gene, the model would account for the properties of *crr* mutants: if no RPr is made, the uptake systems for non-PT sugars would be readily induced, and glucose would not exert inducer exclusion on them. The model would also explain the phenotype of mutants either totally devoid of Enzyme I or still containing a small residual activity of this enzyme: in the former, RPr could not be phosphorylated at all, and in the latter the rate of formation of RPr~P would be so much less than the rate of transfer of phosphate to PT sugars that, in the presence of such sugars, RPr would remain overwhelmingly in the non-phosphorylated state. Moreover, if it is RPr~P that is required to activate adenyl cyclase (Peterkofsky & Gazdar, 1975), cAMP would be expected to reverse inducer exclusion in 'leaky' mutants in Enzyme I: this is observed. But it remains a puzzle that mutants with a deletion in *ptsI* and *crr*, which on the model would not be able to form an active adenyl cyclase, appear not to contain unusually low levels of cAMP and appear to contain adenyl cyclase that, in toluene-treated cells, is readily activated by PEP. Moreover, if component III of the glucose-specific III/II-B' system is the product of the *crr* gene (Kundig, 1974), and is thus identical with RPr, it is difficult to understand why mutants containing Enzyme I but lacking the *crr*$^+$ gene product are not obviously impaired in the formation of inducible enzymes as well as in the utilisation of PT sugars. Clearly, much requires to be learned of the mechanisms through which the PT-system exerts its many effects on the carbohydrate metabolism of *E. coli* and of other micro-organisms that contain this important system of phosphate transfer.

REFERENCES

ADHYA, S. & ECHOLS, H. (1966). Glucose effect and the galactose enzymes oɪ *Escherichia coli*: correlation between glucose inhibition of induction and inducer transport. *Journal of Bacteriology*, **92**, 601–8.

AMARAL, D. & KORNBERG, H. L. (1975). Regulation of fructose uptake by glucose in *Escherichia coli*. *Journal of General Microbiology*, **90**, 157–68.

ASENSIO, C., AVIGAD, G. & HORECKER, B. L. (1963). Preferential galactose utilization in a mutant strain of *E. coli*. *Archives of Biochemistry and Biophysics*, **103**, 299–309.

ASHWORTH, J. M. & KORNBERG, H. L. (1966). The anaplerotic fixation of carbon dioxide by *Escherichia coli*. *Proceedings of the Royal Society of London, Series B*, **165**, 179–88.

BACHMANN, B. J., LOW, K. B. & TAYLOR, A. L. (1976). Recalibrated linkage map of *Escherichia coli* K12. *Bacteriological Reviews*, **40**, 116–67.

BERKOWITZ, D. (1971). D-Mannitol utilization in *Salmonella typhimurium*. *Journal of Bacteriology*, **105**, 232–40.

BERMAN, M., ZWAIG, N. & LIN, E. C. C. (1970). Suppression of a pleiotropic mutant affecting glycerol dissimilation. *Biochemical and Biophysical Research Communications*, **38**, 272–5.

BERNLOHR, R. W., HADDOX, M. R. & GOLDBERG, N. D. (1974). Cyclic guanosine 3':5'-monophosphate in *Escherichia coli* and *Bacillus licheniformis*. *Journal of Biological Chemistry*, **249**, 4329–31.

BONIFACE, J. & KOCH, A. L. (1967). The interaction between permeases as a tool to find their relationship on the membrane. *Biochimica et Biophysica Acta*, **135**, 757–70.

BOURD, G. I., ERLAGAEVA, R. S., BOLSHAKOVA, T. N. & GERSHANOVITCH, V. N. (1975). Glucose catabolite repression in *Escherichia coli* K12 mutants defective in methyl-α-D-glucoside transport. *European Journal o Biochemistry*, **53**, 419–27.

BRICE, C. B. & KORNBERG, H. L. (1967). Location of a gene specifying phosphopyruvate synthase activity on the genome of *Escherichia coli* K12. *Proceedings of the Royal Society of London, Series B*, **168**, 281–92.

BRODA, P. (1967). The formation of Hfr strains in *Escherichia coli* K12. *Genetical Research*, **9**, 35–47.

CÁNOVAS, J. L. & KORNBERG, H. L. (1966). Properties and regulation of phosphopyruvate carboxylase activity in *Escherichia coli*. *Proceedings of The Royal Society of London, Series B*, **165**, 189–205.

CHAMBERS, D. A. & ZUBAY, G. (1969). The stimulatory effect of cyclic adenosine 3′,5′-monophosphate on DNA-directed synthesis of β-galactosidase in a cell-free system. *Proceedings of the National Academy of Sciences, USA*, **63**, 118–22.

COHN, M. & HORIBATA, K. (1959). Inhibition by glucose of the induced synthesis of the β-galactoside-enzyme system of *Escherichia coli*. Analysis of maintenance. *Journal of Bacteriology*, **78**, 601–12.

COOPER, R. A. & KORNBERG, H. L. (1967). The direct synthesis of phosphoenol pyruvate by *Escherichia coli*. *Proceedings of the Royal Society of London, Series B*, **168**, 263–80.

CORDARO, J. C., ANDERSON, R. P., GROGAN, E. W., WENZEL, D. J., ENGLER, M. & ROSEMAN, S. (1974). Promoter-like mutant affecting HPr and Enzyme I of the phosphoenol pyruvate:sugar phosphotransferase system in *Salmonella typhimurium*. *Journal of Bacteriology*, **120**, 245–52.

CORDARO, J. C. & ROSEMAN, S. (1972). Deletion mapping of the genes coding for HPr and Enzyme I of the phosphoenolpyruvate:sugar phosphotransferase system in *Salmonella typhimurium*. *Journal of Bacteriology*, **112**, 17–29.

CURTIS, S. J. & EPSTEIN, W. (1970). Two constitutive P-HPr:glucose phosphotransferases in *Escherichia coli* K12. *Federation Proceedings*, **30**, 1123.

CURTIS, S. J. & EPSTEIN, W. (1975). Phosphorylation of D-glucose in *Escherichia coli* mutants defective in glucosephosphotransferase, mannosephosphotransferase and glucokinase. *Journal of Bacteriology*, **122**, 1189–99.

EPPS, H. M. R. & GALE, E. F. (1942). The influence of the presence of glucose during growth on the enzymic activities of *Escherichia coli*: comparison of the effect with that produced by fermentation acids. *Biochemical Journal*, **36**, 619–23.

EPSTEIN, W. & CURTIS, S. J. (1972). Genetics of the phosphotransferase system. In *Role of Membranes in Secretory Processes*, ed. L. Bolis, R. D. Keynes & W. Wilbrandt, pp. 98–112. North-Holland: Amsterdam.

FERENCI, T. & KORNBERG, H. L. (1971a). Pathway of fructose utilization by *Escherichia coli*. *FEBS Letters*, **13**, 127–30.

FERENCI, T. & KORNBERG, H. L. (1971b). Role of fructose 1,6-diphosphatase in fructose utilization. *FEBS Letters*, **14**, 360–3.

FERENCI, T. & KORNBERG, H. L. (1973). The utilization of fructose by *Escherichia coli*. Properties of a mutant defective in fructose 1-phosphate kinase activity. *Biochemical Journal*, **132**, 341–7.

FERENCI, T. & KORNBERG, H. L. (1974). The role of phosphotransferase-mediated syntheses of fructose 1-phosphate and fructose 6-phosphate in the growth of *Escherichia coli* on fructose. *Proceedings of the Royal Society London, Series B*, **187**, 105–19.

FRAENKEL, D. G. (1968). The phosphoenolpyruvate-initiated pathway of fructose metabolism in *Escherichia coli*. *Journal of Biological Chemistry*, **243**, 6458–64.

HAGGERTY, D. M. & SCHLEIF, R. F. (1975). Kinetics of the onset of catabolite repression in *Escherichia coli* as determined by *lac* messenger ribonucleic acid initiations and intracellular cyclic adenosine 3,5'-monophosphate levels. *Journal of Bacteriology*, **123**, 946–53.

HAMILTON, W. A. (1975). Energy coupling in microbial transport. *Advances in Microbial Physiology*, **12**, 1–53.

HANSON, T. E. & ANDERSON, R. L. (1968). Phosphoenol pyruvate-dependent formation of D-fructose 1-phosphate by a four-component phosphotransferase system. *Proceedings of the National Academy of Sciences, USA*, **61**, 269–76.

HARWOOD, J. P., GAZDAR, C., PRASAD, C., PETERKOFSKY, A., CURTIS, S. J. & EPSTEIN, W. (1976). Involvement of the glucose Enzymes II of the sugar phosphotransferase system in the regulation of adenylate cyclase by glucose in *Escherichia coli*. *Journal of Biological Chemistry*, **251**, 2462–8.

HARWOOD, J. P. & PETERKOFSKY, A. (1975). Glucose-sensitive adenylate cyclase in toluene-treated cells of *Escherichia coli* B. *Journal of Biological Chemistry*, **250**, 4656–62.

HERBERT, D. & KORNBERG, H. L. (1976). Glucose transport as rate-limiting step in the growth of *Escherichia coli* on glucose. *Biochemical Journal*, **156**, 477–80.

JONES-MORTIMER, M. C. & KORNBERG, H. L. (1974). Genetic control of inducer exclusion by *Escherichia coli*. *FEBS Letters*, **48**, 93–5.

JONES-MORTIMER, M. C. & KORNBERG, H. L. (1976). Order of genes adjacent to *ptsX* on the *E. coli* genome. *Proceedings of the Royal Society of London, Series B*, **193**, 313–15.

KEPES, A. (1960). Etudes cinétiques sur la galactoside-perméase d'*Escherichia coli*. *Biochimica et Biophysica Acta*, **40**, 70–84.

KESSLER, D. P. & RICKENBERG, H. V. (1963). The competitive inhibition of α-methylglucoside uptake in *Escherichia coli*. *Biochemical and Biophysical Research Communications*, **10**, 482–7.

KOCH, A. L. (1964). The role of permease in transport. *Biochimica et Biophysica Acta*, **79**, 177–200.

KOCH, A. L. (1971). Local and non-local interactions of fluxes mediated by the glucose and galactose permeases of *Escherichia coli*. *Biochimica et Biophysica Acta*, **249**, 197–215.

KORNBERG, H. L. (1972). Nature and regulation of hexose uptake by *Escherichia coli*. In *The Molecular Basis of Biological Transport*, ed. J. F. Woessner Jr & F. Huijing, Miami Winter Symposia, vol. 3, pp. 157–80. New York & London: Academic Press.

KORNBERG, H. L. (1973). Fine control of sugar uptake by *Escherichia coli*. *Symposia of the Society for Experimental Biology*, **27**, 175–93.

KORNBERG, H. L. & JONES-MORTIMER, M. C. (1975). *PtsX*: a gene involved in the uptake of glucose and fructose by *Escherichia coli. FEBS Letters*, **51**, 1–4.

KORNBERG, H. L. & MALCOVATI, M. (1973). Control *in situ* of the pyruvate kinase activity of *Escherichia coli. FEBS Letters*, **32**, 257–9.

KORNBERG, H. L. & REEVES, R. E. (1972*a*). Correlation between hexose transport and phosphotransferase activity in *Escherichia coli. Biochemical Journal*, **126**, 1241–3.

KORNBERG, H. L. & REEVES, R. E. (1972*b*). Inducible phosphoenolpyruvate-dependent hexose phosphotransferase activities in *Escherichia coli. Biochemical Journal*, **128**, 1339–44.

KORNBERG, H. L. & RIORDAN, C. L. (1976). Uptake of galactose into *Escherichia coli* by facilitated diffusion. *Journal of General Microbiology*, **94**, 75–89.

KORNBERG, H. L. & SMITH, J. (1970). Role of phosphofructokinase in the utilization of glucose by *Escherichia coli. Nature, London*, **227**, 44–6.

KORNBERG, H. L. & SMITH, J. (1972). Genetic control of glucose uptake by *Escherichia coli. FEBS Letters*, **20**, 270–2.

KORNBERG, H. L. & SOUTAR, A. K. (1973). Utilization of gluconate by *Escherichia coli*. Induction of gluconate kinase and 6-phosphogluconate dehydratase activities. *Biochemical Journal*, **134**, 489–98.

KUNDIG, W. (1974). Molecular interactions in the bacterial phosphoenol-pyruvate-phosphotransferase system (PTS). *Journal of Supramolecular Structure*, **2**, 695–714.

KUNDIG, W., GHOSH, S. & ROSEMAN, S. (1964). Phosphate bound to histidine in a protein as an intermediate in a novel phosphotransferase system. *Proceedings of the National Academy of Sciences, USA*, **52**, 1067–74.

KUNDIG, W. & ROSEMAN, S. (1971*a*). Sugar transport. Isolation of a phosphotransferase system from *Escherichia coli. Journal of Biological Chemistry*, **246**, 1393–406.

KUNDIG, W. & ROSEMAN, S. (1971*b*). Sugar transport. Characterization of constitutive membrane-bound enzymes II of the *Escherichia coli* phosphotransferase system. *Journal of Biological Chemistry*, **246**, 1407–18.

LENGELER, J. (1966). Untersuchungen zum Glukose-Effekt bei der Synthese der Galaktose-Enzyme von *Escherichia coli. Zeitschrift für Vererbungslehre*, **98**, 203–29.

LENGELER, J. (1975*a*). Mutations affecting transport of the hexitols D mannitol, D glucitol and galactitol in *Escherichia coli* K12: isolation and mapping. *Journal of Bacteriology*, **124**, 26–38.

LENGELER, J. (1975*b*). Nature and properties of hexitol transport systems in *Escherichia coli. Journal of Bacteriology*, **124**, 38–47.

LIS, J. F. & SCHLEIF, R. (1973). Different cyclic AMP requirements for induction of the arabinose and lactose operons of *Escherichia coli. Journal of Molecular Biology*, **79**, 149–62.

MCFALL, E. & MANDELSTAM, J. (1963). Specific metabolic repression of three induced enzymes in *Escherichia coli. Biochemical Journal*, **89**, 391–8.

MCGINNIS, J. F. & PAIGEN, K. (1969). Catabolite inhibition: a general phenomenon in the control of carbohydrate utilization. *Journal of Bacteriology*, **100**, 902–13.

McGinnis, J. F. & Paigen, K. (1973). Site of catabolite inhibition of carbohydrate metabolism. *Journal of Bacteriology*, **114**, 885–7.

Magasanik, B. (1961). Catabolite repression. *Cold Spring Harbor Symposia on Quantitative Biology*, **26**, 249–56.

Magasanik, B. (1970). Glucose effects: inducer exclusion and repression. In *The Lactose Operon*, ed. J. R. Beckwith & D. Zipser, pp. 189–219. Madison: Cold Spring Harbor Laboratory.

Makman, R. S. & Sutherland, E. W. (1965). Adenosine 3′,5′-phosphate in *Escherichia coli*. *Journal of Biological Chemistry*, **240**, 1309–14.

Malcovati, M. & Kornberg, H. L. (1969). Two types of pyruvate kinase in *Escherichia coli* K12. *Biochimica et Biophysica Acta*, **178**, 420–3.

Monod, J. (1942). *Recherches sur la croissance des cultures bacteriénnes*. Paris: Hermann et Cie.

Monod, J. (1956). Remarks on the mechanism of enzyme induction. In *Enzymes: Units of Biological Structure and Function*, ed. O. H. Gaebler, pp. 7–28. New York & London: Academic Press.

Morgan, M. J. & Kornberg, H. L. (1967). Effect of pyruvate on hexose metabolism by *Escherichia coli*. *Biochemical Journal*, **103**, 57P.

Morgan, M. J. & Kornberg, H. L. (1969). Regulation of sugar accumulation by *Escherichia coli*. *FEBS Letters*, **3**, 53–6.

Paigen, K. (1966). Phenomenon of transient repression in *Escherichia coli*. *Journal of Bacteriology*, **91**, 1201–9.

Pastan, I. & Perlman, R. L. (1969). Repression of β-galactosidase synthesis by glucose in phosphotransferase mutants of *Escherichia coli*. *Journal of Biological Chemistry*, **244**, 5836–42.

Perlman, R. L. & Pastan, I. (1968). Cyclic 3′,5′-AMP: stimulation of β-galactosidase and tryptophanase synthesis in *E. coli*. *Biochemical and Biophysical Research Communications*, **30**, 656–64.

Perlman, R. L. & Pastan, I. (1969). Pleiotropic deficiency of carbohydrate utilization in an adenyl cyclase deficient mutant of *Escherichia coli*. *Biochemical and Biophysical Research Communications*, **37**, 151–7.

Peterkofsky, A. & Gazdar, C. (1974). Glucose inhibition of adenylate cyclase in intact cells of *Escherichia coli* B. *Proceedings of the National Academy of Sciences, USA*, **71**, 2324–8.

Peterkofsky, A. & Gazdar, C. (1975). Interaction of Enzyme I of the phosphoenolpyruvate:sugar phosphotransferase system with adenylate cyclase of *Escherichia coli*. *Proceedings of the National Academy of Sciences, USA*, **72**, 2920–4.

Postma, P. W. (1976). Involvement of the phosphotransferase system in galactose transport in *Salmonella typhimurium*. *FEBS Letters*, **61**, 49–53.

Roseman, S. (1969). The transport of carbohydrates by a bacterial phosphotransferase system. *Journal of General Physiology*, **54**, 138s–80s.

Roseman, S. (1972). A bacterial phosphotransferase system and its role in sugar transport. In *The Molecular Basis of Biological Transport, ed.* J. F. Woessner Jr & F. Huijing, Miami Winter Symposia, vol. 3, pp. 181–218. New York & London: Academic Press.

Roseman, S. (1975). The bacterial phosphoenolpyruvate:sugar phosphotransferase system. *Ciba Foundation Symposia*, **31** (new series), 225–41.

SAIER, M. H., FEUCHT, B. U. & HOFSTADTER, L. J. (1976). Regulation of carbohydrate uptake and adenylate cyclase by the Enzymes II of the phosphoenolpyruvate:sugar phosphotransferase system in *Escherichia coli. Journal of Biological Chemistry*, **251**, 883–92.

SAIER, M. H., FEUCHT, B. U. & MCCAMAN, M. T. (1975). Regulation of intracellular adenosine cyclic 3':5'-monophosphate levels in *Escherichia coli* and *Salmonella typhimurium*. Evidence for energy-dependent excretion of the cyclic nucleotide. *Journal of Biological Chemistry*, **250**, 7593–601.

SAIER, M. H. & ROSEMAN, S. (1972). Inducer exclusion and repression of enzyme synthesis in mutants of *Salmonella typhimurium* defective in enzyme I of the phosphoenolpyruvate:sugar phosphotransferase system. *Journal of Biological Chemistry*, **247**, 972–5.

SAIER, M. H., SIMONI, R. D. & ROSEMAN, S. (1970). The physiological behaviour of Enzyme I and heat-stable protein mutants of a bacterial phosphotransferase system. *Journal of Biological Chemistry*, **245**, 5870–3.

SAIER, M. H. & STILES, C. D. (1975). Regulation of bacterial metabolism. In *Molecular Dynamics in Biological Membranes*, pp. 99–105, Heidelberg Science Library. New York, Heidelberg & Berlin: Springer Verlag.

SETLOW, P. (1973). Inability to detect cyclic AMP in vegetative or sporulating cells or dormant spores of *Bacillus megaterium. Biochemical and Biophysical Research Communications*, **52**, 365–72.

TAO, M. & LIPMANN, F. (1969). Isolation of adenyl cyclase from *Escherichia coli. Proceedings of the National Academy of Sciences, USA*, **63**, 86–92.

TYLER, B., LOOMIS, W. F. & MAGASANIK, B. (1967). Transient repression of the *lac* operon. *Journal of Bacteriology*, **94**, 2001–11.

ULLMANN, A. (1974). Are cyclic AMP effects related to real physiological phenomena? *Biochemical and Biophysical Research Communications*, **57**, 348–52.

ULLMANN, A. & MONOD, J. (1968). Cyclic AMP as an antagonist of catabolite repression in *Escherichia coli. FEBS Letters*, **2**, 57–60.

UMBARGER, H. E. (1961). Feedback control by end product inhibition. *Cold Spring Harbor Symposia on Quantitative Biology*, **26**, 301–12.

VINOPAL, R. T. & FRAENKEL, D. G. (1974). Phenotypic suppression of phosphofructokinase mutations in *Escherichia coli* by constitutive expression of the glyoxalate shunt. *Journal of Bacteriology*, **118**, 1090–100.

WANG, R. J. & MORSE, M. L. (1968). Carbohydrate accumulation and metabolism in *Escherichia coli*. Description of pleiotropic mutants. *Journal of Molecular Biology*, **32**, 59–66.

WANG, R. J., MORSE, H. G. & MORSE, M. L. (1969). Carbohydrate accumulation and metabolism in *Escherichia coli*: the close linkage and chromosomal location of *ctr* mutations. *Journal of Bacteriology*, **98**, 605–10.

WANG, R. J., MORSE, H. G. & MORSE, M. L. (1970). Carbohydrate accumulation and metabolism in *Escherichia coli*: characteristics of the reversions of *ctr* mutations. *Journal of Bacteriology*, **104**, 1318–24.

WINKLER, H. H. & WILSON, T. H. (1967). Inhibition of β-galactoside transport by substrates of the glucose transport system in *Escherichia coli. Biochimica et Biophysica Acta*, **135**, 1030–51.

MICROBIAL METABOLIC REGULATION BY ADENINE NUCLEOTIDE POOLS

CHRISTOPHER J. KNOWLES

Biological Laboratory, University of Kent at Canterbury, Canterbury CT2 7NJ, UK

INTRODUCTION

The crucial role of ATP as the 'coinage' of energy in living organisms has been clear since 1941, when Lipmann's profound review was published. However, it wasn't until 1961 that an attempt was made to measure the ATP content of a growing micro-organism (Franzen & Binkley, 1961). Though the development of the luciferin–luciferase system has recently facilitated estimation of low concentrations of adenine nucleotides (Strehler & McElroy, 1957), there are no methodological reasons for this strangely extended dormant period.

Slow sampling techniques were used in many of the estimates of intracellular ATP content made during the 1960s, and it was not appreciated until the end of this decade that the total ATP pool of growing micro-organisms turns over many times per second (Harrison & Maitra, 1969; Holms, Hamilton & Robertson, 1972; Miović & Gibson, 1973). It is therefore only during the last eight or nine years that realistic measurements have been made of intracellular ATP, ADP and AMP contents of growing micro-organisms. During this period considerable impetus has been given to such studies by the provocative concept of metabolic regulation by adenine nucleotides proposed by Atkinson (1968a).

It is the aim of this review to discuss the current knowledge of adenine nucleotide pools in growing micro-organisms. The initial section is a critique of the energy charge concept, such that the reader may assess its relevance to metabolic regulation by adenine nucleotide pools in micro-organisms, which will be discussed in the subsequent sections.

ADENYLATE ENERGY CHARGE

The concept

Dietary materials are catabolised by living organisms to a limited number of small molecules. These are interconverted, as necessary, and serve as precursors for synthesis of the macromolecular constituents of the organism via the anabolic or biosynthetic sequences.

However, catabolism and anabolism are not linked only by these building-block molecules. Net synthesis of ATP from ADP occurs during catabolism. The ATP is used to drive endergonic biosynthetic pathways, and is converted to ADP and AMP. The adenine nucleotides are also interconverted by the action of adenylate kinase (Noda, 1973).

Adenine nucleotides (ATP, ADP and AMP) therefore act as metabolic 'energy' mediators and are ideally placed to regulate the whole metabolic economy of the organism. Not only is ATP produced and consumed in many individual enzymatic steps of metabolism, but these and other reactions are often regulated by AMP, ADP or ATP in addition to specific feedback control by end-products of the particular pathways.

For example, phosphofructokinase is a key regulatory enzyme in the glycolytic pathway of glucose catabolism. The enzyme from *Escherichia coli* is inhibited by one of the products of glycolysis, phosphoenolpyruvate, and activated by ADP (Blangy, Buc & Monod, 1968). Moreover, since ATP is a substrate for this enzyme, its activity is determined by the relative concentrations of the ATP and ADP present.

In general, catabolic sequences contain regulatory enzymes that are activated by ADP or AMP, or inhibited by ATP: degradation of growth substrates therefore proceeds maximally only when ATP needs to be regenerated. Conversely, many regulatory enzymes of ATP-consuming biosynthetic pathways are activated by ATP or inhibited by ADP or AMP.

Thus, whilst the activities of many regulatory enzymes depend on the concentrations of ATP, ADP or AMP, or on the relative concentrations of ATP:ADP or ATP:AMP it is clearly naive to consider only the concentrations of the individual adenine nucleotides or the ratios of only two of them. In the intracellular environment all three are present and a change in concentration of any one is necessarily reflected in changes in concentration of both of the other adenine nucleotides.

Atkinson & Walton (1967) introduced the concept of 'adenylate energy charge'. This theory was later discussed in more detail (Atkinson, 1968a) and has since been reviewed several times (Atkinson 1968b, 1969, 1970, 1971). Bomsel & Pradet (1968) independently developed a similar theory that stresses the regulatory importance of adenylate kinase.

Adenylate energy charge (EC), where

$$EC = \frac{[ATP] + \frac{1}{2}[ADP]}{[ATP] + [ADP] + [AMP]}$$

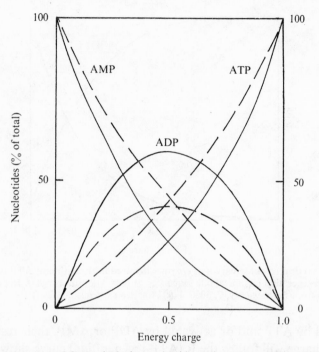

Fig 1. The proportion of the total adenine nucleotide pool present as ATP, ADP and AMP at different energy charge values, with the adenylate kinase equilibrium constant (K_{app}) in the region of 0.2 to 1.2. Solid line, $K_{app} = 0.2$; dashed line, $K_{app} = 1.2$. (From J.-L. Bomsel & A. Pradet (1968). *Biochimica et Biophysica Acta*, **162**, 230–42.)

is a linear function from 0 (all AMP) to 1 (all ATP) of the proportion of the total adenylate pool that contains anhydride-bound phosphate of high free energy of hydrolysis. (NB: ATP contains two 'high-energy' anhydride-bound phosphate residues, ADP one and AMP none; energy charge is thus a measure of *half* the anhydride-bound phosphate of high free energy of hydrolysis.)

At intermediate values of energy charge the relative concentrations of each of the adenine nucleotides depend on the equilibrium position of the adenylate kinase reaction. Fig. 1 shows the effect of variation of energy charge on the relative concentrations of each of the adenine nucleotides, assuming values in the range of 0.2–1.2 for the apparent equilibrium constant of adenylate kinase ($K_{app} = [ATP] \cdot [AMP]/[ADP]^2$). Thus conversion of ATP to ADP, when the system is poised at moderate energy charge values, results in an overall increase in AMP concentrations with little effect on the ADP concentration.

Since ATP-regenerating (catabolic) pathways of metabolism are

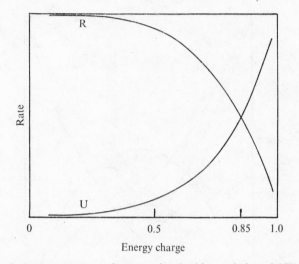

Fig. 2. Theoretical response curves of enzymes involved in regulation of ATP-utilising (U) and ATP-regenerating (R) metabolic sequences. (From Atkinson (1968a). Reprinted with permission from *Biochemistry*, **7**, 4030–4. © American Chemical Society.)

inhibited by ATP and/or activated by ADP or AMP, their response to energy charge will follow the R (ATP-regenerating) curve shown in Fig. 2. Similarly ATP-activated, ADP or AMP-inhibited biosynthetic pathways will follow the U (ATP-utilising) curve shown in Fig. 2.

The portion of these curves that represents the greatest change in enzyme activity with respect to change in energy charge is in the region of 0.6–1.0, and they intersect at an energy charge of about 0.8–0.9. Variations from an energy charge of 0.8–0.9 will therefore adjust the metabolism of the organism and re-establish the energy charge in this region. Decreases in energy charge will 'switch-off' ATP-utilising pathways and activate ATP-regenerating sequences, thus restoring the steady state. This sequence of events will prevent either an excessive supply or a shortage of intermediary metabolites from occurring and will lead to metabolic stabilisation.

Enzymes, of course, cannot respond to an abstract concept like energy charge, however attractive it might be on theoretical grounds. As has already been pointed out, an enzyme will respond to changes in concentration of one of the nucleotides or the relative concentrations of two or, probably rarely, all three. This regulatory effect need not always be due to the presence of a regulatory site on the enzyme distinct from the active site. For enzymes that catalyse reactions involving conversions of adenine nucleotides (e.g. kinases) the response to energy charge will

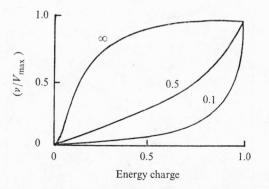

Fig. 3. Theoretical response curves of attainment of maximal initial velocity (v/V_{max} at differing energy charge values for ATP-utilising kinases with various relative affinities for substrate and product. Michaelis constant for the substrate is set at 0.2 mM and the ratio of K_{ADP} to K_{ATP} varied as indicated. Total adenine nucleotide concentration is 5 mM and K_{app} (adenylate kinase equilibrium) is assumed to be 0.8. (Redrawn from Purich & Fromm, (1973). *Journal of Biological Chemistry*, **248**, 461–6.)

depend on the relative affinity of the catalytic site for the substrate and product adenine nucleotides (Fig. 3).

Enzymes that respond to energy charge usually will be subject to feedback regulation by products of the specific metabolic pathway. How then are energy charge effects modified by feedback effectors and vice versa? Atkinson (1968*a*) has predicted that response curves of the types shown in Fig. 4 will occur for ATP-utilising and ATP-regenerating pathways. In both cases, at a given energy charge, the enzyme activity is lowered by an increasing concentration of an inhibitory feedback effector. Thus, the relative flux through any one of many competing metabolic pathways depends on the concentration of the product (i.e. the feedback inhibitor) of that pathway at any instant. Such response modifications mean that the intersection point on the U- and R-type plots (Fig. 2) can no longer be regarded as a single point, but rather as a complex region of overlap. Because the total regulatory effect comes from the interaction between many individual metabolic pathways of the type given in Fig. 4, optimal flexibility of metabolic regulation will be possible: the cell will be extremely responsive to even very minor shifts in energy charge and rapidly act to reassert the maximal possible metabolic activity.

Assessment of the validity of the concept

The energy charge concept is of necessity a simplified representation of metabolic control by adenine nucleotides. Like any other process of

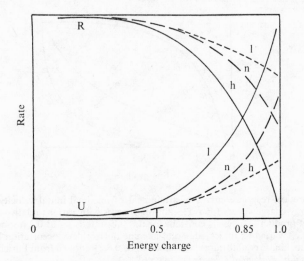

Fig. 4. Computed response curves of regulatory enzymes of ATP-utilising (U) and ATP-regenerating (R) sequences, as modified by high (h), normal (n) and low (l) concentrations of feedback modifiers. (Redrawn from Atkinson (1968a). Reproduced with permission from *Biochemistry*, **7**, 4030–4. © American Chemical Society.)

scientific reduction, it is difficult to determine whether it is a useful unifying hypothesis or whether it is an unrealistic oversimplification. Even where energy charge values of 0.8–0.9 have been obtained in growing micro-organisms (see below) and in higher organisms (c.f. Bomsel & Pradet, 1968; Chapman, Fall & Atkinson, 1971), it is difficult to assess whether energy charge regulation is responsible for metabolic homeostasis or whether these values are a consequence of such homeostasis.

Measurements of the responses of many regulatory enzymes to energy charge have shown that they exhibit the appropriate R- and U-type responses and that their activities are modified by feedback effectors as shown in Fig. 4. With a few exceptions, it is not possible in this review to discuss the properties of individual enzymes that are responsive to energy charge; the reader should consult the following papers for further information: Klungsøyr, Hageman, Fall & Atkinson, 1968; Shen, Fall, Walton & Atkinson, 1968; LeJohn & Stevenson, 1970; Villar-Palasi & Wei, 1970; Chulavatnatol & Atkinson, 1973a, b; Preiss, 1973; Chapman & Atkinson, 1973; Preiss, Greenberg & Sabraw, 1975.

In investigating the properties of isolated enzymes it is important that the assay conditions used should represent as closely as possible the intracellular concentrations of substrate, product, feedback effectors and total adenine nucleotides, as well as pH, Mg^{2+} content, etc. Moreover, it is usually more appropriate to consider the variations in affinity of a regulatory enzyme for its substrate(s) than variations of its maximal activity. In some of the references quoted above insufficient attention has been paid to these points though, where tested, the values of substrate affinities have been found to vary as expected. Purich & Fromm (1972, 1973) have confirmed that the sensitivities of several enzymes to energy charge are strongly dependent on the experimental conditions of pH and Mg^{2+}, substrate and product concentration, etc.; nonetheless their studies showed that over a wide range of conditions the responses were in the correct direction and only the degree of sensitivity changed.

Of great importance in assessing the role of energy charge is the level of available Mg^{2+} relative to the concentrations of the adenine nucleotides. Many enzymes, particularly kinases, respond to $Mg\ ATP^{2-}$ and $Mg\ ADP^-$ rather than to ATP^{4-} and ADP^{3-}; these enzymes are either unresponsive to the free nucleotides or ATP^{4-} or ADP^{3-} act as competitive inhibitors of $MgATP^{2-}$ and $MgADP^-$ (Melchior, 1965; Lowry & Passonneau, 1966; Larsson-Raznikiewicz, 1967). Therefore, when there is a greater concentration of adenine nucleotides than available Mg^{2+} an increase in energy charge could result in an inhibition of ATP-utilising systems rather than the expected stimulation. This has been shown to occur in studies of isolated rat brain hexokinase (Purich & Fromm, 1972).

Unfortunately, in most of the other studies quoted above with isolated enzymes an excess of Mg^{2+} has been used and the effect of variation of Mg^{2+} concentration has not been tested (cf. Shen & Atkinson, 1970).

In several types of mammalian cells the available Mg^{2+} concentration is less than the concentration of the total adenine nucleotide pool (see Purich & Fromm, 1972, 1973 for references). On the other hand, in *E. coli* the total intracellular adenine nucleotide concentration is in the range of 1–5 mM (see following sections) whilst the Mg^{2+} concentration is 20–40 mM (Lusk & Kennedy, 1969; Silver, 1969), though much of the Mg^{2+} could be in a bound form (Hurwitz & Rosano, 1967). *Bacillus subtilis* has also been reported to accumulate high concentrations of Mg^{2+} (Schribner, Eisenstadt & Silver, 1974). Growing *Chromatium* strain D contains about 20 mM Mg^{2+} (Miović & Gibson, 1973).

In addition to Mg^{2+}, other divalent metals (particularly Mn^{2+} and Ca^{2+}) are able to satisfy the requirements of many enzyme processes and also can complex with adenine nucleotides (Mildvan, 1970). Both

9

E. coli and *B. subtilis* have been reported to accumulate Mn^{2+} (Silver & Kralovic, 1969; Eisenstadt, Fisher, Der & Silver, 1973) but not Ca^{2+} (Silver, Toth & Schribner, 1975).

The energy charge concept requires there to be sufficient adenylate kinase present in living cells for it to catalyse a rapid return of the adenine nucleotides to equilibrium following changes in concentration of any of the adenine nucleotides. Adenylate kinase is, in fact, ubiquitously present in living cells (Noda, 1973) and its essential role is illustrated by the dramatic metabolic consequences of shifting mutants of *E. coli* containing a thermolabile adenylate kinase to non-permissive temperatures (Glaser, Nulty & Vagelos, 1975).

There is a Mg^{2+} requirement for the adenylate kinase reaction, which catalyses phosphate transfer between the adenine nucleotides according to the equation:

$$MgATP^{2-}+AMP^{2-} \rightleftharpoons MgADP^{-}+ADP^{3-}$$

Other divalent metals, notably Mn^{2+} and Ca^{2+}, can replace the Mg^{2+} (Noda, 1973).

Blair (1970) has pointed out that:

In a closed system it is obvious that as ATP is hydrolysed to ADP or AMP, the magnesium ion concentration (as distinct from the total magnesium concentration) must rise, as ADP and AMP form much weaker complexes with magnesium than does ATP; as the magnesium ion concentration changes, so complex formation with the three nucleotides will adjust to different extents, and so the apparent equilibrium constant of the reaction catalysed by adenylate kinase will change.

Thus the apparent equilibrium constant of the adenylate kinase reaction ($K_{app} = [ATP_{total}] \cdot [AMP_{total}]/[ADP_{total}]^2$) will vary depending on the prevailing energy charge and the Mg^{2+} : total adenine nucleotide ratio. At a true equilibrium constant ($K_{true} = [ATP^{4-}] \cdot [AMP^{2-}]/[ADP^{3-}]^2$) of 0.311 (or $K_{true} = [MgATP^{2-}] \cdot [AMP^{2-}]/[MgADP^{-}] \cdot [ADP^{3-}] = 5.31$), Blair (1970) has calculated that the K_{app} varies within the limits of 1.2 to 0.7, with an energy charge in the range of 0.96 to 0.52 and at a fixed total adenine nucleotide concentration of 5 mM and total Mg^{2+} concentration of 4–6 mM. The consequences of such changes in K_{app} can be judged crudely from a glance at Fig. 1 (cf. also Fig. 1 of Purich & Fromm, 1972).

It is evident that a modification of the energy charge concept is required to take into account the problems associated with changes in the Mg^{2+} : total adenine nucleotide ratio, the greater affinity of ATP than ADP or AMP for Mg^{2+}, and the greater total Mg^{2+} binding to the adenine nucleotides at higher energy charge values.

However, such considerations do not necessarily invalidate the concept in its present form for cells operating within relatively narrow limits of energy charge, Mg^{2+} and total adenine nucleotide content. This is particularly true when there is sufficient Mg^{2+} to saturate the adenine nucleotide pool, conceivably the case in bacteria. Indeed, these factors could well have a sharpening effect on regulation by adenine nucleotide pools (Blair, 1970, has discussed this point in greater detail). Nonetheless, it is clear that greater attention in both in-vivo and in-vitro studies should, in future, be paid to the levels of Mg^{2+} and other divalent metals.

ATP contains three phosphate residues, ADP two and AMP one, and a fall in energy charge causes an increase in intracellular inorganic orthophosphate content. The activities of many enzymes are affected by phosphate, and the effect of changes in cellular phosphate content with energy charge could be of considerable importance. The energy charge concept, therefore, may require modification to take this factor also into account.

A further major problem in assessing the role of energy charge in eukaryotic organisms is the question of metabolic compartmentation. Respiratory activity of mitochondria is controlled by the availability of ADP and orthophosphate (or substrate supply) within the mitochondrial matrix. The entry and exit of ADP and ATP occur via an adenine nucleotide translocase, which catalyses a 1:1 exchange of endogenous and exogenous ATP and ADP (Klingenberg, 1970). The translocase is highly specific for ATP and ADP; AMP is not transported, and the intra- and extramitochondrial AMP pools are separate. Furthermore, adenylate kinase is located in the intermembrane space and not within the matrix (Heldt & Schwalbach, 1967; Noda, 1973). The intramitochondrial AMP is phosphorylated to ADP by GTP–AMP transphosphorylase and substrate-level phosphorylation (Heldt & Schwalbach, 1967). In controlled mitochondria the ATP:ADP ratio is lower than the extramitochondrial ratio, and presumably the internal energy charge is also lower (Heldt, Klingenberg & Milovancev, 1972). The consequences of this are difficult to assess in terms of metabolic regulation by energy charge, but it has been claimed that rat liver citrate synthase, which in-vitro experiments show to be sensitive to energy charge (Atkinson, 1968b), is insensitive to energy charge in intact mitochondria (Olson & Williamson, 1971).

At 'physiological' energy charge values ($\simeq 0.8$–0.9) a small change in energy charge has only a slight effect on the ATP pool, but a much larger effect on the AMP pool. For example, using the data of Blair

(1970), a change in energy charge from 0.92 to 0.80 causes only a 20% decrease in the ATP pool but a five-fold increase in the AMP level and a change in the ATP:AMP ratio from about 45:1 to about 8:1. This 'amplification' effect means that very small changes in energy charge have profound effects on the activity of enzymes regulated by adenine nucleotides. In the growing cell the energy charge must be essentially invariant, or the change so small as to be detectable only by very sensitive assay systems. Lowry, Carter, Ward & Glaser (1971) have objected to the energy charge concept on the grounds that it is insensitive, with small changes in energy charge disguising large changes in ratios of the nucleotides, as pointed out above. However, Dietzler, Lais & Leckie (1974a) have pointed out that this is akin to stating that the pH scale is an insensitive measure of proton concentration.

Despite the objections listed above to the concept in its present form, it is obvious that the state of the intracellular adenine nucleotide pools has a profound regulatory effect on the metabolism of the cell, and that energy charge is a very convenient expression to describe the state of the pools. However, it should be remembered also that energy charge is a unitless parameter and, as a consequence, there are significant limitations to its usefulness. For example, without extra information, a knowledge of the intracellular energy charge tells us neither the total adenine nucleotide pool size nor the turnover of ATP, both of which could vary enormously without any change in energy charge.

ASSAY OF INTRACELLULAR ADENINE NUCLEOTIDE POOLS

The total adenine nucleotide pools of growing micro-organisms are of the order of 5mM, or less (see below). As a result, the total adenine nucleotide concentration in dilute suspensions of micro-organisms is very low and extremely sensitive assays are necessary for its determination. Three generally accepted assays have been developed: (a) the luciferin–luciferase system (Strehler & McElroy, 1957; Cole, Wimpenny & Hughes, 1967; Chapman et al., 1971; Lundin & Thore, 1975); (b) the fluorimetric–enzymatic method (Estabrook & Maitra, 1962; Williamson & Corkey, 1969; Lowry et al., 1971); and (c) thin-layer chromatography of ^{32}P-labelled nucleotides (Randerath & Randerath, 1964, 1967; Nazar & Wong, 1972). The first two methods are probably more sensitive and give quantitatively better results; the last method has the advantage that other pyrimidine and purine nucleotides can be assayed simultaneously. In the luciferase and fluorimetric methods AMP and

ADP have to be converted to ADP and ATP, respectively, before assay. The energy charge in growing cells is high, and the concentrations of ADP and, especially, AMP are very low. Consequently, the accuracies of assays of ADP and AMP concentrations are somewhat less than for ATP estimates. This has only a small effect on determinations of energy charge values, but leads to greater errors in estimates of K_{app} for the adenylate kinase reaction.

As already stated, the ATP pool of growing micro-organisms turns over very rapidly (Holms et al., 1972) and sampling techniques should have ideally, a quenching time of, at worst, only a few tenths of a second. In practice it is probable that the conditions in the growth medium are usually maintained during the few seconds required to remove a sample from the growth medium and pipette it into a quenching agent. However, it is dangerous to rely on such 'slow' methods of sampling, and reports of adenine nucleotide pools in micro-organisms estimated after, e.g. filtration or centrifugation, must be considered unreliable.

Harrison & Maitra (1969) developed a method for rapid run-off and quenching of bacteria from continuous cultures. Dr Niven, working in the author's laboratory, has devised a simple, spring-loaded syringe technique of sampling (D. F. Niven & C. J. Knowles, unpublished observations). The syringe is partially filled with a quenching agent, a dry, sterile needle attached, and inserted into a growing culture via a serum cap fitted onto a port in the culture vessel. A simple twist of the syringe plunger causes an extremely rapid spring-driven withdrawal of a fixed volume of the culture, with a quenching time of a few milliseconds.

Several methods have been used for extraction of adenine nucleotides and quenching of metabolic activity, e.g. cold perchloric, trichloroacetic or sulphuric acids, hot buffer, ethanol, butanol or chloroform. In utilising any of these methods attention should be given to ensuring that: (a) metabolic activity is rapidly and irreversibly inactivated, as residual ATPase, phosphatase, adenylate kinase activity, etc. will lead to incorrect estimates of the nucleotide composition; (b) all the ATP, ADP and AMP are extracted; and (c) the quenching solution used does not interfere, or interferes in a quantifiably reproducible manner, with the nucleotide assay system. Lundin & Thore (1975) have thoroughly examined and discussed these problems with respect to the luciferase assay system.

BACTERIAL ADENINE NUCLEOTIDE POOLS
AND ENERGY CHARGE

ATP content during growth

The ATP pools (expressed as μmol per g protein, dry weight or nitrogen) of a wide variety of bacteria growing in batch culture have been reported to be constant throughout the logarithmic growth phase (Chapman et al., 1971; Bächi & Ettlinger, 1973; Holms et al., 1972; Miović & Gibson, 1973; Dietzler et al., 1974a; Gadkari & Stolp, 1975). However, Forrest (1965) claimed that the ATP pool of Streptococcus faecalis decreased throughout the logarithmic growth phase. Cole et al., (1967) found that the ATP pool of E. coli was either in balance, decreased or increased during growth depending on the culture conditions, but later work by Chapman et al. (1971), Dietzler et al. (1974a) and Holms et al. (1972) has shown that there is a constant ATP pool during growth of E. coli. Eigener (1975) reported that the ATP content of the chemoautotroph Nitrobacter winogradskyi increased during logarithmic and linear growth phases. Knowles & Smith (1970) suggested that the ATP pool of Azotobacter vinelandii increases during the growth phase; this is erroneous due to underestimation of ATP during the early logarithmic phase, and the pool is in fact relatively constant at 4–6 μmol per g dry wt (C. J. Knowles & L. Smith, unpublished observations). The ATP content of B. subtilis increases during vegetative growth in nutrient media (Klofat, Picciolo, Chappelle & Freese, 1969) but is constant during vegetative growth in a glucose-minimal medium (Hutchison & Hanson, 1974).

Franzen & Binkley (1961) reported that the ATP content of E. coli (as μmol per g dry wt), is independent of growth rate, but that both cell mass and ATP level (as μmol ATP per cell) increase with increases in growth rate. Smith & Maaløe (1964) obtained similar results for Salmonella typhimurium growing at different rates. Therefore, according to these workers, the intracellular ATP molarity is essentially independent of growth rate. In contrast, Neidhardt & Fraenkel (1961) and Bagnara & Finch (1973) found that the ATP contents (as μmol per g protein or dry wt) of Aerobacter (Klebsiella) aerogenes and E. coli vary with the growth rate.

Miović & Gibson (1971, 1973) have studied the effect of growth rate on the ATP content of Chromatium strain D cultured photoheterotrophically. In their earlier paper they reported that the ATP pool (as μmol per g protein) was lower in slowly growing cells, but suggested that this could have reflected a higher relative protein content in the slower growing cells rather than a decrease in concentration of the ATP pool. In their later paper, employing improved experimental techniques, they showed that the ATP pool (as μmol per g protein) was independent of growth rate. Since the volume of slowly growing cells was less than that of rapidly growing cells, the intracellular ATP molarity was slightly higher in the former cells.

In the experiments mentioned in the previous two paragraphs the variations of growth rate were attained by changes in composition of the medium or changes in light intensity. However, determinations of the effect of growth rate on metabolite levels are probably better assessed using continuous cultures. Harrison & Maitra (1969) found the ATP pool of K. aerogenes growing in aerobic glucose-limited continuous culture to be constant at dilution rates of 0.1 to 0.5 h^{-1}. At slower growth rates the ATP pools were lower, probably due to losses of viability. The ATP pool of Selenomonas ruminantium has also been found to be constant when grown anaerobically in continuous culture at dilution rates of 0.065–0.11 h^{-1} (Hobson & Summers, 1972). Swedes, Sedo & Atkinson (1975) have shown, perhaps not surprisingly, that the ATP content of a mutant of E. coli strain K12 auxotrophic for adenine increases from 3 to 11.7 μmol per g protein as the growth rate increases from 0.13 to 0.62 h^{-1} during adenine-limited continuous culture.

Table 1 shows ATP contents, ATP:ADP ratios, values of energy charge, and K_{app} for the adenylate kinase reaction of a variety of bacteria during logarithmic growth in batch culture or in continuous culture. Information about the adenine nucleotide pools of bacteria estimated following 'slow' quenching of metabolic activity has been omitted. Some of these data have previously been tabulated by Chapman et al. (1971).

The ATP contents of aerobically or anaerobically growing heterotrophic or photosynthetic bacteria are remarkably similar and usually in the range of 2–10 μmol per g dry wt, which is equivalent to 0.1–0.5% of the total dry weight. Roberts et al. (1957) have pointed out that small molecular weight metabolites represent about 8% of the dry weight of E. coli. Assuming a similar figure for other bacteria as well, ATP is present as 1–6% of all the small molecules present in bacteria. Winkler & Wilson (1966) have calculated that the cytoplasmic volume of E. coli is 2.7 μl per mg dry wt. If it is assumed that the intracellular water space of other organisms is in the region of 2–4 μl per mg dry wt, intracellular ATP concentrations of bacteria vary in the range of about 0.5–5.0 mM during growth.

An exceptionally low ATP pool of 0.4–1.2 μmol per g dry wt has

Table 1. *The adenine nucleotide pools of growing bacteria*[a]

Organism	Comments	Energy charge	ATP content (μmol per g dry wt)	ATP:ADP ratio	K_{app}[b]	Reference
Acetobacter aceti	—	0.87	7.6	—	—	Bächi & Ettlinger (1973)
Azotobacter vinelandii	—	—	4–6	—	—	Knowles & Smith (1970, and unpublished observations)
A. vinelandii	—	0.8–0.9	—	—	—	Liao & Atkinson (1971)
Bacillus subtilis	—	~0.7	0.4	—	—	Hutchison & Hanson (1974)
B. subtilis	—	0.79–0.86	3.6	—	—	Setlow & Kornberg (1970a)
Bdellovibrio bacteriovorus	—	—	6–10[c]	—	—	Gadkari & Stolp (1975)
Beneckea natriegens	—	0.85–0.9	5–8	—	—	D. F. Niven & C. J. Knowles, unpublished observations
Chromatium strain D	Anaerobic, photoheterotrophic growth					Miović & Gibson (1971)
	(a) high-light	0.76	5.2[c]	2.6	1.3	
	(b) medium-light	0.82	4.1[c]	2.9	0.84	
	(c) low-light	0.83	3.1[c]	3.8	1.7	
Chromatium strain D	Anaerobic, photoheterotrophic growth					Miović & Gibson (1973)
	(a) high-light	0.81	2.2	3.8	1.9	
	(b) low-light	0.85	2.3	4.6	1.7	
Clostridium acetobutylicum	Anaerobic growth	—	3.7	1.54	—	O'Brien & Morris (1971)
Clostridium kluyveri	Anaerobic growth					Decker & Pfitzer (1972)
	(a) ethanol, acetate, bicarbonate medium	0.79	5.48	2.67	0.96	
	(b) crotonate, bicarbonate medium	0.76	7.01	2.38	1.03	

Escherichia coli (strain B)	Various carbon sources and growth rates	0.90–0.93	3.0–6.6	5.9–10.7	0.22–2.2	Franzen & Binkley (1961)
E. coli (strain B)	Aerobic and anaerobic growth	0.8	3.5–7.0	—	—	Chapman, Fall & Atkinson (1971)
E. coli (strain K12)	Aerobic growth	—	6.6–13.5	—	—	Cole, Wimpenny & Hughes (1967)
	Anaerobic growth	—	7.4–9.4	—	—	
	Growth in continuous culture (a) aerobic	—	5.4–7.5	—	—	
	(b) anaerobic	—	3.0–3.5	—	—	
E. coli (strain K12, Hfr 139)	—	0.77–0.86	5.4–6.6	3.0–3.3	0.54–0.57	Lowry, Carter, Ward & Glaser (1971)
E. coli (strain K12, ABLA Leu$^-$)	At 0.4 doublings h^{-1} to	—	5	~5	—	Bagnara & Finch (1973)
	1.0 doubling h^{-1}	—	11	~5	—	
E. coli (strain K12, W4597)		0.74	3.84[c]	1.6	0.4	Dietzler, Lais & Leckie (1974a)
E. coli (strain K12, PCO294) (adenine auxotroph)	Aerobic growth in continuous culture at $D = 0.62\ h^{-1}$	0.89	5.9[c]	—	—	Swedes, Sedo & Atkinson (1975)
	to $D = 0.13\ h^{-1}$ and in batch culture	0.89	1.5[c]	—	—	
		0.90	5.0[c]	—	—	
E. coli (strain ML308)	Various carbon sources and growth rates	—	4.5–7.5	—	—	Holms, Hamilton & Robertson (1972)
Klebsiella (Aerobacter) aerogenes	In continuous culture: glucose-limited, aerobic	0.8–0.84	6.3–9.0[d]	2.7–3.0	0.53–0.59	Harrison & Maitra (1969)
	glucose-limited, anaerobic	0.65	3.7	0.9	0.27	
	nitrogen-limited, aerobic	0.71	4.5	1.7	0.62	

Table 1. Continued

Organism	Comments	Energy charge	ATP content (μmol per g dry wt)	ATP:ADP ratio	K_{app}[b]	Reference
K. aerogenes	—	0.8–0.9	—	—	—	Wiebe & Bancroft (1975)
Myxococcus xanthus	—	0.8–0.87	0.85	6.7	3.7	Hanson & Dworkin (1974)
Nitrobacter winogradskyi	Chemoautotrophic growth					Eigener (1975)
	(a) linear growth phase	0.54–0.60	0.43–0.44[c]	0.30–0.34	0.28–0.32 }	
	(b) logarithmic growth phase	0.36–0.37	0.81–1.23[c]	0.53–0.83	0.30–0.36 }	
Pseudomonas aeruginosa	—	0.8–0.9	—	—	—	Wiebe & Bancroft (1975)
Rickettsia prowazeki	Freshly isolated	0.75	1.45	1.7	0.4	Winkler (1976)
Selenomonas ruminantium	Anaerobic growth in continuous culture ($D = 0.065$–0.11 h^{-1})	—	2.2	—	—	Hobson & Summers (1972)
Zymomonas mobilis	Anaerobic growth	—	3–8	—	—	Lazdunski & Belaich (1972)
Seven species of marine bacteria	Continuous culture ($D = 0.2$ h^{-1})	—	Mean 5.3 (range 2.4–8.8)	—	—	Hamilton & Holm-Hansen (1967)
Five species of bacteria	Logarithmic growth in a microcosm of sand, kaolin and ground terrestrial litter, containing 60% water	—	Mean 4.4 (range 3.2–5.6)	—	— }	Ausmus (1973)
Six species of actinomycetes		—	Mean 7.0 (range 3.6–10.0)	—	— }	

[a] Values taken for aerobic, heterotrophic growth during logarithmic phase, unless otherwise indicated.

[b] Apparent equilibrium constant of the adenylate kinase reaction $\left(K_{app} = \dfrac{[ATP][AMP]}{[ADP]^2} \right)$.

[c] Values derived from original data, assuming nitrogen and protein contents to be 10% and 50% of the dry weight respectively.

[d] ATP pool essentially constant at $D = 0.1$ to 0.5 h^{-1}.

been found in growing *N. winogradskyi* (Eigener, 1975). However, it cannot be assumed from this that chemoautotrophs in general have small ATP pools; harvested *Thiobacillus* species strain C contains an ATP pool of up to 4.7 μmol per g dry wt (Kelly & Syrett, 1966).

Energy charge during growth

From a consideration of the energy charge concept, energy charge values of 0.8–0.9 would be expected to occur in micro-organisms during balanced growth, either in logarithmic growth in batch culture or in continuous culture. Table 1 shows energy charge values found in several bacteria during growth. In almost all instances values of 0.7–0.9, and usually 0.8–0.9, have been obtained. Only when unusual methods of quenching metabolic activity as, for example, in Eigener's (1975) experiments with *N. winogradskyi*, or when 'slow' quenching has been used (e.g. Fanica-Gaignier, Clement-Metral & Kamen, 1971; Schmidt & Kamen, 1971), have lower energy charge values been reported.

In heterotrophic bacteria ATP is generated from catabolism of organic substrates, which also provide biosynthetic building-block molecules. In chemoautotrophic and photosynthetic bacteria ATP generation (by inorganic oxidations and cyclic photophosphorylation respectively) is not directly linked to substrate utilisation (by carbon dioxide fixation or catabolism of organic compounds). It is, therefore, of considerable interest to see whether energy charge values similar to those found in heterotrophic bacteria occur during growth of these organisms. Unfortunately there have not been any satisfactory estimates of energy charge in growing chemoautotrophic bacteria. Furthermore, as far as the author is aware, only in the case of *Chromatium* (Miović & Gibson, 1971, 1973), where values of 0.76–0.85 have been found, has energy charge been estimated in a growing photosynthetic micro-organism using rapid quenching techniques.

What happens to the adenine nucleotide pools and energy charge when there is a sudden change in growth conditions, such as deprivation of a growth substrate? Enzymes that are responsive to energy charge *in vitro* are, within limits, less sensitive to variations of adenine nucleotide concentrations than to the ratios of the nucleotides (ATP:ADP or ATP:AMP). Micro-organisms, therefore, might be expected to respond to metabolic stress by reduction of the adenine nucleotide pool size in order to maintain the energy charge. This could presumably be attained either by excretion of ADP or, especially, AMP or by enzymatic breakdown of AMP within the cell.

Chromatium excretes no ATP, some ADP, and up to 50% of the intracellular AMP during growth (Miović & Gibson, 1973). However, *E. coli* only excretes nucleotides (mainly AMP) when starved for extended periods (Chapman *et al.*, 1971). Moreover, *Beneckea natriegens* does not excrete adenine nucleotides into the medium during growth or for several hours after exhaustion of growth substrates (Niven, Collins & Knowles, 1976), and *Acetobacter aceti* excretes less than 10% of its total adenine nucleotides during growth (Bächi & Ettlinger, 1973).

In-vitro experiments of Schramm & Leung (1973) have shown that AMP nucleosidase (AMP $\longrightarrow$ adenine+ribose-5-phosphate) from *A. vinelandii* is sharply activated as the energy charge declines from 0.9 to 0.7, and that the enzyme is inhibited by orthophosphate. These data, together with results from in-vivo experiments with resting cell suspensions (Schramm & Leung, 1973; Schramm & Lazorik, 1975), suggest that this enzyme acts transiently to remove AMP when the intracellular energy charge falls, thus buffering the system, but that excess removal of AMP and total depletion of the adenine nucleotides is prevented by the build-up of orthophosphate. AMP nucleosidase has been detected also in several other bacteria but not in any eukaryotic organisms (Schramm & Leung, 1973). AMP removal in eukaryotes appears to be by the action of adenylate deaminase (AMP+H_2O $\longrightarrow$ IMP +NH_3), which responds in a similar manner to variations in energy charge (Chapman & Atkinson, 1973).

Strong support for regulation by AMP nucleosidase comes from studies on the changes in adenine nucleotide pools in resting cell suspensions of the strict aerobe *A. vinelandii* following a transition from aerobic to anaerobic conditions (Schramm & Leung, 1973). Immediately after anaerobiosis the ATP pool halves and the ADP and AMP pools double, though, since the AMP plus ADP content is lower than the ATP content, there must be a net loss of adenine nucleotides. The adenine pool remains constant on anaerobiosis, but the hypoxanthine content increases ten-fold. Therefore, AMP nucleosidase converts AMP to adenine, which is then removed as hypoxanthine by the action of adenase.

Exhaustion of adenine from the culture medium during growth of an adenine-requiring mutant of *E. coli* causes a rapid 90% drop in the total adenine nucleotide pool size but only a small decrease, from 0.9 to 0.7, of the energy charge. Furthermore, the energy charge only drops when the adenine nucleotide pool has decreased by 70% (Swedes *et al.*, 1973).

Control of energy charge by removal of AMP does not always seem to occur. Shifts of *Chromatium* growing photoheterotrophically from high to low intensities of illumination result in only small transient decreases in ATP content and energy charge, and hence maintenance of the total nucleotide content (Miović & Gibson, 1973). Shifts to complete darkness cause decreases in the ATP pool and a lowering of energy charge from about 0.8 to about 0.5 but no overall loss of nucleotides; even in the first few seconds after the change in conditions there is no evidence for transient losses of the nucleotides (Miović & Gibson, 1971, 1973). Experiments with several harvested, washed and resuspended bacteria suddenly exposed to conditions of metabolic stress have shown only a redistribution in concentrations of the individual adenine nucleotides rather than losses of the total adenine nucleotide pools (Gibson & Morita, 1967; Schön, 1969; Roberton & Wolfe, 1970; Schön & Bachofen, 1970; Van der Beek & Stouthamer, 1973; Goldenbaum, Keyser & White, 1975).

Adenine nucleotide content and energy charge during the stationary phase

Reserve polymer formation

When growth is limited by a shortage of an essential nutrient (e.g. nitrogen) other than a utilisable carbon source, many organotrophic micro-organisms accumulate energy reserve polymers (glycogen, poly-phosphate and poly-β-hydroxybutyrate) (Dawes & Senior, 1973). It might also be expected that many chemoautotrophic and phototrophic bacteria will form reserve polymers when they have a source of energy (inorganic compounds and light respectively) and carbon, but are unable to grow due to some other limiting factor. Dawes & Senior (1973) have pointed out that such conditions should generate a high energy charge, which could act as a signal to stimulate formation of reserve compounds. On the other hand, under conditions of starvation when the energy charge is lower, the reserve materials are broken down, thereby enabling the organism to survive for extended periods. The energy charge level could again act as a signal, this time for breakdown of the reserve polymers.

ATP is utilised in the formation of glycogen and polyphosphates, and there have been many reports of regulation by adenine nucleotides of enzymes involved in formation of reserve polymers (Dawes & Senior, 1973). It is therefore likely that energy charge regulation does in fact occur. However, only in the case of glycogen formation by E. coli has energy charge regulation of reserve polymer formation been studied in any detail.

Two enzymes, ADP-glucose pyrophosphorylase and α-1,4-glucan-4-glucosyl-transferase, are involved in bacterial glycogen formation. The former enzyme is activated by fructose diphosphate, reduced nicotinamide adenine dinuclotide phosphate (NADPH) and pyridoxal phosphate, and inhibited by AMP, ADP and orthophosphate (Preiss, 1973; Preiss *et al.*, 1975). This enzyme has a particularly sharp U-type response curve (Fig. 2) to energy charge: in the presence of fructose diphosphate its activity at an energy charge of 0.8 is only 18% of the activity at an energy charge of 1.0.

Dietzler, Leckie & Lais (1973) have noted that on cessation of growth due to nitrogen exhaustion (in the presence of excess glucose) glycogen formation by *E. coli* strain K12 is stimulated four-fold without a change in the intracellular content of the enzymes involved in glycogen formation. At the onset of the stationary phase, however, the ATP pool increases by 38% and the energy charge rises from 0.74 to 0.87 (Dietzler *et al.*, 1974a). The effect of this increase in energy charge, which is enough to stimulate greatly ADP-glucose pyrophosphorylase activity and hence increase massively the rate of glycogen deposition, is partially offset by a 77% decrease in the fructose diphosphate concentration. Comparison with data from in-vitro experiments confirms that these changes should result in about a four-fold stimulation of activity of glycogen formation. Recent work has confirmed the correlation between energy charge and fructose diphosphate levels with the rate of glycogen synthesis by *E. coli* K12 (Dietzler, Lais, Magnani & Leckie, 1974b; Dietzler *et al.*, 1975).

Unlike *E. coli* strain K12 (Dietzler *et al.*, 1973), nitrogen limitation of *E. coli* strain ML308 growing in a glucose-minimal medium does not cause an increase in the ATP pool (Holms *et al.*, 1972). Nitrogen limitation of *E. coli* strain B growing in a glucose-minimal medium also does not lead to an increase in the adenine nucleotide content or to a higher energy charge (Chapman *et al.*, 1971). It remains to be seen whether nitrogen limitation of *E. coli* strains B and ML308 causes an increase in the rate of glycogen formation.

In contrast to *E. coli*, nitrogen limitation of *B. natriegens* growing in succinate- or glucose-containing minimal media causes a massive increase in ATP content (from 5–7 μmol per g dry wt up to at least 20 μmol per g dry wt), coupled with maintenance of the energy charge at about 0.9; the ATP level increases linearly with time until the organism runs out of carbon (Niven, Collins & Knowles, 1976). The effect of nitrogen limitation on reserve polymer synthesis by *B. natriegens* is not known.

Starvation

Stabilisation of energy charge at the expense of the adenine nucleotide pool size, by the action of AMP nucleosidase or adenylate deaminase, can be only a transient phenomenon, in order to minimise the effects of fluctuations in substrate supply. There must be other mechanisms available to limit ATP utilisation when there is long-term deprivation of energy sources (Dawes, 1976). On starvation a reduction of the energy charge might be expected to occur (possibly after a short period of stabilisation at the expense of pool size), causing inhibition of energy-utilising biosynthetic processes. Growth will therefore stop, metabolic activity will be reduced, and scarce energy reserves will be conserved for use in essential maintenance reactions (which are presumably less sensitive to decreases in energy charge). In addition, the reduction in energy charge should cause breakdown of energy reserve polymers. As a result the organism will remain viable during extended periods of starvation. Death will presumably occur only when the organism is no longer able to regenerate ATP from reserve polymers and other endogenous sources.

Unfortunately, there is only a limited amount of information available to check whether these predictions are valid. Chapman *et al.* (1971) showed that exhaustion of glucose, as the sole carbon and energy source, from the growth medium causes the intracellular energy charge of *E. coli* to fall from about 0.8 to 0.6-0.5. The latter value is maintained for 60–80 h despite a gradual reduction in the total intracellular adenine nucleotide pool size. A rapid decrease in energy charge then occurs, coincident with a loss of viability. The energy charge of *Chromatium* also falls to 0.5–0.6 when there is no energy supply (i.e. no light), but the effects of long periods of darkness on viability, adenine nucleotide pool size and energy charge have not been tested (Miović & Gibson, 1973). In *Peptoccocus prévotii*, which does not form reserve polymers, there is a rapid drop in adenine nucleotide content immediately after exhaustion of growth substrate, followed by an increase (as RNA is degraded to replenish the pool) and then by a further reduction in nucleotide content (Montague & Dawes, 1974). During this sequence of events the energy charge falls only gradually from about 0.8 to 0.5 and viability is maintained. However, when the adenine nucleotide pool size has been greatly reduced there is a further sharp reduction in energy charge and a concomitant loss of viability.

On harvesting, washing and starving, some bacteria maintain their adenine nucleotide pools at about their growth levels as well as exhibiting high energy charge values under appropriate conditions (Gibson &

Morita, 1967; Schön, 1969; Schön & Bachofen, 1970; Knowles & Smith, 1970; Van der Beek & Stouthamer, 1973). On the other hand not all bacteria retain their adenine nucleotide pools on harvesting (Cole et al., 1967; Chapman et al., 1971; Eigener, 1975; D. F. Niven & C. J. Knowles, unpublished observations). The reason for this could be differences between bacterial species in their contents of reserve polymers, functioning of AMP nucleosidase, requirement for energy of maintenance, etc., as well as differences in experimental technique. So far little attention has been paid to these factors; because studies on the adenine nucleotide pools of intact micro-organisms are potentially extremely useful in assessing, for example, the efficiencies of oxidative phosphorylation and photophosphorylation, it is extremely important that the effects on the pool size of harvesting bacteria be assessed carefully as a preliminary to such experiments.

Adenine nucleotide content and energy change during sporulation and germination

Following vegetative growth in *Bacillus* species there is a period of several hours in which commitment to sporulation occurs, but before refractile spores appear. During this period there is an obligatory requirement for a functioning citric acid cycle and mutants of *B. subtilis* deficient in any one of several citric acid cycle enzymes have been shown not to sporulate (Freese et al., 1969; Klofat et al., 1969; Yousten & Hanson, 1972). In nutrient media the ATP pool drastically decreases in size in the mutants during the post-exponential growth period, but there is only a smaller, transient decrease in ATP content of the wild-type. Addition of citric acid cycle compounds that precede the blockage has no effect, but cycle compounds that follow the deficient enzyme cause an increase in the ATP pool, though sporulation is not restored. Moreover, a high content of ATP is not essential as: (*a*) at least one citric acid cycle mutant of *B. subtilis* is available that sporulates well, but which has a low ATP content in the period after completion of growth (Yousten & Hanson, 1972); and (*b*) growth of the wild-type in a minimal medium results in a low ATP pool after completion of growth yet there is massive sporulation (Hutchison & Hanson, 1974).

Resuspension of citric acid cycle mutants in spent broth obtained from growth of the wild-type (plus a post-blockage citric acid cycle compound) causes restoration of sporulation (Yousten & Hanson, 1972). Since spent broth from growth of *E. coli* also causes restoration of sporulation, it is probable that the restoration is due to supply of some nutrient essential for sporulation plus removal of an inhibitor formed in the original medium, which in wild-type cells is removed as a consequence of the action of the citric acid cycle. Sporulation by this resuspension technique occurs only with mutants deficient in enzymes from the first half of the citric acid cycle and not with mutants deficient in the later part of the cycle.

B. *subtilis* grown in glucose- or phosphate-limited minimal media has a constant ATP pool and has a fixed energy charge of about 0.7 during exponential growth (Hutchison & Hanson, 1974). During the post-exponential phase the ATP content is much lower, and the energy charge drops to 0.3 for several hours, before rising to 0.6–0.7 due to the further loss of adenine nucleotides immediately before refractile forespores are seen. Setlow & Kornberg (1970a) have shown that dormant spores contain only slightly lower levels of adenine nucleotides than vegetative cells, but energy charge is only about 0.1 in spores.

Both glucose and phosphate limitation derepress the citric acid cycle and sporulation of bacilli, but tryptophan or sulphate limitation increases the severity of catabolite repression. The strain of B. *subtilis* used in the studies of Hutchison & Hanson (1974) is auxotrophic for tryptophan; when the medium is limited in both tryptophan and phosphate the ATP content is much less reduced during the post-exponential phase, the energy charge remains at 0.6–0.8, and there is no refractile forespore formation.

Hutchison & Hanson (1974) and Hanson (1975) have reasoned that for spore formation to occur there must be a transient or long-term decrease in ATP level or energy charge, which acts as a signal during the stationary phase and that ATP, energy charge, or possibly other nucleotide triphosphates should be considered to be effectors of the genes involved in sporulation of bacilli.

The requirement for an active citric acid cycle coupled with a lowered energy charge during the stationary phase possibly can be explained by Ohné's recent (1975) observations. She has shown that many of the citric acid cycle enzymes of B. *subtilis* are inactivated at high energy charge values (R-type enzymes, Fig. 2).

The breaking of dormancy of B. *subtilis* spores requires neither energy nor macromolecular synthesis, but ATP is needed immediately afterwards as RNA and protein synthesis commence (Setlow & Kornberg, 1970a, b; Setlow, 1975). How then do dormant spores generate ATP following breaking of dormancy?

Setlow & Kornberg (1970a) have shown that there are three stages of ATP formation during germination. In stage I (0–5 min) the ATP level rises 100-fold, in stage II (5–15 min) it is approximately constant, and in stage III (15–50 min) there is a further seven- to ten-fold increase in ATP content. In stages I and II ATP is formed from endogenous sources since the rise and stabilisation of the ATP pool can occur in carbon-free media. During stage I ATP is formed mainly from 3-phospho-D-glyceric acid (PGA), the major phosphorylated low molecular weight metabolite of spores. This occurs via substrate level phosphorylation as ATP synthesis is unaffected by anaerobiosis or by cyanide, but is inhibited by fluoride (an inhibitor of enolase). During stage II the ATP pool decays in the presence of cyanide or when there is anaerobiosis; presumably it is maintained by respiration and oxidative phosphorylation (cf. Wilkinson & Ellar, 1975). Fluoride does not inhibit ATP formation in stage II, and PGA is probably broken down via phosphoglycerate kinase, gluconeogenesis and the pentose phosphate pathway. In stage III there is de-novo synthesis of ATP and other nucleotides, catalysed by oxidation of exogenous sources of energy. Unlike commitment to sporulation, germination of *B. subtilis* appears not to require a functioning citric acid cycle, since stages I to III are not inhibited by fluoroacetate (an inhibitor of aconitase).

The precise energy charge values that occur during germination of *B. subtilis* are unknown, but Setlow & Kornberg (1970*a*) have measured the ATP:total adenine nucleotide ratio. Within 2 min of breaking dormancy the ratio is about 0.6. The ratio is stable until 10 min, when it gradually rises, in the presence of an exogenous carbon source, until it is about 0.9 at 20 min. This increase in energy charge during the stage II–stage III transition period, at a time when there is a switch over to obligatory use of exogenous carbon sources and the de-novo synthesis of ATP, is clearly of major regulatory importance and merits closer examination: presumably the increase in energy charge activates many biosynthetic sequences.

Unlike the situation in bacilli, energy charge remains relatively constant during the life cycle of *Myxococcus xanthus* (Hanson & Dworkin, 1974). The energy charge is 0.85 ± 0.02 in vegetative cells and dormant myxospores, and during myxospore formation and germination. The adenine nucleotide pools increase slightly during the initial stages of myxospore formation, followed by a decrease to about the original value after they become refractile spheres. No changes in adenine nucleotide pool size occurs during germination.

Adenine nucleotides and macromolecular synthesis

ATP is frequently used only indirectly in anabolic sequences, especially during macromolecular synthesis. Other nucleoside triphosphates (GTP,

UTP or CTP) are used, and in the process converted to the corresponding mono- or diphosphates. The triphosphates are resynthesised by phosphotransfer from ATP by the action of nucleoside diphosphate kinase (Parks & Agarwal, 1973):

$$ATP + XDP \rightleftharpoons ADP + XTP$$

Thompson & Atkinson (1971) have suggested that the use of nucleoside triphosphates other than ATP ('indirect coupling') enables organisms to preferentially suspend synthesis of macromolecules under adverse conditions, thus conserving scarce resources for use in essential maintainance processes. Bovine liver nucleoside diphosphate kinase has been shown to have a particularly sharp U-type response (Fig. 2) to energy charge (Thompson & Atkinson, 1971). Therefore, under adverse conditions, a small decrease in energy charge has a dramatic effect on the activity of nucleoside diphosphate kinase and the rate of nucleoside triphosphate synthesis is sharply curtailed, thus amplifying the effect of lowered energy charge and drastically inhibiting macromolecular synthesis.

It is also reasonable to suppose that when ATP is consumed directly during macromolecular synthesis the process will be inactivated at low energy charge values. For example, Brenner, De Lorenzo & Ames (1970) have shown that in *S. typhimurium* the initial step in protein synthesis, catalysed by aminoacyl transfer ribonucleic acid synthetases, is inactivated by ADP and AMP and consequently responsive to energy charge.

A major role of ATP is its incorporation, with other nucleoside triphosphates (GTP, UTP and CTP), into RNA. In this context ATP has a dual function: not only is it incorporated into RNA, but it is converted to ADP or AMP at various steps during the biosynthesis of the nucleoside monophosphates as well as in the formation of the other nucleoside triphosphates. Energy charge has been claimed to regulate these sequences, not only by the regulation of nucleoside diphosphate kinase (see above), but also by the inhibition of phosphoribosyl pyrophosphate synthetase (ATP + D-ribose-phosphate $\longrightarrow$ AMP + P-ribose-PP) at low energy charge values (Klungsøyr *et al.*, 1968).

The intracellular content of stable RNA in *E. coli* is proportional to the growth rate, as are the contents of the nucleoside triphosphates (Beck, Ingraham, Maaløe & Neuhard, 1973; Bagnara & Finch, 1973) and P-ribose-PP (Bagnara & Finch, 1973). Abrupt increases in growth rate, due to an enrichment of the medium, or other effects that induce an increase in the rate of RNA synthesis, cause a transient drop in the intracellular levels of the nucleoside triphosphates but a transient increase in the P-ribose-PP content. The increased demand for RNA precursors presumably causes the initial decrease in the nucleoside triphosphate contents and their lower levels in turn reduce feedback inhibition of P-ribose-PP synthetase, thus increasing the P-ribose-PP content. Bagnara & Finch (1973) have reasoned that the lowered ATP content and the decreased ATP:ADP ratio, but with increased P-ribose-PP synthesis during this period, is evidence against control of the synthetase by the adenine nucleotide pool. However, they did not measure the AMP content, and the energy charge could have been maintained despite the lower ATP content (particularly when the increase in rate of RNA synthesis was due to enrichment of the medium; Beck *et al.*, 1973).

It is not yet clear how bacteria regulate RNA synthesis, though it is certain that there is no simple relationship between the size of the nucleoside triphosphate pools and the rate of RNA synthesis. In experiments where there is a sudden shift-up or shift down of the rate of RNA synthesis the increases or decreases in nucleoside triphosphate levels are relatively slow in comparison with the changes in rate of RNA synthesis (Nazar & Wong, 1972; Erlich, Gallant & Lazzarini, 1975). Further consideration of the regulation of RNA synthesis is outside the scope of this review: the reader should consult the references given above and Cashel (1975).

Permeability to adenine nucleotides

Healthy bacteria, like other micro-organisms, are usually considered to be impermeable to adenine nucleotides. However, there is some evidence available to suggest that this is not always the case. The 'leakiness' of some bacteria under certain conditions has already been discussed. In addition, exogenous ATP has been claimed to affect bacteriochlorophyll synthesis in photosynthetic bacteria, either due to its effect on the ATP pool size or on energy charge (Schmidt & Kamen, 1971; Zilinsky, Sojka & Gest, 1971).

The obligate intracellular parasite *Rickettsia prowazeki* has recently been shown to have a carrier-mediated transport system for the 1:1 exchange of ATP and ADP, presumably for ATP uptake from the nucleotide-rich intracellular fluid of the host (Winkler, 1976).

ADENINE NUCLEOTIDE POOLS OF EUKARYOTIC MICRO-ORGANISMS

At present we have a very inadequate knowledge of the adenine nucleotide content of eukaryotic micro-organisms. As far as the reviewer is aware, there have been no studies on the intracellular distribution of these nucleotides within growing eukaryotic micro-organisms. Therefore, information is available only about the *average* composition of the adenine nucleotide pools.

ATP can be generated both within mitochondria (citric acid cycle plus respiration) and in the cytoplasm (glycolysis). Adenylate kinase, though present in the functional cytoplasmic compartment, is not found within the mitochondrial matrix (Noda, 1973) and an adenine nucleotide translocase for ATP:ADP exchange connects the intra- and extra-mitochondrial adenine nucleotide pools (Klingenberg, 1970). It is therefore essential that an understanding is gained about both the mitochondrial and the cytoplasmic adenine nucleotide pools of growing eukaryotic micro-organisms before we can begin to comprehand metabolic regulation by, or of, these nucleotides. In the absence of such information, measurements of the gross energy charge of growing eukaryotes are of only limited usefulness.

Table 2 shows the ATP content and energy charge of growing eukaryotic micro-organisms in which measurements have been made using reasonably rapid sampling and quenching methods. In addition there have been several other papers published recording the adenine nucleotide contents of eukaryotes, that have involved freezing, filtration or centrifugation stages in the sampling process, and which have resulted in lower, probably unrealistic, values of adenine nucleotide content and energy charge (e.g. Kahn & Blum, 1966; Woolley & Jones, 1970; Akbar, Rickard & Moss, 1974).

It can be seen from Table 2 that the ATP contents of eukaryotic micro-organisms in the logarithmic growth phase are about 2.4–8.9 μmol per g dry wt, and hence remarkably similar to the ATP contents of growing bacteria (Table 1). With one or two exceptions the energy charge values are in the expected region of 0.8–0.93. Holm-Hansen (1970) has shown that the ATP content per cell is proportional to cell size in a wide variety of algae.

Anaerobic or aerobic growth of yeast on glucose-containing media causes repression of mitochondrial formation (Lloyd, 1974, pp. 39–49) and fermentation or partial fermentation of the glucose to ethanol. Under aerobic conditions ethanol is then used as an oxidative growth substrate, following derepression of the mitochondrial enzymes. Ball & Atkinson (1975) have shown that there is a similar adenine nucleotide content and energy charge (0.8–0.9) in *Saccharomyces cerevisiae* during anaerobic or aerobic growth on glucose, or during aerobic growth on ethanol. In contrast, Weibel, Mor & Fiechter (1974) have shown that during aerobic growth on glucose that *S. cerevisiae* has an energy

Table 2. *The adenine nucleotide pools of growing eukaryotic micro-organisms*[a]

Organisms	Comments	Energy charge	ATP content (μmol per g dry wt)	ATP:ADP ratio	K_{app}[c]	Reference
Crithidia fasciculata	Asynchronous growth	0.61–0.66	—	—	—	Edwards, Statham & Lloyd (1975)
	Synchronous growth	0.75 & 0.66–0.47[d]	—	~0.1–0.2	—	
Neurospora crassa	Values vary throughout logarithmic growth phase	0.67–0.84	2.8–5.3	~1–10	0.52–3.1	Slayman (1973)
N. crassa (poky *f* mutant)	Respires partially via non-phosphorylating alternate oxidase	—	6.6–7.4	—	—	Slayman, Rees, Orchard & Slayman (1975)
N. crassa (strain *bd*)	Exhibits circadian growth rhythm; values given are for optimal growth on solid media	0.93	8.9	3.5	~0.93	Delmer & Brody (1975)
Polytoma uvella	Mid-logarithmic phase	0.89	—	7.3	3.1	Mangat (1971)
	Late-logarithmic phase	0.87	—	6.3	3.6	
Saccharomyces cerevisiae	Aerobic growth	0.85	—	—	—	Kopperschläger, von Baehr & Hofmann (1967)
	Anaerobic growth	0.81	—	—	—	
S. cerevisiae	Aerobic growth, glucose-containing medium	—	~7.5	—	—	Bailey & Parks (1972)
S. cerevisiae wild-type (A31)	Aerobic growth, glucose-containing medium	0.85	6.2	3.6	0.84	Talwalkar & Lester (1973)
Respiratory deficient mutant (RDA 3)		0.88	5.6	5.5	1.64	

S. cerevisiae[e]	Aerobic growth, glucose containing medium					Weibel, Mor & Fiechter (1974)
	(a) glucose-oxidising phase	0.59	~4	~0.7	~0.26	
	(b) ethanol-oxidising phase	0.78	~8	~2.2	~0.62	
S. cerevisiae	Anaerobic growth, glucose-containing medium	0.8	4.8	—	—	Kudrna & Edlin (1975)
S. cerevisiae	Aerobic growth, glucose-containing medium (glucose and ethanol utilising phases)	0.9	—	—	—	Ball & Atkinson (1975)
	Aerobic growth, ethanol-containing medium	0.84	—	—	—	
Schizosaccharomyces pombe	Respiratory deficient mutant	—	4.8–7.2[b]	—	—	Foury & Goffeau (1975)
Thirty species of marine and freshwater algae (seven phyla)		—	2.4–3.1	—	—	Holm-Hansen (1970)
Eight species of fungi	Logarithmic growth in a microcosm of sand, kaolin, and ground terrestrial litter, containing 60% water	—	Mean 4.4 (range 3.2–5.6)	—	—	Ausmus (1973)
Six species of algae		—	Mean 7.0 (range 3.6–10.0)	—	—	

[a] Values taken from logarithmic growth phase unless otherwise stated.

[b] Value derived on the assumption that protein content is 50% of the dry wt.

[c] Apparent equilibrium constant of the adenylate kinase reaction $\left(K_{APP} = \dfrac{[ATP][AMP]}{[ADP]^2}\right)$.

[d] Energy charge, adenine nucleotide content and respiratory activity oscillate throughout the growth cycle (cf. text).

[e] Values derived from original diagram.

charge of about 0.59, but during the second, ethanol-oxidising, growth phase the energy charge rises to about 0.78.

In growing *Neurospora crassa* there seems to be a complex interrelation between the adenine nucleotide pools (Slayman, 1973). At the start of logarithmic growth the rate of respiration attains a maximum and then gradually falls throughout the remainder of the growth phase. The ADP and AMP pools are maximal shortly afterwards, but the ATP pool is constant during the early logarithmic growth period, doubles at about the mid-logarithmic phase and then slowly declines. As a result the energy charge oscillates between 0.67 and 0.84 during growth.

Edwards, Statham & Lloyd (1975) studied adenine nucleotide content during asynchronous and synchronous growth of the trypanosomatid, *Crithidia fasciculata*, which forms a single large mitochondrion (kinetoplast). Their results are unusual in that the ADP pool is much larger than the ATP pool, and hence the values for energy charge, ATP:ADP ratio and K_{app} (adenylate kinase reaction) are all comparatively low. In asynchronous culture the ATP content stays constant during growth, whilst the ADP and AMP pools rise, causing the energy charge to fall from 0.66 to 0.61. In synchronous culture the overall adenine nucleotide content doubles during the cell cycle, as do the ADP and AMP contents, but the ATP pool size does not follow the increase in dry weight. Respiration, energy charge (0.47–0.66) and adenine nucleotide content all oscillate during the growth cycle, showing five maxima per cycle. With one exception there is a correlation between the maxima of respiratory activity, and ATP and AMP pool sizes, with the ADP pool size at a maximum with a 36° phase retardation. DNA synthesis (S-phase) in the nucleus and mitochondrion occurs concurrently in synchronous cultures of *C. fasciculata*; there is no obvious perturbation of the adenine nucleotides during the S-phase period.

Nuclei of the polynucleate organism *Physarum polycephalum* divide synchronously in the absence of cell division (Chin & Bernstein, 1968). During interphase the ATP pool is high but falls drastically during early mitosis (prophase) and recovers somewhat during the latter stages of mitosis.

Very little is known about the effect of starvation on the adenine nucleotide composition of eukaryotic micro-organisms. Ball & Atkinson (1975) showed that *S. cerevisiae*, following anaerobic growth on a medium containing high concentrations of glucose and yeast extract, can maintain the adenine nucleotide pool at the growth level for over 24 h with no loss of viability, but with a gradual decrease in energy charge to less than 0.1. After anaerobic growth on a medium containing limiting glucose and yeast extract there is an immediate decrease in ATP content and energy charge (to about 0.5) though with little decrease in the total adenine nucleotide content. Following aerobic growth on glucose, and after the ethanol-oxidising growth phase, the total adenine nucleotide content and energy charge (0.9) are maintained for at least 16 h. After nitrogen exhaustion from an ammonia-limited, glucose-containing medium the energy charge of aerobic cultures remains at about 0.75, and the total adenine nucleotide pool shows a slight increase.

Starvation of *Polytoma uvella* causes a sharp, five-fold decrease in the adenine nucleotide content and a decrease in energy charge from the growth value of 0.89 to 0.63 (Mangat, 1971).

There is also a paucity of information on the effect of sporulation and germination on the adenine nucleotide pools of eukaryotes. Sporulation of *Acanthamoeba castellanii* causes a 15% drop in the nucleotide pool and a decrease in energy charge from the logarithmic growth value of 0.83 to as low as 0.58 (Jantzen, 1974). Spores of *Phycomyces blakesleeanus* also contain a significant amount of ATP and other nucleotides, and during germination the ATP pool increases in stages, whilst the AMP and ADP pools decrease slightly (Furch, 1974).

Band mutant *bd* of *N. crassa* exhibits a 22-h circadian rhythm (Delmer & Brody, 1975). This biological clock is expressed as a conidiation (spore-forming) rhythm. During growth on solid media there is a rhythmic cycle of energy charge values, with a minimum of 0.65 during conidiation and a maximum of 0.95 during the mid-band period (optimal growth phase). The change in energy charge is due to a cyclical five-fold change in the AMP content, with the ATP and ADP pools relatively constant. Therefore adenylate kinase does not appear to keep the adenylates in equilibrium at all stages of the circadian cycle. Bearing in mind that there is no adenylate kinase activity within the mitochondrial matrix, the authors have speculated that the rhythmic changes in energy charge are due to uncoupling of oxidative phosphorylation and increases in the mitochondrial AMP content. However, it is not clear whether these changes are the cause or a consequence of the conidiation rhythm.

Unlike many bacteria, harvesting and washing of eukaryotic microorganisms does not appear to reduce the adenine nucleotide content, and under appropriate conditions resting cell suspensions can maintain high energy charge values (Betz & Chance, 1965; Barwell & Brunt, 1969; Somlo, 1970, 1971; Barwell & Hess, 1971; Maitra, 1971; Slayman, 1973; Talwalkar & Lester, 1973; Ball & Atkinson, 1975; Slayman, Rees, Orchard & Slayman, 1975).

CONCLUDING REMARKS

It is, in retrospect, not surprising that growing micro-organisms closely control their adenine nucleotide content and the proportion of the total pool present as ATP, ADP and AMP. Nor is it surprising that they have evolved sensitive mechanisms for protecting the adenine nucleotide pools against fluctuations in nutritional status and other environmental factors.

Adenylate energy charge is a convenient parameter for expressing the state of the adenine nucleotide pool; together with a knowledge of the adenylate kinase equilibrium position, it tells us the exact adenine nucleotide composition of the cell. Moreover, studies of the effect of adenine nucleotides on the activities of isolated enzymes can be assessed using different applied energy charge values as well as ATP:ADP or ATP:AMP ratios, since they relate the in-vitro conditions to the adenylate-kinase-catalysed equilibrium of the nucleotides within the cell.

There are, however, severe limitations to the usefulness of the energy charge concept, as have already been discussed. In particular, a knowledge of the energy charge tells us nothing about the turnover of ATP and the total adenine nucleotide pool of an organism, which can vary widely at a fixed energy charge value. Furthermore, in eukaryotes there is the problem of physical compartmentation of the adenine nucleotide pools within the cell, as against possible kinetic compartmentation in bacteria. There is usually no adenylate kinase activity within the intramitochondrial compartment, and hence the energy charge concept applies directly only to the cytosol. To comprehend metabolic regulation by adenine nucleotide pools of eukaryotes, therefore, we need to know the states of both the intra- and extramitochondrial adenine nucleotide pools. Measurements of the intracellular distributions of adenine nucleotides within growing micro-organisms are likely to present considerable technical problems, and so far nothing is known about the individual intracellular pools. However, in liver cells isolated from rats fed normally the intramitochondrial ATP:ADP ratio is considerably lower than the cytosolic ratio (Siess & Wieland, 1976).

Because of their central position linking the ATP-producing catabolic pathways and the ATP-utilising biosynthetic sequences, the adenine nucleotide pools of living cells are ideally suited to act as 'overall' regulators of metabolic activity. Catabolic pathways are also often oxidative, and biosynthetic pathways reductive. It is probable, therefore, that cells additionally maintain the status of their 'reduction balance' during growth as a further central regulator of metabolic activity.

It is a pity that the state of the pyridine nucleotide pools during growth of micro-organisms has so far received only scant attention, despite the wide range of enzymes known to be regulated by one or more of the pyridine nucleotides.

This review has discussed exclusively the non-cyclic adenine nucleotide (ATP, ADP and AMP) contents of growing micro-organisms. Adenosine 3′:5′-cyclic monophosphate (cyclic AMP) also acts as a regulatory molecule in many, though probably not all, micro-organisms (Rickenberg, 1974). For example, growth of E. coli on substrates such as glucose, on which it can grow rapidly, results in a depression of the cyclic AMP pool and an antithetical increase in the guanosine 3′:5′-cyclic monophosphate (cyclic GMP) pool (Makman & Sutherland, 1965; Buettner, Spitz & Rickenberg, 1973; Bernlohr, Haddox & Goldberg, 1974; Epstein, Rothman-Denes & Hesse, 1975). Depression of the cyclic AMP level causes transient and catabolite repression of many enzymes, such as the structural genes of the lactose operon, which can be overcome by addition of exogenous cyclic AMP together with a specific inducer molecule. Thus whilst the *activities* of many regulatory enzymes are controlled by the prevailing ATP:ADP or ATP:AMP ratios, the *formation* of several enzymes are affected by the cellular cyclic AMP level. It is therefore of importance to determine in what way, if any, these two central regulatory systems are interrelated.

The intracellular level of cyclic AMP is probably mediated by the effect of the growth carbon source on the phosphotransferase system (Kornberg, this volume), which in turn controls adenylate cyclase activity for cyclic AMP formation from ATP (Abou-Sabé & Mento, 1975; Peterkofsky & Gazdar, 1975). Since formation of cyclic AMP is via an ATP-utilising pathway (adenylate cyclase) and its breakdown (by cyclic AMP phosphodiesterase) is an AMP-forming pathway, it is possible that the cyclic AMP content of micro-organisms could be sensitive also to energy charge; growth on substrates that promote rapid growth and ATP formation could slightly raise the energy charge and inhibit cyclic AMP formation or stimulate its breakdown. There is currently no evidence available in support of these speculations. On the other hand, a functional interrelation between the non-cyclic adenine nucleotides and cyclic AMP has recently been shown by the effect of the intracellular cyclic AMP content of E. coli on the formation of enzyme systems involved in ATP-regeneration (citric acid cycle enzymes and cytochromes, respiratory activity and possibly the efficiency of oxidative phosphorylation) (Broman, Dobrogosz & White, 1974; Hemfling & Mainzer, 1975; Takahashi, 1975).

I thank Drs Robert Freedman, Donald Niven and Chris Pogson for their critical reading of the manuscript, and Barbara Kerry and Sadie Loughlin for their excellent typing of the manuscript.

REFERENCES

ABOU-SABÉ, M. & MENTO, S. (1975). On the regulation of cyclic AMP level in bacteria. II. In vitro regulation of adenylate cyclase activity. Solubilization and reconstruction of a functional membrane-bound adenylate cyclase system responsive to regulation by glucose. *Biochimica et Biophysica Acta*, **385**, 294–304.

AKBAR, M. D., RICKARD, P. A. D. & MOSS, F. J. (1974). Response of the adenosine phosphate pool level to changes in the catabolite pattern of *Saccharomyces cerevisiae*. *Biotechnology and Bioengineering*, **16**, 455–74.

ATKINSON, D. E. (1968a). The energy charge of the adenylate pool as a regulatory parameter. Interaction with feedback modifiers. *Biochemistry*, **7**, 4030–4.

ATKINSON, D. E. (1968b). Citrate and the citrate cycle in the regulation of energy metabolism. *Biochemical Society Symposia*, **27**, 23–40.

ATKINSON, D. E. (1969). Regulation of enzyme function. *Annual Review of Microbiology*, **23**, 47–68.

ATKINSON, D. E. (1970). Enzymes as control elements in metabolic regulation. In *The Enzymes*, ed. P. D. Boyer, 3rd edn, vol. 1, pp. 461–89. London & New York: Academic Press.

ATKINSON, D. E. (1971). Adenine nucleotides as stoichiometric coupling agents in metabolism and as regulatory modifiers: the adenylate energy charge. In *Metabolic Pathways*, ed. D. M. Greenberg, 3rd edn, vol. 5, pp. 1–21. London & New York: Academic Press.

ATKINSON, D. E. & WALTON, G. M. (1967). Adenosine triphosphate conservation in metabolic regulation. Rat liver citrate cleavage enzyme. *Journal of Biological Chemistry*, **242**, 3239–41.

AUSMUS, B. S. (1973). The use of the ATP assay in terrestrial decomposition studies. *Bulletin of Ecological Research Communications, Stockholm*, **17**, 223–34.

BÄCHI, B. & ETTLINGER, L. (1973). Influence of glucose on adenine nucleotide levels and energy charge in *Acetobacter aceti*. *Archiv für Mikrobiologie*, **93**, 155–64.

BAGNARA, A. S. & FINCH, L. R. (1973). Relationships between intracellular contents of nucleotides and 5-phosphoribosyl-1-pyrophosphate in *Escherichia coli*. *European Journal of Biochemistry*, **36**, 422–7.

BAILEY, R. B. & PARKS, L. W. (1972). Response of the intracellular adenosine triphosphate pool of *Saccharomyces cerevisiae* to growth inhibition induced by excess L-methionine. *Journal of Bacteriology*, **111**, 542–6.

BALL, W. J. JR & ATKINSON, D. E. (1975). Adenylate energy charge in *Saccharomyces cerevisiae* during starvation. *Journal of Bacteriology*, **121**, 975–82.

BARWELL, C. J. & BRUNT, R. V. (1969). The regulation of aerobic polysaccharide synthesis in resting cells of *Saccharomyces cerevisiae*. *Archiv für Mikrobiologie*, **66**, 59–62.

BARWELL, C. J. & HESS, B. (1971). Regulation of pyruvate kinase during glyconeogenesis in *Saccharomyces cerevisiae*. *FEBS Letters*, **19**, 1–4.

BECK, C., INGRAHAM, J., MAALØE, O. & NEUHARD, J. (1973). Relationship between the concentration of nucleoside triphosphate and the rate of synthesis of RNA. *Journal of Molecular Biology*, **78**, 117–21.

BERNLOHR, R. W., HADDOX, M. K. & GOLDBERG, N. D. (1974). Cyclic guanosine 3′:5′-monophosphate in *Escherichia coli* and *Bacillus licheniformis*. *Journal of Biological Chemistry*, **249**, 4329–31.

BETZ, A. & CHANCE, B. (1965). Phase relationship of glycolytic intermediates in yeast cells with oscillatory metabolic control. *Archives of Biochemistry and Biophysics*, **109**, 585–94.

BLAIR, J. McD. (1970). Magnesium, potassium, and the adenylate kinase equilibrium. Magnesium as a feedback signal from the adenine nucleotide pool. *European Journal of Biochemistry*, **13**, 384–90.

BLANGY, D., BUC, H. & MONOD, J. (1968). Kinetics of the allosteric interactions of phosphofructokinase from *Escherichia coli*. *Journal of Molecular Biology*, **31**, 13–35.

BOMSEL, J.-L. & PRADET, A. (1968). Study of adenosine 5′-mono, di-, and triphosphates in plant tissues. IV. Regulation of the level of nucleotides, *in vivo*, by adenylate kinase: theoretical and experimental study. *Biochimica et Biophysica Acta*, **162**, 230–42.

BRENNER, M., DE LORENZO, F. & AMES, B. N. (1970). Energy charge and protein synthesis. Control of aminoacyl transfer ribonucleic acid synthetases. *Journal of Biological Chemistry*, **245**, 450–2.

BROMAN, R. L., DOBROGOSZ, W. J. & WHITE, D. C. (1974). Stimulation of cytochrome synthesis in *Escherichia coli* by cyclic AMP. *Archives of Biochemistry and Biophysics*, **162**, 595–601.

BUETTNER, M. J., SPITZ, E. & RICKENBERG, H. V. (1973). Cyclic adenosine 3′,5′-monophosphate in *Escherichia coli*. *Journal of Bacteriology*, **114**, 1068–73.

CASHEL, M. (1975). Regulation of bacterial ppGpp and pppGpp. *Annual Review of Microbiology*, **29**, 301–18.

CHAPMAN, A. G. & ATKINSON, D. E. (1973). Stabilization of adenylate energy charge by the adenylate deaminase reaction. *Journal of Biological Chemistry*, **248**, 8309–12.

CHAPMAN, A. G., FALL, L. & ATKINSON, D. E. (1971). Adenylate energy charge in *Escherichia coli* during growth and starvation. *Journal of Bacteriology*, **108**, 1072–86.

CHIN, B. & BERNSTEIN, I. A. (1968). Adenosine triphosphate and synchronous mitosis in *Physarum polycephalum*. *Journal of Bacteriology*, **96**, 330–7.

CHULAVATNATOL, M. & ATKINSON, D. E. (1973a). Phosphoenolpyruvate synthetase from *Escherichia coli*. Effects of adenylate energy charge and modifier concentrations. *Journal of Biological Chemistry*, **248**, 2712–15.

CHULAVATNATOL, M. & ATKINSON, D. E. (1973b). Kinetic competition *in vitro* between phosphoenolpyruvate synthetase and the pyruvate dehydrogenase complex from *Escherichia coli*. *Journal of Biological Chemistry*, **248**, 2716–21.

COLE, H. A., WIMPENNY, J. W. T. & HUGHES, D. E. (1967). The ATP pool in *Escherichia coli*. I. Measurement of the pool using a modified luciferase assay. *Biochimica et Biophysica Acta*, **143**, 445–53.

DAWES, E. A. (1976). Endogenous metabolism and the survival of starved prokaryotes. In *The Survival of Vegetative Microbes*, ed. T. R. G. Gray & J. R. Postgate, *Society for General Microbiology Symposium*, **26**, 19–53. London: Cambridge University Press.

DAWES, E. A. & SENIOR, P. J. (1973). The role and regulation of energy reserve polymers in micro-organisms. *Advances in Microbial Physiology*, **10**, 136–266.

DECKER, K. & PFITZER, S. (1972). Determination of steady-state concentrations of adenine nucleotides in growing *C. kluyveri* cells by biosynthetic labelling. *Analytical Biochemistry*, **50**, 529–39.

DELMER, D. P. & BRODY, S. (1975). Circadian rhythms in *Neurospora crassa*: oscillation in the level of an adenine nucleotide. *Journal of Bacteriology*, **121**, 548–53.

DIETZLER, D. N., LAIS, C. J. & LECKIE, M. P. (1974a). Simultaneous increases of the adenylate energy charge and the rate of glycogen synthesis in nitrogen-starved *Escherichia coli* W4597 (K). *Archives of Biochemistry and Biophysics*, **160**, 14–25.

DIETZLER, D. N., LAIS, C. J., MAGNANI, J. L. & LECKIE, M. P. (1974b). Maintenance of the energy charge in the presence of large decreases in the total adenylate pool of *Escherichia coli* and concurrent changes in glucose-6-P, fructose-P$_2$ and glycogen synthesis. *Biochemical and Biophysical Research Communications*, **60**, 875–81.

DIETZLER, D. N., LECKIE, M. P. & LAIS, C. J. (1973). Rates of glycogen synthesis and cellular levels of ATP and FDP during exponential growth and the nitrogen-limited stationary phase of *Escherichia coli* W4597 (K). *Archives of Biochemistry and Biophysics*, **156**, 684–93.

DIETZLER, D. N., LECKIE, M. P., MAGNANI, J. L., SUGHRUE, M. J. & BERNSTEIN, P. E. (1975). Evidence for the coordinate control of glycogen synthesis, glucose utilization, and glycolysis in *Escherichia coli*. II. Quantitative correlation of the inhibition of glycogen synthesis and the stimulation of glucose utilization by 2,4-dinitrophenol with the effects on the cellular levels of glucose 6-phosphate, fructose 1,6-diphosphate and total adenylates. *Journal of Biological Chemistry*, **250**, 7194–203.

EDWARDS, C., STATHAM, M. & LLOYD, D. (1975). The preparation of large-scale synchronous cultures of the trypanosomatid, *Crithida fasciculata*, by cell-size selection: changes in respiration and adenylate charge through the cell-cycle. *Journal of General Microbiology*, **88**, 141–52.

EIGENER, U. (1975). Adenine nucleotide pool variations in intact *Nitrobacter winogradskyi* cells. *Archives of Microbiology*, **102**, 233–40.

EIGENER, U. & BOCK, E. (1975). Study of the regulation of oxidation and CO$_2$ assimilation in intact *Nitrobacter winogradskyi* cells. *Archives of Microbiology*, **102**, 241–6.

EISENSTADT, E., FISHER, S., DER, C.-L. & SILVER, S. (1973). Manganese transport in *Bacillus subtilis* W23 during growth and sporulation. *Journal of Bacteriology*, **113**, 1363–72.

EPSTEIN, W., ROTHMAN-DENES, L. B. & HESSE, J. (1975). Adenosine 3':5'-cyclic monophosphate as mediator of catabolite repression in *Escherichia coli*. *Proceedings of the National Academy of Sciences, USA*, 72, 2300–4.

ERLICH, H., GALLANT, J. & LAZZARINI, R. A. (1975). Synthesis and turnover of ribosomal ribonucleic acid in guanine-starved cells of *Escherichia coli*. *Journal of Biological Chemistry*, 250, 3057–61.

ESTABROOK, R. W. & MAITRA, R. B. (1962). A fluorimetric method for the quantitative micro-analysis of adenine and pyridine nucleotides. *Analytical Biochemistry*, 3, 369–82.

FANICA-GAIGNIER, M., CLEMENT-METRAL, J. & KAMEN, M. D. (1971). Adenine nucleotide levels and photopigment synthesis in a growing photosynthetic bacterium. *Biochimica et Biophysica Acta*, 226, 135–43.

FORREST, W. W. (1965). Adenosine triphosphate pool during growth cycle in *Streptococcus faecalis*. *Journal of Bacteriology*, 90, 1013–16.

FOURY, F. & GOFFEAU, A. (1975). Stimulation of active uptake of nucleosides and amino acids by cyclic adenosine 3':5'-monophosphate in the yeast *Schizosaccharomyces pombe*. *Journal of Biological Chemistry*, 250, 2354–62.

FRANZEN, J. S. & BINKLEY, S. B. (1961). Comparison of acid-soluble nucleotides in *Escherichia coli* at different growth rates. *Journal of Biological Chemistry*, 236, 515–19.

FREESE, E., FORTNAGEL, P., SCHMITT, R., KLOFAT, W., CHAPPELLE, E. & PICCIOLA, G. (1969). Biochemical genetics of initial sporulation stages. In *Spores IV*, ed. L. L. Campbell & H. O. Halvorson, pp. 82–101. Washington DC: American Society for Microbiology.

FURCH, B. (1974). Changes in free nucleotide pattern of spores from *Phycomyces* in relation to heat-induced germination. *Archives of Microbiology*, 98, 77–84.

GADKARI, D. & STOLP, H. (1975). Energy metabolism of *Bdellovibrio bacteriovorus*. I. Energy production, ATP pool, energy charge. *Archives of Microbiology*, 102, 179–85.

GIBSON, J. & MORITA, S. (1967). Changes in adenine nucleotides of intact *Chromatium* D produced by illumination. *Journal of Bacteriology*, 93, 1544–50.

GLASER, M., NULTY, W. & VAGELOS, P. R. (1975). Role of adenylate kinase in the regulation of macromolecular biosynthesis in a putative mutant of *Escherichia coli* defective in membrane phospholipid biosynthesis. *Journal of Bacteriology*, 123, 128–36.

GOLDENBAUM, P. E., KEYSER, P. D. & WHITE, D. C. (1975). Role of vitamin K-2 in the organization and function of *Staphylococcus aureus* membranes. *Journal of Bacteriology*, 121, 442–9.

HAMILTON, R. D. & HOLM-HANSEN, O. (1967). Adenosine triphosphate content of marine bacteria. *Limnology and Oceanography*, 12, 319–24.

HANSON, C. W. & DWORKIN, M. (1974). Intracellular and extracellular nucleotides and related compounds during the development of *Myxococcus xanthus*. *Journal of Bacteriology*, 118, 486–96.

HANSON, R. S. (1975). Role of small molecules in regulation of gene expression and sporogenesis in Bacilli. In *Spores VI*, ed. P. Gerhardt,

R. N. Costilow & H. L. Sadoff, pp. 318–26. Washington DC: American Society for Microbiology.

HARRISON, D. E. F. & MAITRA, P. K. (1969). Control of respiration and metabolism in growing *Klebsiella aerogenes*. The role of adenine nucleotides. *Biochemical Journal*, 112, 647–56.

HELDT, H. W., KLINGENBERG, M. & MILOVANCEV, M. (1972). Differences between the ATP/ADP ratios in the mitochondrial matrix and in the extramitochondrial space. *European Journal of Biochemistry*, 30, 434–4.

HELDT, H. W. & SCHWALBACH, K. (1967). The participation of GTP-AMP-P transferase in substrate level phosphate transfer in rat liver mitochondria. *European Journal of Biochemistry*, 1, 199–206.

HEMPFLING, W. P. & MAINZER, S. E. (1975). Effects of varying the carbon source limiting growth on yield and maintenance characteristics of *Escherichia coli* in continuous culture. *Journal of Bacteriology*, 123, 1076–87.

HOBSON, P. N. & SUMMERS, R. (1972). ATP pool and growth yield in *Selenomonas ruminantium*. *Journal of General Microbiology*, 70, 351–60.

HOLM-HANSEN, O. (1970). ATP levels in algal cells as influenced by environmental conditions. *Plant and Cell Physiology*, 11, 689–700.

HOLMS, W. H., HAMILTON, I. D. & ROBERTSON, A. G. (1972). The rate of turnover of the adenosine triphosphate pool of *Escherichia coli* growing aerobically in simple defined media. *Archiv für Mikrobiologie*, 83, 95–109.

HURWITZ, C. & ROSANO, C. L. (1967). The intracellular concentration of bound and unbound magnesium ions in *Escherichia coli*. *Journal of Biological Chemistry*, 242, 3719–22.

HUTCHISON, K. W. & HANSON, R. S. (1974). Adenine nucleotide changes associated with the initiation of sporulation in *Bacillus subtilis*. *Journal of Bacteriology*, 119, 70–5.

JANTZEN, H. (1974). Das Adenosinphosphat-system während Wachstum und Entwicklung von *Acanthamoeba castellanii*. *Archives of Microbiology*, 101, 391–9.

KAHN, V. & BLUM, J. J. (1966). The intracellular nucleotides of *Astasia longa*. *Archives of Biochemistry and Biophysics*, 113, 315–23.

KELLY, D. P. & SYRETT, P. J. (1966). Energy coupling during sulphur compound oxidation by *Thiobacillus* sp. strain C. *Journal of General Microbiology*, 43, 109–18.

KLINGENBERG, M. (1970). Metabolite transport in mitochondria: an example for intracellular membrane function. In *Essays in Biochemistry*, ed. P. N. Campbell & F. Dickens, vol. 6, pp. 119–59. New York & London: Academic Press.

KLOFAT, W., PICCIOLO, G., CHAPPELLE, E. W. & FREESE, E. (1969). Production of adenosine triphosphate in normal cells and sporulation mutants of *Bacillus subtilis*. *Journal of Biological Chemistry*, 244, 3270–6.

KLUNGSØYR, L., HAGEMEN, J. H., FALL, L. & ATKINSON, D. E. (1968). Interaction between energy charge and product feedback in the regulation of biosynthetic enzymes. Aspartokinase, phosphoribosyladenosine triphosphate synthetase and phosphoribosyl pyrophosphate synthetase. *Biochemistry*, 7, 4035–40.

KNOWLES, C. J. & SMITH, L. (1970). Measurements of ATP levels of intact *Azotobacter vinelandii* under different conditions. *Biochimica et Biophysica Acta*, 197, 152–60.

KOPPERSCHLÄGER, G., VON BAEHR, M.-L. & HOFMANN, E. (1967). Zur Regulation des mehrphasigen Verlaufes des aeroben und anaeroben Glucoseverbrauches in Hefezellen. *Acta Biologica et Medica Germanica*, 19, 691–704.

KUDRNA, R. & EDLIN, G. (1975). Nucleotide pools and regulation of ribonucleic acid synthesis in yeast. *Journal of Bacteriology*, 121, 740–2.

LARSSON-RAZNIKIEWICZ, M. (1967). Kinetic studies of the reaction catalyzed by phosphoglycerate kinase. II. The kinetic relationships between 3-phosphoglycerate $MgATP^{2-}$ and activating metal ion. *Biochimica et Biophysica Acta*, 132, 33–40.

LAZDUNSKI, A. & BELAICH, J. P. (1972). Uncoupling in bacterial growth: ATP pool variation in *Zymomonas mobilis* cells in relation to different uncoupling conditions of growth. *Journal of General Microbiology*, 70, 187–97.

LEJOHN, H. B. & STEVENSON, R. M. (1970). Multiple regulatory processes in nicotinamide adenine dinucleotide-specific glutamic dehydrogenases. *Journal of Biological Chemistry*, 245, 3890–900.

LIAO, C. L. & ATKINSON, D. E. (1971). Regulation at the phosphoenolpyruvate branchpoint in *Azotobacter vinelandii*:pyruvate kinase. *Journal of Bacteriology*, 106, 37–44.

LIPMANN, F. (1941). Metabolic generation and utilization of phosphate bond energy. In *Advances in Enzymology*, ed. F. F. Nord & C. H. Werkman, vol. 1, pp. 99–162. New York: Interscience Publishers Inc.

LLOYD, D. (1974). The mitochondria of microorganisms. New York & London: Academic Press.

LOWRY, O. H., CARTER, J., WARD, J. B. & GLASER, L. (1971). The effect of carbon and nitrogen sources on the level of metabolic intermediates in *Escherichia coli*. *Journal of Biological Chemistry*, 246, 6511–21.

LOWRY, O. H. & PASSONNEAU, J. V. (1966). Kinetic evidence for multiple binding sites on phosphofructokinase. *Journal of Biological Chemistry*, 241, 2268–79.

LUNDIN, A. & THORE, A. (1975). Comparison of methods for extraction of bacterial adenine nucleotides determined by firefly assay. *Applied Microbiology*, 30, 713–21.

LUSK, J. E. & KENNEDY, E. P. (1969). Magnesium transport in *Escherichia coli*. *Journal of Biological Chemistry*, 244, 1653–5.

MAITRA, P. K. (1971). Glucose and fructose metabolism in a phosphoglucoisomeraseless mutant of *Saccharomyces cerevisiae*. *Journal of Bacteriology*, 107, 759–69.

MAKMAN, R. S. & SUTHERLAND, E. W. (1965). Adenosine 3′,5′-phosphate in *Escherichia coli*. *Journal of Biological Chemistry*, 240, 1309–14.

MANGAT, B. S. (1971). Changes in the free nucleotide pool during growth in cultures of *Polytoma uvella*. *Canadian Journal of Biochemistry*, 49, 811–15.

MELCHIOR, J. B. (1965). The role of metal ions in the pyruvic kinase reaction. *Biochemistry*, 4, 1518–25.

MILDVAN, A. S. (1970). Metals in enzyme catalysis. In *The Enzymes*, ed. P. D. Boyer, 3rd edn, vol. 2, pp. 445–536. New York & London: Academic Press.

MIOVIĆ, M. L. & GIBSON, J. (1971). Nucleotide pools in growing *Chromatium* strain D. *Journal of Bacteriology*, **108**, 954–6.

MIOVIĆ, M. L. & GIBSON, J. (1973). Nucleotide pools and adenylate energy charge in balanced and unbalanced growth of *Chromatium*. *Journal of Bacteriology*, **114**, 86–95.

MONTAGUE, M. D. & DAWES, E. A. (1974). The survival of *Peptococcus prévotii* in relation to the adenylate energy charge. *Journal of General Microbiology*, **80**, 291–9.

NAZAR, R. N. & WONG, J. T.-F. (1972). Nucleotide changes and the regulation of ribonucleic acid accumulation during growth rate shifts in *Escherichia coli*. *Journal of Biological Chemistry*, **247**, 790–7.

NEIDHARDT, F. C. & FRAENKEL, D. G. (1961). Metabolic regulation of RNA synthesis in bacteria. *Cold Spring Harbor Symposia on Quantitative Biology*, **26**, 63–74.

NIVEN, D. F., COLLINS, P. A. & KNOWLES, C. J. (1976). Adenylate energy charge during batch culture of *Beneckea natriegens*. *Journal of General Microbiology*, in press.

NODA, L. (1973). Adenylate kinase. In *The Enzymes*, ed. P. D. Boyer, 3rd edn, vol. 8, pp. 279–305. New York & London: Academic Press.

O'BRIEN, R. W. & MORRIS, J. G. (1971). Oxygen and the growth and metabolism of *Clostridium acetobutylicum*. *Journal of General Microbiology*, **68**, 307–18.

OHNÉ, M. (1975). Regulation of the dicarboxylic acid part of the citric acid cycle of *Bacillus subtilis*. *Journal of Bacteriology*, **122**, 224–34.

OLSON, M. S. & WILLIAMSON, J. R. (1971). Regulation of citrate synthesis in isolated rat liver mitochondria. *Journal of Biological Chemistry*, **246**, 7794–803.

PARKS, R. E. & AGARWAL, R. P. (1973). Nucleoside diphosphokinases. In *The Enzymes*, ed. P. D. Boyer, 3rd edn, vol. 8, pp. 307–33. New York & London: Academic Press.

PETERKOFSKY, A. & GAZDAR, C. (1975). Interaction of Enzyme I of the phosphoenolpyruvate: sugar phosphotransferase system with adenylate cyclase of *Escherichia coli*. *Proceedings of the National Academy of Sciences, USA*, **72**, 2920–4.

PREISS, J. (1973). Adenosine diphosphoryl glucose pyrophosphorylase. In *The Enzymes*, ed. P. D. Boyer, 3rd edn, vol. 8, pp. 73–119. New York & London: Academic Press.

PREISS, J., GREENBERG, E. & SABRAW, A. (1975). Biosynthesis of bacterial glycogen. Kinetic studies of a glucose-1-phosphate adenyltransferase (EC 2.7.7.27) from a glycogen-deficient mutant of *Escherichia coli* B. *Journal of Biological Chemistry*, **250**, 7631–8.

PURICH, D. L. & FROMM, H. J. (1972). Studies on factors influencing enzyme responses to adenylate energy charge. *Journal of Biological Chemistry*, **247**, 249–55.

PURICH, D. L. & FROMM, H. J. (1973). Additional factors influencing enzyme responses to the adenylate energy charge. *Journal of Biological Chemistry*, **248**, 461–6.

RANDERATH, E. & RANDERATH, K. (1964). Resolution of complex nucleotide mixtures by two dimensional anion-exchange thin-layer chromatography. *Journal of Chromatography*, **15**, 126–9.

RANDERATH, K. & RANDERATH, E. (1967). Thin-layer separation methods for nucleic acid derivatives. In *Methods in Enzymology*, ed. L. Grossman & K. Moldave, vol. 12A, pp. 323–47. New York & London: Academic Press.

RICKENBERG, H. V. (1974). Cyclic AMP in prokaryotes. *Annual Review of Microbiology*, **28**, 353–69.

ROBERTS, R. B., COWIE, D. B., ABELSON, P. H., BOLTON, E. T. & BRITTEN, R. J. (1957). *Studies of biosynthesis in* Escherichia coli. *Carnegie Institute of Washington, Publication 607.* Washington, DC.

ROBERTSON, A. M. & WOLFE, R. S. (1970). Adenosine triphosphate pools in *Methanobacterium*. *Journal of Bacteriology*, **102**, 43–51.

SCHÖN, G. (1969). Der Einfluss der Kulturbedingungen auf den ATP-, ADP- und AMP-Spiegel bei *Rhodospirillum rubrum*. *Archiv für Mikrobiologie*, **66**, 348–64.

SCHÖN, G. & BACHOFEN, R. (1970). Der Einfluss des Sauerstoffpartialdrucks und der Lichtintensität auf den ATP-spiegel in Zellen von Athiorhodaceae. *Archiv für Mikrobiologie*, **73**, 34–46.

SCHMIDT, G. L. & KAMEN, M. D. (1971). Control of chlorophyll synthesis in *Chromatium vinosum*. *Archiv für Mikrobiologie*, **76**, 51–64.

SCHRAMM, V. L. & LAZORIK, F. C. (1975). The pathway of adenylate catabolism in *Azotobacter vinelandii*. Evidence for adenosine monophosphate nucleosidase as the regulatory enzyme. *Journal of Biological Chemistry*, **250**, 1801–8.

SCHRAMM, V. L. & LEUNG, H. (1973). Regulation of adenosine monophosphate levels as a function of adenosine triphosphate and inorganic phosphate. A proposed metabolic role for adenosine monophosphate nucleosidase from *Azotobacter vinelandii*. *Journal of Biological Chemistry*, **248**, 8313–15.

SCHRIBNER, H., EISENSTADT, E. & SILVER, S. (1974). Magnesium transport in *Bacillus subtilis* W23 during growth and sporulation. *Journal of Bacteriology*, **117**, 1224–30.

SETLOW, P. (1975). Energy and small-molecule metabolism during germination of *Bacillus* spores. In *Spores VI*, ed. P. Gerhardt, R. N. Costilow & H. L. Sadoff, pp. 443–50. Washington DC: American Society for Microbiology.

SETLOW, P. & KORNBERG, A. (1970a). Biochemical studies of bacterial sporulation and germination. XXII. Energy metabolism in early stages of germination of *Bacillus megaterium* spores. *Journal of Biological Chemistry*, **245**, 3637–44.

SETLOW, P. & KORNBERG, A. (1970b). Biochemical studies of bacterial sporulation and germination. XXIII. Nucleotide metabolism during spore germination. *Journal of Biological Chemistry*, **245**, 3645–52.

SHEN, L. C. & ATKINSON, D. E. (1970). Regulation of adenosine diphosphate glucose synthase from *Escherichia coli*. Interactions of adenylate energy charge and modifier concentrations. *Journal of Biological Chemistry*, **245**, 3996–4000.

SHEN, L. C., FALL, L., WALTON, G. M. & ATKINSON, D. E. (1968). Interaction between energy charge and metabolite modulation in the regulation of enzymes of amphibolic sequences. Phosphofructokinase and pyruvate dehydrogenase. *Biochemistry*, **7**, 4041–5.

SIESS, E. A. & WIELAND, O. H. (1976). Phosphorylation state of cytosolic and mitochondrial adenine nucleotides and of pyruvate dehydrogenase in isolated rat liver cells. *Biochemical Journal*, **156**, 91–102.

SILVER, S. (1969). Active transport of magnesium in *Escherichia coli*. *Proceedings of the National Academy of Sciences, USA*, **62**, 764–71.

SILVER, S. & KRALOVIC, M. L. (1969). Manganese accumulation by *Escherichia coli*: evidence for a specific transport system. *Biochemical and Biophysical Research Communications*, **34**, 640–5.

SILVER, S., TOTH, K. & SCHRIBNER, H. (1975). Facilitated transport of calcium by cells and subcellular membranes of *Bacillus subtilis* and *Escherichia coli*. *Journal of Bacteriology*, **122**, 880–5.

SLAYMAN, C. L. (1973). Adenine nucleotide levels in *Neurospora*, as influenced by conditions of growth and by metabolic inhibitors. *Journal of Bacteriology*, **114**, 752–66.

SLAYMAN, C. W., REES, D. C., ORCHARD, P. P. & SLAYMAN, C. L. (1975). Generation of adenosine triphosphate in cytochrome-deficient mutants of *Neurospora*. *Journal of Biological Chemistry*, **250**, 396–408

SMITH, R. C. & MAALØE, O. (1964). Effect of growth rate on the acid-soluble nucleotide composition of *Salmonella typhimurium*. *Biochimica et Biophysica Acta*, **86**, 229–34.

SOMLO, M. (1970). Respiration-linked ATP formation by an 'oxidative phosphorylation mutant' of yeast. *Archives of Biochemistry and Biophysics*, **136**, 122–33.

SOMLO, M. (1971). Oxidative phosphorylation in intact cells of normal and op_1 mutant yeast undergoing respiratory adaptation. *Biochimie*, **53**, 819–30.

STREHLER, B. L. & MCELROY, W. D. (1957). Assay of adenosine triphosphate. In *Methods in Enzymology*, ed. S. P. Colowick & N. O. Kaplan, vol. 3, pp. 871–3. New York & London: Academic Press.

SWEDES, J. S., SEDO, R. J. & ATKINSON, D. E. (1975). Relation of growth and protein synthesis to the adenylate energy charge in an adenine-requiring mutant of *Escherichia coli*. *Journal of Biological Chemistry*, **250**, 6930–8.

TAKAHASHI, Y. (1975). Effect of glucose and cyclic adenosine 3′,5′-monophosphate on the synthesis of succinate dehydrogenase and isocitrate lyase in *Escherichia coli*. *Journal of Biochemistry*, **78**, 1097–100.

TALWALKAR, R. T. & LESTER, R. L. (1973). The response of diphosphoinositide and triphosphoinositide to perturbations of the adenylate energy charge in cells of *Saccharomyces cerevisiae*. *Biochimica et Biophysica Acta*, **306**, 412–21.

THOMPSON, F. M. & ATKINSON, D. E. (1971). Response of nucleoside diphosphate kinase to the adenylate energy charge. *Biochemical and Biophysical Research Communications*, **45**, 1581–5.

VAN DER BEEK, E. G. & STOUTHAMER, A. H. (1973). Oxidative phosphorylation in intact bacteria. *Archives of Microbiology*, **89**, 327–39.

VILLAR-PALASI, C. & WEI, S. H. (1970). Conversion of glycogen phosphorylase *b* to *a* by non-activated phosphorylase *b* kinase: an *in vitro* model of the mechanism of increase in phosphorylase *a* activity with muscle contraction. *Proceedings of the National Academy of Sciences, USA*, **67**, 345–50.

WEIBEL, K. E., MOR, J.-R. & FIECHTER, A. (1974). Rapid sampling of yeast cells and automated assays of adenylate, citrate, pyruvate and glucose-6-phosphate pools. *Analytical Biochemistry*, **58**, 208–16.

WIEBE, W. J. & BANCROFT, K. (1975). Use of the adenylate energy charge ratio to measure growth state of natural microbial communities. *Proceedings of the National Academy of Sciences, USA*, **72**, 2112–15.

WILKINSON, B. J. & ELLAR, D. J. (1975). Morphogenesis of the membrane-bound electron-transport system in sporulating *Bacillus megaterium* KM. *European Journal of Biochemistry*, **55**, 131–9.

WILLIAMSON J. R. & CORKEY, B. E. (1969). Assays of intermediates of the citric acid cycle and related compounds by fluorometric enzyme methods. In *Methods in Enzymology*, ed. J. M. Lowenstein, vol. 13, pp. 434–513. New York & London: Academic Press.

WINKLER, H. H. (1976). Rickettsial permeability. An ADP–ATP transport system. *Journal of Biological Chemistry*, **251**, 389–96.

WINKLER, H. H. & WILSON, T. H. (1966). The role of energy coupling in the transport of β-galactosides by *Escherichia coli*. *Journal of Biological Chemistry*, **241**, 2200–11.

WOOLLEY, D. E. & JONES, P. C. T. (1970). Correlation of adenosine phosphate levels with cellular morphology in *Dictyostelium discoideum* amoebae. *Journal of Cellular Physiology*, **76**, 191–6.

YOUSTEN, A. & HANSON, R. S. (1972). Sporulation of tricarboxylic acid cycle mutants of *Bacillus subtilis*. *Journal of Bacteriology*, **109**, 886–94.

ZILINSKY, J. W., SOJKA, G. A. & GEST, H. (1971). Energy charge regulation in photosynthetic bacteria. *Biochemical and Biophysical Research Communications*, **42**, 955–61.

ENERGETIC ASPECTS OF THE GROWTH
OF MICRO-ORGANISMS

A. H. STOUTHAMER

Biological Laboratory, Free University,
de Boelelaan 1087, The Netherlands

List of symbols

Symbol	Specification	Units
Y_{glu}	Molar growth yield for glucose	g dry weight per mol glucose
Y_{ATP}	Molar growth yield for ATP	g dry weight per mol ATP
Y_O	Growth yield per g atom O	g dry weight per g atom O
Y_{O_2}	Growth yield per mol O_2	g dry weight per mol O_2
Y_{glu}^{MAX}	Molar growth yield for glucose corrected for energy of maintenance	g dry weight per mol glucose
Y_{succ}^{MAX}	Molar growth yield for succinate corrected for energy of maintenance	g dry weight per mol succinate
Y_{ATP}^{MAX}	Molar growth yield for ATP corrected for energy of maintenance	g dry weight per mol ATP
$Y_{O_2}^{MAX}$	Growth yield per mol O_2 corrected for energy of maintenance	g dry weight per mol O_2
$Y_{el.acc}^{MAX}$	Growth yield per mol of electron acceptor (oxygen, nitrate or nitrite) corrected for energy of maintenance	g dry weight per mol electron acceptor
Y_{el}^{MAX}	Growth yield per g equivalent of electrons transferred to oxygen, nitrate or nitrite corrected for energy of maintenance	g dry weight per g equivalent of electrons
m_s	Maintenance coefficient	mol glucose per g dry weight h^{-1}
m_{succ}	Maintenance coefficient	mol succinate per g dry weight h^{-1}
m_e	Maintenance coefficient	mol ATP per g dry weight h^{-1}
m_o	Maintenance coefficient	mol O_2 per g dry weight h^{-1}
m_g	Growth-rate-dependent energy requirement for purposes other than formation of new cell material	mol ATP per g dry weight h^{-1}

List of symbols – contd.

Symbol	Specification	Units
$(Y_{ATP}^{MAX})_{theor}$	Theoretical ATP requirement for the formation of biomass of a given composition	g dry weight per mol ATP
μ	Specific growth rate	h^{-1}
q_{ATP}	Specific rate of ATP production	mol ATP per g dry weight h^{-1}
q_{sub}	Specific rate of substrate consumption	mol substrate per g dry weight h^{-1}
$P/2e^-$	Moles of ATP formed per electron pair transferred to oxygen, nitrate or nitrite	mol ATP per electron pair
H^+/O	Number of protons ejected per atom O	number of H^+ per atom O

Y_{ATP} FOR VARIOUS MICRO-ORGANISMS

In 1960 Bauchop & Elsden concluded that the amount of growth of a micro-organism was directly proportional to the amount of ATP that could be obtained from the degradation of the energy source in the medium. They introduced the coefficient Y_{ATP}, which was defined as the number of g dry weight produced per mol of ATP. Under anaerobic conditions the ATP yield for substrate breakdown can be calculated exactly, since the catabolic pathways for anaerobic breakdown of substrates are known, and consequently Y_{ATP} can be determined for micro-organisms growing anaerobically. Anaerobic experiments are therefore ideally suited to identify the environmental factors which influence Y_{ATP}. The subject has been reviewed by Stouthamer (1969, 1976a), Payne (1970), Decker, Jungermann & Thauer (1970), Forrest & Walker (1971) and Stouthamer & Bettenhaussen (1973).

Originally it was thought that Y_{ATP} was a biological constant independent of the nature of the organism and of the environment. It therefore seemed justified to utilise the mean value of Y_{ATP} calculated from the results obtained with a number of organisms for the prediction of the ATP yield in another organism for a process for which the ATP production is not known. However recent studies indicate that Y_{ATP} is not a constant for different micro-organisms (Stouthamer & Bettenhaussen, 1973). The values obtained for several micro-organisms growing anaerobically invarious media with glucose are given in Table 1. Y_{ATP} can vary from 4.7 for *Zymomonas mobilis* (Bélaich, Bélaich & Simonpietri, 1972) to 20.9 for *Lactobacillus casei* (de Vries, Kapteijn, van der Beek & Stouthamer, 1970). Utilisation of $Y_{ATP} = 10.5$ for the estimation of

Table 1. Y_{ATP} *for a number of organisms growing anaerobically in various media with glucose*

Organism	Medium[a]	Y_{ATP}	Reference
Streptococcus faecalis	c	10.9	Bauchop & Elsden (1960); Sokatch & Gunsalus (1957); Forrest & Walker (1965); Beck & Shugart (1966)
Streptococcus agalactiae	c	9.3	Mickelson (1972)
Streptococcus pyogenes	s	9.8	Davies, Karush & Rudd (1968)
Lactobacillus plantarum	c	10.2	Oxenburgh & Snoswell (1965)
Lactobacillus casei	c	20.9	de Vries, Kapteijn, van der Beek & Stouthamer (1970)
Bifidobacterium bifidum	c	13.1	de Vries & Stouthamer (1968)
Saccharomyces cerevisiae	c	10.2	Bauchop & Elsden (1960)
Saccharomyces rosei	c	11.6	Bulder (1966)
Zymomonas mobilis	c	8.5	Bauchop & Elsden (1960); Bélaich & Senez (1965); Dawes, Ribbons & Rees (1966); Forrest (1967)
	s	6.5	Bélaich, Bélaich & Simonpietri (1972)
	m	4.7	Bélaich *et al.* (1972)
Zymomonas anaerobia	c	5.9	McGill & Dawes (1971)
Sarcina ventriculi	s	11.7	Stephenson & Dawes (1971)
Enterobacter aerogenes	m	10.2	Hadjipetrou, Gerrits, Teulings & Stouthamer (1964)
	c	16.1	A. H. Stouthamer (unpublished results)
Aerobacter cloacae	m	11.9	Hernandez & Johnson (1967a)
Escherichia coli	m	11.2	Stouthamer (1969)
Proteus mirabilis	m	5.5	Stouthamer & Bettenhaussen (1972)
	c	12.6	Stouthamer & Bettenhaussen (1972)
Ruminococcus flavefaciens	c	10.6	Hopgood & Walker (1967)
Actinomyces israeli	c	12.3	Buchanan & Pine (1967)
Clostridium perfringens	c	14.6	Hasan & Hall (1975)

[a] Media: m, minimal; s, synthetic; c, complex.

ATP yields can thus give erroneous results. Furthermore the data in Table 1 show that Y_{ATP} is influenced also by the growth conditions, the values being lower in minimal media than in complex media.

INFLUENCE OF GROWTH RATE AND MAINTENANCE ENERGY ON Y_{ATP}

Growing bacteria, like all living organisms, require a certain amount of energy for maintenance processes. Maintenance energy is required for the turnover of cellular constituents, the preservation of the right ionic composition and intracellular pH of the cell, and the maintenance of a pool of intracellular metabolites against a concentration gradient. Pirt (1965) has derived an equation which relates the molar growth yield and the specific growth rate. In this derivation it is assumed that during growth the consumption of the energy source is partly growth-dependent and partly growth-independent, i.e.:

$$1/Y_{glu} = m_s/\mu + 1/Y_{glu}^{MAX}, \tag{1}$$

where Y_{glu} is the molar growth yield for glucose, m_s is the maintenance coefficient (mol glucose per g dry weight h^{-1}), μ is the specific growth rate, and Y_{glu}^{MAX} is the molar growth yield for glucose corrected for energy of maintenance.

The amount of energy source which is used independently of growth is called by definition the maintenance coefficient. It has been concluded, however, that this growth-independent consumption of energy source is used for more purposes than the true maintenance processes as originally defined (Stouthamer & Bettenhaussen, 1975), which indicates that the use of the term maintenance coefficient for the mathematical parameter m_s in equation (1) may be misleading.

The maintenance coefficient is normally determined by studying the influence of the specific growth rate on the molar growth yield. A double reciprocal plot of estimated values of Y_{glu} against the experimental μ values yields a straight line whose intercept is the reciprocal of Y_{glu}^{MAX} and whose slope is the maintenance coefficient (Pirt, 1965; Stouthamer & Bettenhaussen, 1973). In a number of cases no straight line was obtained (Pirt, 1965; de Vries et al., 1970; Watson, 1970), which was shown to be due to an influence of μ on the fermentation pattern and ATP yield of the organism. Therefore in equation (1) Y_{glu} and Y_{glu}^{MAX} were replaced by Y_{ATP} and Y_{ATP}^{MAX}. By multiplication of this changed equation by μ, equation (2) was obtained (Stouthamer & Bettenhaussen, 1973):

$$q_{ATP} = \mu/Y_{ATP} = \mu/Y_{ATP}^{MAX} + m_e. \tag{2}$$

In this equation Y_{ATP}^{MAX} is the growth yield per mol of ATP corrected for energy of maintenance; q_{ATP} is the specific rate of ATP production (mol ATP per g dry weight h^{-1}); and m_e is the maintenance coefficient (mol ATP per g dry weight h^{-1}).

It is clear from equation (2) that q_{ATP} is a linear function of μ and that Y_{ATP} is dependent on the growth rate. This has indeed been observed for *Lactobacillus casei* (de Vries et al., 1970), *Saccharomyces cerevisiae* (Watson, 1970), *Enterobacter aerogenes* (Stouthamer & Bettenhaussen, 1973, 1975) and *Escherichia coli* (Hempfling & Mainzer, 1975). The Y_{ATP}^{MAX} and m_e values determined for these organisms are shown in Table 2. The results with *E. aerogenes* show that both parameters are dependent on the growth conditions. The same conclusion has been

Table 2. *Growth parameters of some organisms growing anaerobically in a chemostat.*

Organism	Growth-limiting factor	m_e	Y_{ATP}^{MAX}	Reference
Lactobacillus casei	Glucose	1.5	24.3	de Vries *et al.* (1970)
Enterobacter aerogenes	Glucose, minimal medium	6.8(5.4–8.0)[a]	14.0(13.2–14.7)[a]	Stouthamer & Bettenhaussen (1973, 1975)
	Glucose, complex medium	9.9(6–14)	19.9(17.4–23.2)	A. H. Stouthamer (unpublished results)
	Tryptophan	38.7(33.9–41.7)	25.4(21.2–32.2)	A. H. Stouthamer (unpublished results)
	Citrate	2.2(0.4–3.9)	9.0(8.2–9.6)	Hempfling & Mainzer (1975)
Escherichia coli	Glucose	18.9(15.6–21.8)	10.3(8.9–10.7)	Watson (1970)
Saccharomyces cerevisiae	Glucose	0.5	11.0	Rogers & Stewart (1974)
	Glucose	0.25	13.0	Rogers & Stewart (1974)
Saccharomyces cerevisiae (petite)	Glucose	0.7	11.3	Rogers & Stewart (1974)
Candida parapsilosis	Glucose	0.21	12.5	

[a] The numbers given in brackets are the 95% confidence limits calculated by the model of de Kwaadsteniet *et al.* (1976).

reached for m_e from the results of aerobic yield studies with *E. coli* (Hempfling & Mainzer, 1975), which are included in Table 7. A stochastic model for heterotrophic growth of bacteria (de Kwaadsteniet, Jager & Stouthamer, 1976) allows the determination to the 95% confidence limits of Y_{ATP}^{MAX} and m_e, and of Y_{ATP} at various μ values. The results (Table 2) show that the confidence limits of Y_{ATP}^{MAX} and m_e are rather large. Estimates for Y_{ATP} are fairly precise however.

Very different values of the maintenance coefficient for different micro-organisms may be one of the factors for the explanation of the large variation in Y_{ATP} (Stouthamer & Bettenhaussen, 1973).

THEORETICAL Y_{ATP}^{MAX} VALUES

Influence of the carbon source and complexity of the medium

In theoretical calculations of the ATP requirement for the formation of cell material the macromolecular composition of the cells is taken as a base. The ATP requirement for the formation of each cell constituent is subsequently calculated. Such calculations have been performed by Gunsalus & Shuster (1961), Forrest & Walker (1971) and Stouthamer (1973). The results of the calculations of the ATP requirement for the formation of *E. coli* cell material in various media is given in Table 3. The results show that the theoretical Y_{ATP}^{MAX} values are 31.9 for growth with glucose and pre-formed monomers (A) and 28.8 for growth with glucose and inorganic salts (B). These values do not show a large difference, which is due to the small ATP requirement for monomer synthesis from glucose. The ATP requirement for monomer synthesis can be determined from Table 3 as follows. The ATP requirement for protein synthesis from glucose and pre-formed monomers amounts to 191.4×10^{-4} mol g^{-1} (column A), which represents the amount of ATP needed for the polymerisation of the amino acids to protein. Consequently the difference in the amount of ATP needed for protein synthesis between glucose and pre-formed monomers (A) and glucose and inorganic salts (B) represents the ATP requirement for the formation of amino acids from glucose. The ATP requirement for the formation of other monomers can be obtained in a similar way. The results shown in Table 3 demonstrate that the nature of the carbon source has a strong influence on the theoretical Y_{ATP}^{MAX} values. For growth with acetate and inorganic salts Y_{ATP}^{MAX} is 10, and for autotrophic growth Y_{ATP}^{MAX} is only 6.5. The small Y_{ATP}^{MAX} values for growth in these media are due to a larger requirement for monomer formation and for transport processes than during growth with glucose and inorganic salts.

Table 3. *ATP requirement for the formation of microbial cells in various media*

Synthesis of:	ATP requirement (mol×10^{-4} per g cells) in various media[a]					
	A	B	C	D	E	F
Polysaccharide	20.5	20.5	71	51	92	195
Protein	191.4	205.0	339	285	427	907
Lipid	1.5	1.5	27	25	50	172
RNA	37.9	58.6	85	70	101	} 212
DNA	5.8	10.5	16	13	19	
Transport	57.5	52.1	200	200	306	52
Total	314.6	348.2	738	644	995	1538
Theoretical Y_{ATP}^{MAX}	31.9	28.8	13.4	15.4	10.0	6.5

The cell composition was as described by Morowitz (1968) for *E. coli* (polysaccharide 16.6%; protein 52.4%; lipid 9.4%; RNA 15.7% and DNA 3.2%)

[a] Media: A, glucose, amino acids and nucleic acid bases; B–F, inorganic salts plus (B) glucose, (C) lactate, (D) malate, (E) acetate, and (F) carbon dioxide. (Data simplified from Stouthamer (1973) and Harder & Dijken (1976).)

These calculations offer an explanation for two experimental observations: (*a*) in aerobic experiments the Y_O values (g dry weight per g atom oxygen taken up during growth) are smaller for growth with simple substrates than for growth with glucose, e.g. in *E. aerogenes* the Y_O value for growth with glucose is 31.9 (Hadjipetrou *et al.*, 1964), whereas the Y_O value for growth of *Pseudomonas oxalaticus* with formate is only 3.9 (Whitaker & Elsden, 1963); (*b*) addition of amino acids to *Candida utilis* growing with glucose has scarcely any influence on the Y_O value, whereas this addition strongly increases the Y_O value for cells growing with acetate or ethanol (Hernandez & Johnson, 1967*b*). This difference is explained by the higher ATP requirement for amino acid formation during growth with ethanol and acetate than during growth with glucose.

Influence of cell composition

The influence of cell composition on the theoretical Y_{ATP}^{MAX} values is relatively small. As an example it may be mentioned that for *E. aerogenes*, which has a high protein (75%) and a low polysaccharide (2.7%) content (Tempest, Hunter & Sykes, 1965) a theoretical Y_{ATP}^{MAX} value of about 25 can be calculated. It is well known that the composition of microbial cells is influenced by the growth rate and the growth conditions (Herbert, 1961; Neidhardt, 1963), but the influence of these changes on the theoretical Y_{ATP}^{MAX} values is extremely small, unless large

Table 4. *ATP requirement for the formation of cellular macromolecules in a glucose–inorganic salts medium*

Macromolecule	ATP requirement $(\text{mol} \times 10^{-4} \text{ per g macromolecule})$
Polysaccharide	123.6
Protein	391.1
Lipid	14.8
RNA	373.2
DNA	330.0

Calculated from the data in Table 2 (column B).

amounts of storage materials are formed. The data in Table 4 show that the ATP requirement for the formation of polysaccharide and lipid is much smaller than the ATP requirement for the formation of protein, RNA and DNA. If large amounts of polysaccharide and lipid are formed as storage materials the theoretical $Y_{\text{ATP}}^{\text{MAX}}$ values will be much higher. The same applies to the accumulation of poly-β-hydroxy-butyrate and polymetaphosphate. The influence of growth conditions on the formation of storage materials has been discussed by Dawes & Senior (1973).

Influence of the nitrogen source

The molar growth yields for organisms growing with molecular nitrogen as nitrogen source are much smaller than those for the same organism growing with ammonia (Hill, Drozd & Postgate, 1972). This is because of the large ATP requirement of nitrogen fixation. In cell-free extracts utilisation of 12–15 mol ATP per mol nitrogen fixed has been observed (Zumft & Mortenson, 1975). In non-growing cells of a derepressed nitrogen-fixing mutant of *Klebsiella pneumoniae* about 15 mol ATP are used per mol ammonia produced. Theoretical $Y_{\text{ATP}}^{\text{MAX}}$ values of 12.3, 9.7, 7.9 and 6.7 can be calculated if we assume ATP requirements of 6, 9, 12 or 15 mol per mol of ammonia, respectively.

Influence of the carbon assimilation pathway of the substrate

Different assimilation pathways have been found for growth with methane and methanol (Ström, Ferenci & Quayle, 1974; Quayle, 1976; Harder & Dijken, 1976). The $Y_{\text{ATP}}^{\text{MAX}}$ values for growth with methane or methanol using the different carbon assimilation pathways are shown in Table 5, and it can be seen that differences in carbon assimilation pathways have a large influence on the theoretical $Y_{\text{ATP}}^{\text{MAX}}$ value.

Table 5. *Theoretical Y_{ATP}^{MAX} values for micro-organisms with different carbon assimilation pathways for growth with methane and methanol*

Assimilation pathway	Theoretical Y_{ATP}^{MAX}
Ribulose monophosphate cycle, fructose diphosphate aldolase variant	27.3
Ribulose monophosphate cycle, 2-keto-3-deoxy-6-phosphoglucon-ate aldolase variant	19.4
Serine pathway	12.5
Ribulose diphosphate cycle	6.5

Data from Harder & Dijken (1976).

MAINTENANCE ENERGY, THEORETICAL AND EXPERIMENTAL Y_{ATP}^{MAX} VALUES AND ENERGETIC COUPLING

The results show that in most cases a large discrepancy exists between the theoretical (Table 3) and the experimental (Table 2) Y_{ATP}^{MAX} values. Only in glucose-limited cultures of *L. casei* and tryptophan-limited cultures of *E. aerogenes* are the experimental Y_{ATP}^{MAX} values close to the theoretical Y_{ATP}^{MAX} value. The values for Y_{ATP} observed in batch cultures also indicate that in most cases not all the ATP produced in catabolism can be accounted for by the ATP requirement for synthesis of cell material. The presence of energy-requiring processes other than the formation of new cell material, but related to the growth rate, has therefore been proposed (Stouthamer & Bettenhaussen, 1975; Stouthamer, 1976a). The relation between experimental and theoretical Y_{ATP}^{MAX} values can be described by equation (3) (A. H. Stouthamer & J. Vegter, unpublished results):

$$q_{ATP} = \mu / Y_{ATP}^{MAX} + m_e = \mu / (Y_{ATP}^{MAX})_{theor} + m_g \mu + m_e, \qquad (3)$$

where $(Y_{ATP}^{MAX})_{theor}$ is the theoretical Y_{ATP}^{MAX} value and m_g is the growth-rate-dependent energy requirement for purposes other than formation of new cell material. The results in Table 1 suggest that Y_{ATP} is higher in complex media than in minimal media in that the organisms showing high Y_{ATP} values were grown in complex media. High Y_{ATP} values have been reported for a number of *Clostridium* spp. growing in complex media with various substrates (Decker *et al.*, 1970; Stouthamer, 1976a). Very high Y_{ATP} values have been measured for intraperiplasmic growth of *Bdellovibrio bacteriovorus* on *E. coli* as substrate organism, Y_{ATP} value of 18.5 being obtained from single growth cycle experiments and 25.9 from multicycle experiments (Rittenberg & Hespel, 1975). During intraperiplasmic growth the bdellovibrios do not need to synthesise monomers, since these are directly derived from the substrate organism

(Pritchard, Langley & Rittenberg, 1975). Furthermore Y_{ATP}^{MAX} for growth of *E. aerogenes* in complex medium is higher than that for growth in minimal medium (Table 2). These data indicate that during growth in complex media m_g is much lower than during growth in minimal medium.

Senez (1962) introduced the term 'uncoupled growth' to denote that under some conditions the growth yield is much lower than expected on basis of the ATP yield. Uncoupled growth occurs under the following growth conditions: (*a*) in minimal media; (*b*) in the presence of inhibitory compounds (Stouthamer, 1976*a*); (*c*) in media which contain suboptimal amounts of an essential growth factor (Rosenberger & Elsden, 1960; Bélaich *et al.*, 1972): in these cases there is an excess of the energy source in the medium; (*d*) at temperatures above the optimum (Forrest & Walker, 1971; Coulgate & Sundaram, 1975); (*e*) with compounds as carbon and energy sources, with which growth is not energy-limited. The occurrence of this phenomenon has been demonstrated in aerobic yield studies with *E. aerogenes* (Neijssel & Tempest, 1975).

The following factors, which explain these environmental effects on molar growth yields and Y_{ATP} may be mentioned. (*a*) Under the conditions (*a–d*) mentioned above, the molar growth yield, Y_{ATP} and the specific growth rate decrease. Consequently a larger proportion of the energy source is used for maintenance purposes. (*b*) The actual magnitude of m_e and/or of m_g can also be altered.

It has already been concluded that during growth in complex media m_g is much lower than during growth in minimal medium. Neijssel & Tempest (1975) have concluded that the Y_{O_2} values for growth of *E. aerogenes* in minimal medium with reduced compounds such as mannitol or glycerol are lower than the value for growth with glucose, suggesting that not all the extra energy that is released in oxidising the additional reducing equivalents (contained in mannitol or glycerol) is coupled to biomass formation. This means that m_g during growth with mannitol or glycerol is higher than that with glucose. In my opinion the influence of the complexity of the medium and the influence of the nature of the energy source are related phenomena and are both due to the limiting capacity of assimilation during growth of an organism in minimal medium.

These data suggest that there is an imbalance between the rate of ATP production by catabolism and the rate of ATP utilisation by anabolism under many growth conditions. Evidently the capacities of bacteria to regulate cellular processes are insufficient to regulate the rate of catabolism exactly to the needs of anabolism. During growth of *Zymomonas mobilis* in a medium with suboptimal amounts of pantothenate, ATP has been shown to be degraded by an ATPase (Lazdunski & Bélaich, 1972). Part of the ATP used for maintenance purposes (m_e) and for energy-requiring processes other than the formation of new cell material (m_g) is therefore proposed to be degraded by the ATPase.

The conditions which lead to uncoupling under anaerobic conditions give the same effect under aerobic conditions. It therefore seems likely that under aerobic conditions also ATP is degraded by an ATPase when uncoupling occurs. As an alternative it may be suggested that respiration and phosphorylation are not obligately coupled. At the moment no choice can be made between these two possibilities. However the difference does not seem very relevant. By respiration a membrane potential is generated which is the driving force for ATP synthesis (Garland, C. W. Jones, Haddock, this volume). During uncoupled growth either the electrical potential or ATP is dissipated. Both may be brought about by an ATPase regulated by energy charge (Knowles, this volume). In this respect it is worth mentioning that in *E. coli* an ATPase inhibitor has been identified (Nieuwenhuis *et al.*, 1973; Nieuwenhuis, van der Drift, Voet & van Dam, 1974), the dissociation of which is markedly influenced by the ADP/ATP ratio.

In cases where the energy source is in excess there is still another method for the adjustment of the energy production to the needs for growth. Under these conditions in some organisms the substrate is not completely oxidised and all kind of metabolites are excreted, which lowers the ATP production. This phenomenon has been called 'overflow metabolism' (Neijssel & Tempest, 1975).

INFLUENCE OF HYDROGEN ACCEPTORS ON MOLAR GROWTH YIELDS

Influence of hydrogen acceptors in facultative bacteria

In the presence of external hydrogen acceptors the energy source may be degraded further than under anaerobic conditions. Concomitantly the ATP yield in the presence of hydrogen acceptors will be larger than in their absence. However the total ATP yield during growth in the presence of hydrogen acceptors cannot be determined easily, since the $P/2e^-$ ratio for phosphorylation coupled to electron transport to the hydrogen acceptor is unknown. During growth in the presence of hydrogen acceptors larger molar growth yields are expected than during growth in their absence. The effect of hydrogen acceptors on molar growth yields is most easily studied in facultative micro-organisms. Some results are shown in Table 6. It is evident that the highest molar growth yields are obtained with oxygen. The yields with nitrate are lower, partly due to the absence of citric acid cycle activity during anaerobic growth in the presence of nitrate (Forget & Pichinoty, 1967; Wimpenny & Cole, 1967).

From the data in Table 6, $P/2e^-$ ratios were calculated using equation (4) (Stouthamer & Bettenhaussen, 1972):

Total ATP production $= Y_{glu}/Y_{ATP} = 2$ (part of the glucose dissimilated)+(acetate production per mol glucose)+(hydrogen acceptor utilised per mol glucose)$\cdot P/2e^-$. (4)

It must be clearly stated here that the $P/2e^-$ ratio determined from yield studies differs from the more usual meaning of the number of phosphorylation sites for the oxidation of a substrate, e.g. NADH or succinate, by the respiratory chain. This is due to the fact that the $P/2e^-$ determined from yield studies is an average $P/2e^-$ ratio, computed from the total amount of ATP produced by oxidative phosphorylation during the complete oxidation of a substrate, and the number of electron pairs which are donated to the respiratory chain by that substrate. For example, during the complete oxidation of succinate to carbon dioxide via the tricarboxylic acid cycle, 5 NADH and 2 $FADH_2$ are generated. If there are two phosphorylation sites in the respiratory chain the ATP production by oxidative phosphorylation for the complete oxidation of 1 mol succinate will be $5 \times 2 + 2 \times 1 = 12$ mol ATP. Consequently in this case the $P/2e^-$ ratio in the growing culture will be $12/7 = 1.7$. If there are three phosphorylation sites the $P/2e^-$ ratio for complete oxidation of succinate will be 2.7.

With equation (4) high $P/2e^-$ ratios were obtained from the data in Table 6, using the Y_{ATP} value found for cultures growing anaerobically without external hydrogen acceptor. However in the presence of hydrogen acceptors the specific growth rate is larger than in their absence. The results discussed earlier show that there is a large influence of the specific growth rate on Y_{ATP}, larger values of Y_{ATP} being found at higher growth rates due to the utilisation of a relatively smaller part of the energy source for maintenance. Consequently it has been concluded that in previous publications $P/2e^-$ ratios have been over-estimated (Stouthamer & Bettenhaussen, 1973).

The yield studies indicate that in *Streptococcus faecalis* oxidative phosphorylation occurs. In *S. faecalis* 10C1 no cytochromes are present, whereas in *S. faecalis* var. *zymogenes* cytochromes are formed

Table 6. *Molar growth yield for a number of facultative micro-organisms growing with glucose and various hydrogen acceptors*

Organism	Molar growth yield with hydrogen acceptor				Reference
	None	Oxygen	Nitrate	Tetrathionate	
Enterobacter aerogenes	26.1	72.7	45.5	n.r.	Hadjipetrou et al. (1964); Hadjipetrou & Stouthamer (1965) Stouthamer & Bettenhaussen (1972)
Proteus mirabilis	14.0	58.1	30.1	34.8	Kapralek (1972)
Citrobacter freundii	45.0	96.1	65.8	69.7	
Streptococcus faecalis 10C1	21.5	58.2	n.r.	n.r.	Smalley, Jahrling & Van Demark (1968)
Streptococcus faecalis var. *zymogenes*	20.5	40.6; 50.8[a]	n.r.	n.r.	Ritchey & Seeley (1974)

[a] For growth in the presence of haematin.
n.r., not reduced.

during growth with haematin (Ritchey & Seeley, 1974). The yield studies indicate a P/2e⁻ ratio with oxygen of 0.6 for *S. faecalis* 10C1 (Smalley *et al.*, 1968), 0.75 for *S. faecalis* var. *zymogenes* (Ritchey & Seeley, 1974), and about 1.0 for *S. agalactiae* (Mickelson, 1972).

Aerobic yield studies: Y_O and $Y_{O_2}^{MAX}$ values

In the past several approaches have been used to calculate the P/2e⁻ ratio from aerobic yield studies. Surveying the literature it must be concluded that at the moment a chaotic situation exists in this field. This results from the fact that either Y_{ATP} or Y_{ATP}^{MAX} must be known to determine the P/2e⁻ ratio or conversely the P/2e⁻ ratio must be known to determine Y_{ATP}^{MAX}. In this section an attempt will be made to review the present situation and to suggest some possible ways out of the difficulties.

Whitaker & Elsden (1963) introduced the coefficient Y_{O_2} (g dry weight produced per mol O_2 taken up during growth) in which the oxygen uptake is taken as a measure of that part of the glucose that is completely oxidised. In continuous culture studies $Y_{O_2}^{MAX}$ is used in many cases, which is determined by equation (5):

$$1/Y_{O_2} = m_o/\mu + 1/Y_{O_2}^{MAX}. \tag{5}$$

It is evident that equation (5) is a modified form of equation (2). $Y_{O_2}^{MAX}$ is the growth yield per mol O_2 corrected for energy of maintenance, and m_o is the maintenance respiration rate (mol O_2 per g dry weight h⁻¹). Hadjipetrou *et al.* (1964) used the coefficient Y_O, which has the advantage that division by Y_{ATP} gives a measure of the P/2e⁻ ratio where oxygen is the electron acceptor. In such calculations the following equations were used:

$$Y_O/Y_{ATP} = P/2e^-; \tag{6a}$$
$$Y_{O_2}/Y_{ATP} = 2\,P/2e^-; \tag{6b}$$

and

$$Y_{O_2}^{MAX}/Y_{ATP}^{MAX} = 2\,P/2e^-. \tag{6c}$$

Previously P/2e⁻ ratios of about 3 were calculated for batch cultures of *E. aerogenes* (Hadjipetrou *et al.*, 1964), *P. mirabilis* (Stouthamer & Bettenhaussen, 1972) and *C. freundii* (Kapralek, 1972) using equation (6a). However this calculation leads to an over-estimation of the P/2e⁻ ratio for the following reasons:

(a) In these calculations the anaerobic Y_{ATP} value was used. However as outlined earlier in this article, the aerobic Y_{ATP} is higher than the anaerobic Y_{ATP}.

(b) When Y_O is divided by Y_{ATP} the contribution of substrate-level phosphorylation to the total ATP production is neglected. The influence of this factor will be large when the $P/2e^-$ ratio is smaller than 3 or when large amounts of acetate accumulate.

The second disadvantage also applies to the calculation of $P/2e^-$ ratios from continuous culture studies. Growth parameters for chemostat cultures of various micro-organisms, calculated with the aid of the equations (1) and (5) are shown in Table 7. The disadvantage of using $Y_{O_2}^{MAX}$ values can be clearly demonstrated from the comparison of glucose- and sulphate-limited chemostat cultures of *E. aerogenes* (Stouthamer & Bettenhaussen, 1975). The $Y_{O_2}^{MAX}$ values are similar for these cultures, but the Y_{glu}^{MAX} values are markedly different. In sulphate-limited cultures a strong accumulation of acetate was observed. From these data a $P/2e^-$ ratio of about 1.3 was calculated for glucose-limited chemostat cultures and a value of about 0.4 for sulphate-limited cultures (Stouthamer & Bettenhaussen, 1975). At the same growth rate the part of the total ATP production produced by substrate-level phosphorylation amounts to 20% in glucose-limited cultures, as against 77% in sulphate-limited cultures. This example clearly demonstrates the great influence of substrate-level phosphorylation. By using equation (6c) Hempfling & Mainzer (1975) calculated a $P/2e^-$ ratio of about 1 for glucose-limited chemostat cultures of *E. coli*.

Under aerobic conditions the total ATP production may be given by equation (7):

$$q_{ATP} = a\,(1-\beta)\,q_{sub} + q_{O_2} \cdot 2(P/2e^-). \tag{7}$$

In this equation a is the number of ATP molecules formed by substrate phosphorylation during the complete oxidation of the substrate, β is the part of the substrate that is assimilated, and q_{sub} and q_{O_2} are the specific rate of substrate or oxygen consumption respectively. It is evident that from an aerobic chemostat experiment insufficient information is obtained to allow the determination of Y_{ATP}^{MAX}, m_e and $P/2e^-$ with equations (2) and (7). Thus additional information is needed for the determination of these three growth parameters. For this purpose four approaches have been used:

(1) The H^+/O ratio has been taken as a measure of the efficiency of oxidative phosphorylation. According to the chemiosmotic hypothesis two protons are extruded per atom oxygen taken up per phosphorylation site (Garland, this volume). Evidence is available that the H^+/O quotients are indeed an indication of the efficiency of oxidative phosphorylation in bacteria (C. W. Jones, this volume). By substitution of

Table 7. Growth parameters for chemostat cultures of a number of micro-organisms growing with various substrates

Organism	Substrate	Limiting factor[a]	m_0	m_s	$Y_{O_2}^{MAX}$	$Y_{SUCC.}^{MAX}$	Reference
Enterobacter aerogenes	Glucose		1.7	0.3	44.0	68.0	Stouthamer & Bettenhaussen (1975)
	Glucose	Sulphate	4.9	1.2	46.7	42.0	
	Glucose	Ammonium	3.7	6.0	77.8	168	A. H. Stouthamer (unpublished results)
	Lactate		4.2	0.3	20.4	24.5	
	Citrate		1.5	0.3	32.5	54.5	
	Mannitol		2.3	1.9	49.7	76.5	
	Glycerol		3.4	0.8	30	51	Herbert (1958)
Saccharomyces cerevisiae	Glucose		0.6	0.1	34.5	91	von Meyenburg (1969)
Proteus mirabilis	Glucose		1.0	0	51.8	59.2	van der Beek (1976)
Azotobacter vinelandii	Glucose	Oxygen	5.5	0.8	13	46	Nagai & Aiba (1972)
Escherichia coli	Glucose		3.0	0.5	42	95	Schultze & Lipe (1964)
	Glucose*		0.6	0.4	24.7	63.4	
	Glucose+cAMP*		5.1	3.1	72.6	96.6	
	Galactose		1.8	—	26.4	—	
	Mannitol		6.1	—	40.2	—	Hempfling & Mainzer (1975)
	Glycerol		10.0	—	29.8	—	
	Acetate		25.4	—	29.4	—	
	Succinate		12.1	—	22.4	—	
	Glutamate		17.7	—	51.0	—	
Hydrogenomonas eutropha	Fructose		2.4	—	100	—	Drozd & Jones (1974)
Bacillus megaterium M	Glycerol	Ammonium	1.2	—	47.6	—	
	Glycerol		2.7	—	52.0	—	Downs & Jones (1975)
	Glycerol	Oxygen	1.0	—	48.9	—	
D440	Glycerol		0.6	—	53.8	—	
Pseudomonas sp.	Methane		2.8	1.9	19	16	Nagai, Mori & Aiba (1973)

—, not determined; *, calculated with the model of de Kwaadsteniet et al. (1976) from the data in the original paper.

[a] If no growth-limiting factor is indicated the culture is substrate-limited.

$2 P/2e^- = H^+/O$ in equation $(6c)$, equation $(8a)$ is obtained (Brice, Law, Meyer & Jones, 1974; Downs & Jones, 1974):

$$Y_{O_2}^{MAX}/Y_{ATP}^{MAX} = H^+/O. \tag{8a}$$

In later publications equation $(8b)$ was used:

$$Y_{O_2}^{MAX} = N \cdot Y_{ATP}^{MAX}. \tag{8b}$$

In this equation, $N = $ total ATP production (i.e. oxidative plus substrate-level phosphorylation) per molecule oxygen taken up. In the calculation of N, the relation $2P/2e^- = H^+/O$ is also used. With the aid of equations $(8a)$ and $(8b)$ and the data in Table 7, Y_{ATP}^{MAX} values of 19.0 and 17.8 were calculated for batch cultures of *E. coli* and *K. pneumoniae*, respectively (Brice *et al.*, 1974). For two strains of *B. megaterium*, Y_{ATP}^{MAX} values of 12.7 and 10.8 were found (Downs & Jones, 1975). If the same approach is used for *Paracoccus denitrificans* we arrive at Y_{ATP}^{MAX} values of only 5–5.5 for growth with mannitol and gluconate (van Verseveld & Stouthamer, 1976), which seems unacceptable. The objection against the calculation of growth parameters with equations $(8a)$ and $(8b)$ is the uncertainty as to whether two protons are in fact translocated per phosphorylation site for each pair of electrons transferred along the respiratory chain (see later).

(2) Under normal conditions the specific rates of consumption of substrate and oxygen are linear functions of μ. It has been observed (Fig. 1) that for glucose-limited growth of *E. aerogenes* with nitrate as nitrogen source two phases can be differentiated (Stouthamer & Betten-haussen, 1975). At μ values below 0.53 h^{-1} the rate of oxygen consumption was a linear function of μ, and at higher μ values the slope of the curve relating the rate of oxygen uptake and μ became less. At the same μ value acetate production started. The slope of the curve relating the rate of glucose consumption to μ became higher at μ values above 0.53 h^{-1}. The division of this experiment into two parts allows the determination of the growth parameters. The best fit to the experimental results was obtained with the values $Y_{ATP}^{MAX} = 12.4$, $P/2e^- = 1.41$ and $m_e = 5$.

(3) A comparison of aerobic and anaerobic yield data in facultative organisms can make a determination of the $P/2e^-$ ratio possible. With the aid of the stochastic model Y_{ATP}^{MAX}, m_e and Y_{ATP} can be calculated within 95% confidence limits at various μ values for hypothetical $P/2e^-$ ratios in the range from 0 to 3. This is demonstrated in Table 8 for glucose- and citrate-limited chemostat cultures of *E. aerogenes*. Confidence limits for the $P/2e^-$ ratio can be obtained by assuming either

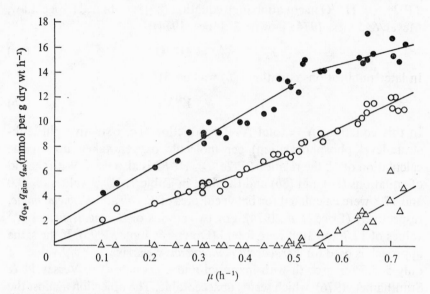

Fig. 1. Effect of specific growth rate on the rates of glucose and oxygen consumption and acetate production by *Enterobacter aerogenes* during glucose-limited aerobic growth with nitrate as nitrogen source. ●——● specific rate of oxygen consumption (q_{O_2}); ○——○ specific rate of glucose consumption (q_{glu}); △——△ specific rate of acetate production (q_{ac}). For those parameters where two lines are given, the first is for $\mu < 0.51$ h^{-1} and the second for $\mu > 0.54$ h^{-1}. Data from Stouthamer & Bettenhaussen, 1975.

$(Y_{ATP}^{MAX})_{aer} = (Y_{ATP}^{MAX})_{anaer}$ or $(Y_{ATP})_{aer} = (Y_{ATP})_{anaer}$. The first assumption seems a logical one, since the anabolic pathways for anaerobic and aerobic growth are the same. It is implicit in the second assumption that both Y_{ATP}^{MAX} and m_e are the same under aerobic and anaerobic conditions, which will not be true in all cases. For both assumptions it is essential that the cultures are energy-limited. If, for instance, anaerobic growth with glucose is energy-limited and aerobic growth not, this might lead to differences in m_g and consequently also in differences in Y_{ATP}^{MAX}. The aerobic experiment with glucose in Table 8 is the same as that which could be differentiated into two parts (Fig. 1) and thus in this case growth is indeed energy-limited. Nitrate was used as a nitrogen source and consequently reducing equivalents are needed for nitrate assimilation, which causes a lower ATP production by oxidative phosphorylation than during growth with ammonia as nitrogen source. The observation that during growth with ammonia the molar growth yields are much higher than during growth with nitrate further confirms that growth in the latter case was indeed energy-limited. The P/2e$^-$ intervals obtained with these assumptions are included in Table 8. The narrowest

Table 8. *Growth parameters for aerobic and anaerobic glucose-
or citrate-limited chemostat cultures of* E. aerogenes

	P/2e$^-$ glucose	95% confidence limits			P/2e$^-$ interval	
		Y_{ATP}^{MAX}	m_e	Y_{ATP} at $\mu = 0.34\,h^{-1}$	A	B
Anaerobic	—	13.2–14.7	5.4–8.0	11.0–11.4		
Aerobic	1.0	14.5–18.6	0.8–6.7	13.5–14.7		
	1.1	13.5–17.3	0.9–7.2	12.6–13.6		
	1.2	12.6–16.2	1.0–7.7	11.8–12.7	1.04–1.43	1.28–1.41
	1.3	11.8–15.2	1.1–8.3	11.0–11.9		
	1.4	11.1–14.3	1.2–8.8	10.4–11.2		
	1.5	10.5–13.5	1.3–9.4	9.8–10.6		
	P/2e$^-$ citrate	Y_{ATP}^{MAX}	m_e	Y_{ATP} at $\mu = 0.22\,h^{-1}$		
Anaerobic	—	8.3–10.6	0.4–3.9	8.1–9.0		
Aerobic	1.0	12.4–17.2	0.6–7.7	10.4–12.3		
	1.1	11.4–15.8	0.6–8.3	9.6–11.3		
	1.2	10.5–14.6	0.6–8.9	8.9–10.5		
	1.3	9.8–13.6	0.6–9.5	8.3–9.8	1.31–1.98	1.24–1.52
	1.4	9.2–12.7	0.7–10.1	7.8–9.2		
	1.5	8.6–11.9	0.7–10.7	7.3–8.6		
	1.6	8.1–11.2	0.7–11.3	6.9–8.2		
	1.7	7.7–10.6	0.7–11.9	6.5–7.7		
	1.8	7.3–10.0	0.7–12.5	6.2–7.3		
	1.9	6.9–9.5	0.7–13.1	5.9–6.9		
	2.0	6.6–9.1	0.8–13.7	5.6–6.6		

The intervals for the growth parameters were calculated with the model of de Kwaadsteniet
et al. (1976). The calculation of the interval for the P/2e$^-$ ratio is based on the assumption
that either (A) the aerobic and anaerobic Y_{ATP}^{MAX} values or (B) the Y_{ATP} values at the mean μ
value of the aerobic experiments are equal. (Unpublished data of R. J. Boender & A. H.
Stouthamer.)

interval is obtained with the assumption of equal Y_{ATP} values under
aerobic and anaerobic conditions. The agreement for the glucose-
limited culture between this approach and that using the division of the
experiment in two parts seems reasonable.

(4) The procedure outlined above cannot be used for obligate aerobic
bacteria and therefore the estimation of growth parameters for these
bacteria is a very difficult task. As a general method, estimates of
growth parameters in facultative and in strictly aerobic bacteria may be
obtained from comparison of substrate-limited and sulphate-limited
growth. In *Candida utilis* (Haddock & Garland, 1971) and in *E. coli*
(Poole & Haddock, 1974, 1975) phosphorylation site 1 is lost during
sulphate-limited growth. As mentioned earlier P/2e$^-$ ratios of 1.3 for
glucose-limited growth and 0.4 for sulphate-limited growth were calculated for *E. aerogenes* (Stouthamer & Bettenhaussen, 1975). The respiratory chain of *E. aerogenes* is shown in Fig. 2(*a*). Phosphorylation sites

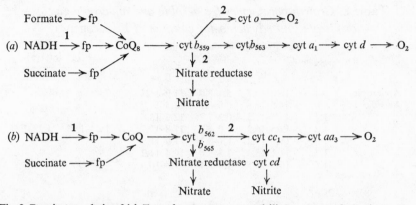

Fig. 2. Respiratory chain of (a) *Enterobacter aerogenes* and (b) *Paracoccus denitrificans*. For data see Stouthamer (1976b) and John & Whatley (1975), respectively. The phosphorylation sites are indicated by 1 and 2.

are thought to be associated with the NADH-ubiquinone oxidoreductase and with the ubiquinol oxidase via cyt o. The yield studies are in accordance with the presence of two phosphorylation sites in the respiratory chain to oxygen. From these data one can conclude that during normal growth ATP generation by oxidative phosphorylation might occur mostly at site 1, the contribution of phosphorylation site 2 being less important. This may be due to branching of the respiratory chain and the absence of a phosphorylation site in the branch with cytochrome d. For phosphorylation coupled to nitrate respiration a $P/2e^-$ ratio of 1.8–1.9 has been calculated (Stouthamer & Bettenhaussen, 1973), indicating the presence of two sites in the respiratory chain to nitrate (Fig. 2a).

This method has also been applied to a strictly aerobic organism; *P. denitrificans* was chosen, since the respiratory chain of this organism is well known (Fig. 2b) and the efficiency of oxidative phosphorylation can easily be determined with cell-free systems (John & Whatley, 1975). In heterotrophically grown cells phosphorylation site 3 is absent (Imai, Asano & Sato, 1967; Knoblock, Ishaque & Aleem, 1971; van Verseveld & Stouthamer, 1976). In *P. denitrificans* phosphorylation site 1 is also lost during sulphate-limited growth (E. M. Meyer, H. W. van Verseveld & A. H. Stouthamer, unpublished results). The results of various chemostat cultures with *P. denitrificans* are included in Table 9. It is remarkable that the m_e values for this organism are very low. Calculation of Y_{ATP}^{MAX}, m_e and $P/2e^-$ ratio can be achieved on the basis of the following assumptions: (a) that Y_{ATP}^{MAX} is a constant for the various growth condi-

Table 9. *Determination of growth parameters for growth of* Paracoccus denitrificans *with succinate*

Limiting factor Hydrogen acceptor	Succinate Oxygen	Sulphate Oxygen	Succinate Nitrate	Sulphate Nitrate	Nitrite Nitrite
Y^{MAX}_{succ}	40.2(36.2–43.6)	37.5(34.2–41.5)	35.2(28.9–44.8)	29.0(24.4–35.5)	38.4(31.7–48.5)
$Y^{MAX}_{el.acc}$	34.2(30.6–38.8)	21.0(18.8–23.7)	18.5(15.8–22.3)	14.3(12.2–17.3)	13.6(12.3–15.2)
Y^{MAX}_{el}	8.6(7.7–9.7)	5.3(4.7–5.9)	3.9(3.2–4.5)	2.9(2.4–3.5)	4.5(4.1–5.1)
m_{succ}	0.6	1.0	0.04	0.6	0.05
$m_{el.acc}$	1.5	1.6	−0.22	0.8	−0.36
Theoretical P/2e$^-$ ratio	1.71	1.0	0.60–1.31	0.60	1.0
Presence of site 1	+	−	±	−	−
P/2e$^-$ rato and interval for $Y^{MAX}_{ATP} = 9.1$	1.70(1.49–1.91)	1.03(0.91–1.15)	0.71(0.58–0.84)	0.52(0.42–0.61)	0.89(0.79–1.00)
$Y^{MAX}_{el}/(P/2e^-)$	5.1	5.1	5.2	5.6	5.1
H^+/O	7.5(6.95–8.05)	3.40(3.23–3.57)	4.46(4.32–4.60)	3.39(3.29–3.48)	3.50(3.39–3.61)

Theoretical P/2e$^-$ ratios were calculated from the composition of the respiratory chain (Fig. 2(b); John & Whatley, 1975); in the respiratory chain to nitrous oxide two phosphorylation sites were assumed to be present. The presence of phosphorylation site 1 was tested by measuring the P/2e$^-$ ratios with NADH and succinate in membrane vesicles. The P/2e$^-$ intervals were calculated using an extension of the model of de Kwaadsteniet *et al.* (1976) by R. J. Boender & A. H. Stouthamer. $Y^{MAX}_{el.acc} = Y^{MAX}_{O_2}$, $Y^{MAX}_{NO_3^-}$ or $Y^{MAX}_{NO_2^-}$. Y^{MAX}_{el} can be calculated from $Y^{MAX}_{el.acc}$ by using the fact that O_2, NO_3^- and NO_2^- accept 4, 5 and 3 electrons respectively. H^+/O ratios were measured by the oxygen pulse method for endogenous substrate. All values in brackets are 95% confidence limits. (Unpublished data of H. W. van Verseveld, E. M. Meijer & A. H. Stouthamer.)

tions and growth under all conditions is energy-limited; and (b) that the $P/2e^-$ ratio for sulphate-limited cultures is 0.7 less than for succinate-limited cultures. In this way we arrive at a Y_{ATP}^{MAX} value of about 9.1 for succinate and $P/2e^-$ ratios of 1.7 and 1.0 for succinate- and sulphate-limited cultures respectively. These $P/2e^-$ ratios are in accordance with the available knowledge on the composition of the respiratory chain and the absence of pohsphorylation site 3. The validity of the assumptions made under (a) can be tested by using growth conditions in which nitrate or nitrite are used as alternative electron acceptors. The $P/2e^-$ ratios for phosphorylation coupled to the reduction of these compounds can also be calculated and differs from the $P/2e^-$ ratio with oxygen as electron acceptor. It was observed that the presence of nitrite in the growth medium causes loss of phosphorylation site 1, which may be due to the toxic effect exerted by nitrite on the organism. The amount of nitrite present or accumulated determines whether the loss of phosphorylation site 1 is complete or not. A good correspondence was observed between the functioning of site 1 phosphorylation and the H^+/O ratio. A constant value 5.1–5.6 was found for $Y_{el.\,acc}^{MAX}/(P/2e^-)$ ratio. In all cases the $P/2e^-$ values calculated from the experimental data are close to the theoretical $P/2e^-$ ratio, indicating that Y_{ATP}^{MAX} for succinate in this organism is independent of the growth condition and that growth with succinate is always energy-limited. Similar observations were made for gluconate. In this case a Y_{ATP}^{MAX} value of 10 was found. From these data we may conclude that in *P. denitrificans* the rate of ATP production may be regulated by adjustment of the number of phosphorylation sites in the respiratory chain. Phosphorylation at site 3 has been detected so far only in autotrophically grown cells (Knobloch *et al.*, 1971). Under sulphate-limited conditions or in the presence of nitrite, phosphorylation site 1 is lost. High H^+/O ratios have been observed for *P. denitrificans* (Scholes & Mitchell, 1970; Lawford, Cox, Garland & Haddock, 1976; Table 9). Generally it has been assumed that H^+/O ratios of about 8 are an indication of the presence of four proton-translocating loops (Scholes & Mitchell, 1970; Jones, Brice, Downs & Drozd, 1975; Lawford *et al.*, 1976; C. W. Jones, this volume). In these studies the presence of phosphorylation site 3 has been assumed without experimental evidence. Due to the absence of site 3 (Imai *et al.*, 1967; Knobloch *et al.*, 1971; van Verseveld & Stouthamer, 1976) in hetero-trophically grown cells this interpretation seems incorrect for *P. denitri-ficans*. On the basis of these data we have concluded that three to four protons might be ejected per oxygen atom taken up and per site in *P. denitrificans* (Meijer, van Verseveld, van der Beek & Stouthamer, unpublished results). Similar findings have recently been reported for mitochondria (Brand, Reynafarje & Lehninger, 1976; Papa, 1976).

In Table 9 we have included the 95% confidence limits of the Y_{sub}^{MAX} and $Y_{O_2}^{MAX}$ values. It is evident that estimates of these parameters are highly inaccurate, as outlined earlier for the Y_{ATP}^{MAX} estimates in anaerobic experiments. Great caution is thus needed in the interpretation of the data in Table 7. This further illustrates the great value of the stochastic model developed by de Kwaadsteniet et al. (1976).

Influence of growth rate on the maintenance energy requirement in aerobic cultures

In equations (1) and (2) the maintenance energy requirement is assumed to be a constant at all growth rates. In aerobic systems a number of deviations have been found. In glucose-limited chemostat cultures of *Azotobacter vinelandii* high values of m_s and m_o were observed and in addition negative values of $Y_{O_2}^{MAX}$ and Y_{sub}^{MAX} were obtained (Nagai & Aiba, 1972). Under the experimental conditions an increase in the specific growth rate led to a sharp decrease in the dissolved oxygen tension. The authors proposed that the amount of energy-uncoupled growth was proportional to the dissolved oxygen tension. This means that the maintenance coefficient was supposed to be inversely related to the growth rate. When nutrients other than the carbon and energy source are the growth-limiting factor and when the maintenance energy requirement is high, deviations from equation (1) and (2) may also occur (Neijssel & Tempest, 1976). This is clearly demonstrated by the results of aerobic ammonium-limited chemostat cultures of *E. aerogenes* (A. H. Stouthamer, unpublished results), which are included in Table 7. The values for Y_{glu}^{MAX} and $Y_{O_2}^{MAX}$ for these cultures are unreasonably high. In these cultures the maintenance energy requirement is high, which is due to uncoupling of growth and energy production caused by the energy source being present in large excess. The explanation for the high Y_{glu}^{MAX} and $Y_{O_2}^{MAX}$ values is that at higher growth rates the amount of energy source used for maintenance purposes is smaller than at lower growth rates. Therefore the amount of energy used for growth may be the same in energy-sufficient and in energy-limited cultures. Thus the high Y_{glu}^{MAX} and $Y_{O_2}^{MAX}$ values are misleading. It has been pointed out that the Y_{ATP}^{MAX} values observed by Stouthamer & Bettenhaussen (1973, 1975) for anaerobic, tryptophan-limited chemostat cultures may be high for similar reasons (Neijssel & Tempest, 1976). Consequently the determination of growth parameters must preferably be performed with substrate-limited cultures.

Influence of the reduction of nitrate and fumarate
on molar growth yields

In principle the calculation of growth parameters for growth conditions under which reduction of nitrate or fumarate occurs, is performed in exactly the same way as outlined in the previous section for aerobic growth. The influence of the reduction of nitrate on energy production in micro-organisms has recently been reviewed by Stouthamer (1976*b*). Some data on nitrate reduction have already been included in the previous section.

Fumarate is also able to act as a terminal hydrogen acceptor (Kröger, this volume). A number of organisms form succinate or propionate as the ultimate fermentation product; reduction of fumarate is one of the intermediate steps in these fermentations. Consequently these organisms are able to form a terminal electron acceptor themselves. The influence of fumarate reduction on molar growth yields has been studied in most detail for *Selenomonas ruminantium* (Hobson & Summers, 1972), *Propionibacterium pentosaceum* (de Vries, van Wijck-Kapteijn & Stouthamer, 1973), *Bacteroides fragilis* (Macy, Probst & Gottschalk, 1975) and *Proteus rettgeri* (Kröger, 1974). The molar growth yield for glucose in continuous cultures of *S. ruminatium* is 62 and that of batch cultures of *P. pentosaceum* is 76.3. It is evident that the molar growth yields of these organisms are much higher than those of other bacteria growing anaerobically with glucose (cf. Table 1). The molar growth yields of these two species for glycerol are of the same order of magnitude as those for glucose in other bacteria growing anaerobically. In *B. fragilis* formation of succinate occurs only when the organism is grown in the presence of haemin (Macy *et al.*, 1975). In the absence of haemin fumarate is one of the ultimate fermentation products. The generation time decreased from 8 h to 2 h in the presence of haemin and the molar growth yield for glucose increased from 17.9 to 47. The formation of cytochromes was also dependent on the presence of haemin in the medium. The explanation for the increase in the molar growth yield of *B. fragilis* in the presence of haemin and the high growth yields of *S. ruminantium* and *P. pentosaceum* is the occurrence of oxidative phosphorylation coupled to anaerobic electron transport in which cytochromes are involved. This aspect is treated in more detail by Kröger (this volume).

Recently another factor involved in the high growth yields of propionibacteria has been mentioned by O'Brien, Bowien & Wood (1975). They have shown that the reaction: fructose-6-phosphate+pyrophosphate →

fructose-1,6-diphosphate+phosphate, plays an important role in the glycolytic system. The pyrophosphate may in part be generated by anabolic reactions, e.g. amino acid activation. The high growth efficiencies of propionibacteria might in part be due to the conservation of the energy of pyrophosphate. The relative importance of this factor is not yet known.

SUMMARY

A wide range of Y_{ATP} values has been found for micro-organisms. Furthermore Y_{ATP} is influenced strongly by the growth conditions. The following factors influence Y_{ATP}: (a) the maintenance coefficient; (b) the specific growth rate; (c) the complexity of the medium; (d) the nature of the carbon source and its pathway of assimilation; (e) the nature of the nitrogen source; (f) the macromolecular composition of the cells; and (g) the presence of energy-requiring processes, other than the formation of new cell material, which are related to the growth rate. It is concluded that during growth in many media an imbalance exists between the rate of ATP production by catabolism and the rate at which ATP can be consumed by anabolism and that in this aspect the capacities of bacteria to regulate cellular processes fall short. Thus during growth in many media ATP seems to be constantly spoiled. The degree of coupling between growth and energy production is reflected in the magnitude of m_e and/or of m_g. Uncoupling between growth and energy production has been demonstrated during growth (a) in minimal media, (b) in the presence of inhibitory compounds, (c) in media which contain suboptimal amounts of an essential growth factor, (d) at temperatures above the maximum, and (e) with substrates as carbon and energy source with which growth is not energy-limited. Furthermore it is concluded that the magnitude of the maintenance coefficient gives no indication of the energy requirement for 'true' maintenance processes. In addition to the factors which influence Y_{ATP} the following factors influence molar growth yields: (a) the pathway for the degradation of the energy source in the medium; (b) the possibility for anaerobic electron transport during fermentation; (c) the availability of and the capability to use external hydrogen acceptors; and (d) the $P/2e^-$ ratio for phosphorylation coupled to electron transport. At the moment a chaotic situation exists in the interpretation of aerobic growth experiments, because either Y_{ATP}^{MAX} must be known to calculate the $P/2e^-$ ratio, or vice versa. Possible methods to determine growth parameters for aerobic cultures are

310   A. H. STOUTHAMER

proposed: (a) comparison of aerobic and anaerobic growth in facultative organisms, and (b) comparison of substrate- and sulphate-limited growth in aerobic organisms.

BAUCHOP, T. & ELSDEN, S. R. (1960). The growth of microorganisms in relation to their energy supply. *Journal of General Microbiology*, **23**, 457–69.
BECK, R. W. & SHUGART, L. R. (1966). Molar growth yields in *Streptococcus faecalis* var. *liquefaciens*. *Journal of Bacteriology*, **92**, 802–3.
VAN DER BEEK, E. G. (1976). Oxidative phosphorylation and electron transport in *Proteus mirabilis*. PhD Thesis, Free University, Amsterdam, The Netherlands.
BÉLAICH, J. P., BÉLAICH, A. & SIMONPIETRI, P. (1972). Uncoupling in bacterial growth: effect of pantothenate starvation on growth of *Zymomonas mobilis*. *Journal of General Microbiology*, **70**, 179–85.
BÉLAICH, J. P. & SENEZ, J. C. (1965). Influence of aeration and of pantothenate on growth yields of *Zymomonas mobilis*. *Journal of Bacteriology*, **89**, 1195–200.
BRAND, M. D., REYNAFARJE, B. & LEHNINGER, A. L. (1976). Stoichiometric relationship between energy-dependent proton ejection and electron transport in mitochondria. *Proceedings of the National Academy of Sciences, USA*, **73**, 437–41.
BRICE, J. M., LAW, J. F., MEYER, D. J. & JONES, C. W. (1974). Energy conservation in *Escherichia coli* and *Klebsiella pneumoniae*. *Biochemical Society Transactions*, **2**, 523–6.
BUCHANAN, B. B. & PINE, L. (1967). Path of glucose breakdown and cell yields of a facultative anaerobe, *Actinomyces naeslandii*. *Journal of General Microbiology*, **46**, 225–36.
BULDER, C. J. E. A. (1966). Lethality in respiratory deficiency and utilization of fermentation energy in petite negative yeasts. *Archiv für Mikrobiologie*, **53**, 189–94.
COULGATE, T. P. & SUNDARAM, T. K. (1975). Energetics of *Bacillus stearothermophilus* growth: molar growth yield and temperature effects on growth efficiency. *Journal of Bacteriology*, **121**, 55–64.
DAVIES, H. C., KARUSH, F. & RUDD, J. H. (1968). Synthesis of M protein by group A hemolytic streptococci in completely synthetic media during steady-state growth. *Journal of Bacteriology*, **95**, 162–8.
DAWES, E. A., RIBBONS, D. W. & REES, D. A. (1966). Sucrose utilization by *Zymomonas mobilis*: formation of a levan. *Biochemical Journal*, **98**, 804–12.
DAWES, E. A. & SENIOR, P. J. (1973). The role and regulation of energy reserve polymers in micro-organisms. *Advances in Microbial Physiology*, **10**, 135–266.
DECKER, K., JUNGERMANN, K. & THAUER, R. K. (1970). Energy production in anaerobic organisms. *Angewandte Chemie Internationale Edition*, **9**, 138–58.

DOWNS, A. J. & JONES, C. W. (1974). Energy conservation in *Bacillus mega-therium*. *Biochemical Society Transactions*, **2**, 526–9.

DOWNS, A. J. & JONES, C. W. (1975). Energy conservation in *Bacillus mega-therium*. *Archives of Microbiology*, **105**, 159–67.

DROZD, J. W. & JONES, C. W. (1974). Oxidative phosphorylation in *Hydro-genomonas eutropha* H16 grown with and without iron. *Biochemical Society Transactions*, **2**, 529–31.

FORGET, P. & PICHINOTY, F. (1967). Le cycle tricarboxylique chez *Aerobacter aerogenes*. *Annales de l'Institut Pasteur*, **112**, 261–90.

FORREST, W. W. (1967). Energies of activation and uncoupled growth in *Streptococcus faecalis* and *Zymomonas mobilis*. *Journal of Bacteriology*, **94**, 1459–63.

FORREST, W. W. & WALKER, D. J. (1965). Synthesis of reserve materials for endogenous metabolism in *Streptococcus faecalis*. *Journal of Bacteriology*, **89**, 1448–52.

FORREST, W. W. & WALKER, D. J. (1971). The generation and utilization of energy during growth. *Advances in Microbial Physiology*, **5**, 213–74.

GUNSALUS, I. C. & SHUSTER, C. W. (1961). Energy-yielding metabolism in bacteria. In *The Bacteria*, ed. I. C. Gunsalus & R. Y. Stanier, vol. 2, pp. 1–58. New York & London: Academic Press.

HADDOCK, B. A. & GARLAND, P. B. (1971). Effect of sulphate-limited growth on mitochondrial electron transfer and energy conservation between reduced nicotinamide-adenine dinucleotide and the cytochromes in *Torulopsis utilis*. *Biochemical Journal*, **124**, 155–70.

HADJIPETROU, L. P., GERRITS, J. P., TEULINGS, F. A. G. & STOUTHAMER, A. H. (1964). Relation between energy production and growth of *Aero-bacter aerogenes*. *Journal of General Microbiology*, **36**, 139–50.

HADJIPETROU, L. P. & STOUTHAMER, A. H. (1965). Energy production during nitrate respiration by *Aerobacter aerogenes*. *Journal of General Micro-biology*, **38**, 29–34.

HARDER, W. & DIJKEN, J. P. VAN (1976). Theoretical considerations on the relation between energy production and growth of methane-utilizing bac-teria. In *Microbial Production and Utilization of Gases (H_2, CH_4, CO)*, ed. H. G. Schlegel, N. Pfennig & G. Gottschalk. Göttingen: E. Goltze Verlag (in press).

HASAN, S. M. & HALL, J. B. (1975). The physiological function of nitrate reduction in *Clostridium perfringens*. *Journal of General Microbiology*, **87**, 120–8.

HEMPFLING, W. P. & MAINZER, S. E. (1975). Effects of varying the carbon source limiting growth on yield and maintenance characteristics of *Escherichia coli* in continuous culture. *Journal of Bacteriology*, **123**, 1076–87.

HERBERT, D. (1958). Some principles of continuous culture. *Symposium of the International Congress of Microbiology*, **6**, 381–96.

HERBERT, D. (1961). The chemical composition of microorganisms as a func-tion of their environment. In *Microbial Reaction to Environment*, ed. G. G. Meynell & H. Gooder, *Society for General Microbiology Sym-posium*, **11**, pp. 391–416. London: Cambridge University Press.

312   A. H. STOUTHAMER

HERNANDEZ, E. & JOHNSON, M. J. (1967a). Anaerobic growth yields of *Aerobacter cloacae* and *Escherichia coli*. *Journal of Bacteriology*, 94, 991–5.

HERNANDEZ, E. & JOHNSON, M. J. (1967b). Energy supply and cell yield in aerobically grown microorganisms. *Journal of Bacteriology*, 94, 996–1001.

HILL, S., DROZD, J. W. & POSTGATE, J. R. (1972). Environmental effects on the growth of nitrogen-fixing bacteria. *Journal of Applied Chemistry and Biotechnology*, 22, 541–58.

HOBSON, P. N. & SUMMERS, R. (1972). ATP pool and growth yield in *Selenomonas ruminantium*. *Journal of General Microbiology*, 70, 351–60.

HOPGOOD, M. F. & WALKER, D. J. (1967). Succinic acid production by rumen bacteria. I. Isolation and metabolism of *Ruminococcus flavefaciens*. *Australian Journal of Biological Sciences*, 20, 165–82.

IMAI, K., ASANO, A. & SATO, R. (1967). Oxidative phosphorylation in *Micrococcus denitrificans*. I. Preparation and properties of phosphorylating membrane fragments. *Biochimica et Biophysica Acta*, 143, 462–76.

JOHN, P. & WHATLEY, F. R. (1975). *Paracoccus denitrificans* and the evolutionary origin of the mitochondrion. *Nature, London*, 254, 495–8.

JONES, C. W., BRICE, J. M., DOWNS, A. J. & DROZD, J. W. (1975). Bacterial respiration-linked proton translocation and its relation to respiratory chain composition. *European Journal of Biochemistry*, 52, 265–71.

KAPRALEK, F. (1972). The physiological role of tetrathionate respiration in growing *Citrobacter*. *Journal of General Microbiology*, 71, 135–9.

KNOBLOCH, K., ISHAQUE, M. & ALEEM, M. I. H. (1971). Oxidative phosphorylation in *Micrococcus denitrificans* under autotrophic growth conditions. *Archiv für Mikrobiologie*, 76, 114–24.

KRÖGER, A. (1974). Electron-transport phosphorylation coupled to fumarate reduction in anaerobically grown *Proteus rettgeri*. *Biochimica et Biophysica Acta*, 347, 273–89.

DE KWAADSTENIET, J. W., JAGER, J. C. & STOUTHAMER, A. H. (1976). A quantitative description of heterotrophic growth in micro-organisms. *Journal of Theoretical Biology*, 57, 103–20.

LAWFORD, H. G., COX, J. C., GARLAND, P. B. & HADDOCK, B. A. (1976). Electron transport of aerobically grown *Paracoccus denitrificans*: kinetic characterization of the membrane-bound cytochromes and the stoichiometry of respiration-driven proton translocation. *FEBS Letters*, 64, 369–74.

LAZDUNSKI, A. & BÉLAICH, J. P. (1972). Uncoupling in bacterial growth: ATP pool variation in *Zymomonas mobilis* cells in relation to different uncoupling conditions of growth. *Journal of General Microbiology*, 70, 187–97.

MCGILL, D. J. & DAWES, E. A. (1971). Glucose and fructose metabolism in *Zymomonas anaerobia*. *Biochemical Journal*, 125, 1059–68.

MACY, J., PROBST, I. & GOTTSCHALK, G. (1975). Evidence for cytochrome involvement in fumarate reduction and adenosine-5'-triphosphate synthesis by *Bacteroides fragilis* grown in the presence of hemin. *Journal of Bacteriology*, 123, 436–42.

VON MEYENBURG, H. K. (1969). Energetics of the budding cycle of *Saccharo-*

myces cerevisiae during glucose-limited aerobic growth. *Archiv für Mikrobiologie*, **66**, 289–303.

MICKELSON, M. N. (1972). Glucose degradation, molar growth yields and evidence for oxidative phosphorylation in *Streptococcus agalactiae*. *Journal of Bacteriology*, **109**, 96–105.

MOROWITZ, H. J. (1968). *Energy Flow in Biology*. New York: Academic Press.

NAGAI, S. & AIBA, S. (1972). Reassessment of maintenance and energy uncoupling in the growth of *Azotobacter vinelandii*. *Journal of General Microbiology*, **73**, 531–8.

NAGAI, S., MORI, T. & AIBA, S. (1973). Investigations of the energetics of methane-utilizing bacteria in methane- and oxygen-limited chemostat cultures. *Journal of Applied Chemistry and Biotechnology*, **23**, 549–62.

NEIDHARDT, F. C. (1963). Effects of environment on the composition of bacterial cells. *Annual Review of Microbiology*, **17**, 61–86.

NEIJSSEL, O. M. & TEMPEST, D. W. (1975). The regulation of carbohydrate metabolism in *Klebsiella aerogenes* NCTC 418 organisms, growing in chemostat culture. *Archives of Microbiology*, **106**, 251–8.

NEIJSSEL, O. M. & TEMPEST, D. W. (1976). Bioenergetic aspects of aerobic growth of *Klebsiella aerogenes* NCTC 418 in carbon-limited and carbon-sufficient chemostat cultures. *Archives of Microbiology*, **107**, 215–21.

NIEUWENHUIS, F. J. R. M., DRIFT, J. A. M. VAN DER, VOET, A. B. & DAM, K. VAN (1974). Evidence for a naturally occurring ATP'ase inhibitor in *Escherichia coli*. *Biochimica et Biophysica Acta*, **368**, 461–3.

NIEUWENHUIS, F. J. R. M., KANNER, B. I., GUTNICK, D. L., POSTMA, P. W. & DAM, K. VAN (1973). Energy conservation in membranes of mutants of *Escherichia coli* defective in oxidative phosphorylation. *Biochimica et Biophysica Acta*, **325**, 62–71.

O'BRIEN, W. E., BOWIEN, S. & WOOD, H. G. (1975). Isolation and characterization of a pyrophosphate-dependent phosphofructokinase from *Propionibacterium shermanii*. *Journal of Biological Chemistry*, **250**, 8690–5.

OXENBURGH, M. S. & SNOSWELL, A. M. (1965). Use of molar growth yields in the evaluation of energy-producing pathways in *Lactobacillus plantarum*. *Journal of Bacteriology*, **89**, 913–14.

PAPA, S. (1976). Proton translocation reactions in the respiratory chains. *Biochimica et Biophysica Acta*, **456**, 39–84.

PAYNE, W. J. (1970). Energy yields and growth of heterotrophs. *Annual Review of Microbiology*, **24**, 17–52.

PIRT, S. J. (1965). The maintenance energy of bacteria in growing cultures. *Proceedings of the Royal Society of London, Series B*, **163**, 224–31.

POOLE, R. K. & HADDOCK, B. A. (1974). Effects of sulphate-limited growth in continuous culture on the efficiency of energy conservation in *Escherichia coli*. *Biochemical Society Transactions*, **2**, 941–4.

POOLE, R. K. & HADDOCK, B. A. (1975). Effects of sulphate-limited growth in continuous culture on the electron transport chain and energy conservation in *Escherichia coli* K12. *Biochemical Journal*, **152**, 537–46.

PRITCHARD, M. A., LANGLEY, D. & RITTENBERG, S. C. (1975). Effects of methotrexate on intraperiplasmic growth of *Bdellovibrio bacteriovorus*. *Journal of Bacteriology*, **121**, 1131–6.

QUAYLE, J. R. (1976). Mechanisms of C1-oxidation by methene utilisers and growth of methane-utilizing bacteria. In *Microbial Production and Utilization of Gases* (*H₂*, *CH₄*, *CO*), ed. H. G. Schlegel, N. Pfennig & G. Gottschalk. Göttingen: E. Goltze Verlag. (In press.)

RITCHEY, T. W. & SEELEY, H. W. (1974). Cytochromes in *Streptococcus faecalis* var. *zymogenes* grown in a haematin-containing medium. *Journal of General Microbiology*, **85**, 220–8.

RITTENBERG, S. C. & HESPELL, R. B. (1975). Energy efficiency of intraperiplasmic growth of *Bdellovibrio bacteriovorus*. *Journal of Bacteriology*, **121**, 1158–65.

ROGERS, P. J. & STEWART, P. R. (1974). Energetic efficiency and maintenance energy characteristics of *Saccharomyces cerevisiae* (wild type and petite) and *Candida parapsilosis* grown aerobically and micro-aerobically in continuous culture. *Archives of Microbiology*, **99**, 25–46.

ROSENBERGER, R. F. & ELSDEN, S. R. (1960). The yields of *Streptococcus faecalis* grown in continuous culture. *Journal of General Microbiology*, **22**, 727–39.

SCHOLES, P. & MITCHELL, P. (1970). Respiration-driven proton translocation in *Micrococcus denitrificans*. *Journal of Bioenergetics*, **1**, 309–23.

SCHULTZE, K. L. & LIPE, R. S. (1964). Relationships between substrate concentration, growth rate and respiration rate of *Escherichia coli* in continuous culture. *Archiv für Mikrobiologie*, **48**, 1–20.

SENEZ, J. C. (1962). Some considerations on the energetics of bacterial growth. *Bacteriological Reviews*, **26**, 95–107.

SMALLEY, A. J., JAHRLING, P. & VAN DEMARK, P. J. (1968). Molar growth yields as evidence for oxidative phosphorylation in *Streptococcus faecalis* strain 10C1. *Journal of Bacteriology*, **96**, 1595–600.

SOKATCH, J. T. & GUNSALUS, I. C. (1957). Aldonic acid metabolism. I. Pathway of carbon in an inducible gluconate fermentation by *Streptococcus faecalis*. *Journal of Bacteriology*, **73**, 452–60.

STEPHENSON, M. P. & DAWES, E. A. (1971). Pyruvic acid and formic acid metabolism in *Sarcina ventriculi* and the role of ferredoxin. *Journal of General Microbiology*, **69**, 331–43.

STOUTHAMER, A. H. (1969). Determination and significance of molar growth yields. In *Methods in Microbiology*, ed. J. R. Norris & D. W. Ribbons, vol. 1, pp. 629–63. New York & London: Academic Press.

STOUTHAMER, A. H. (1973). A theoretical study on the amount of ATP required for synthesis of microbial cell material. *Antonie van Leeuwenhoek*, **39**, 545–65.

STOUTHAMER, A. H. (1976a). *Yield Studies in Micro-organisms*. Durham, England: Meadowfield Press Ltd.

STOUTHAMER, A. H. (1976b). Biochemistry and genetics of nitrate reductase in bacteria. *Advances in Microbial Physiology*, **14**, 315–75.

STOUTHAMER, A. H. & BETTENHAUSEN, C. W. (1972). Influence of hydrogen acceptors on growth and energy production of *Proteus mirabilis*. *Antonie van Leeuwenhoek*, **38**, 81–90.

STOUTHAMER, A. H. & BETTENHAUSSEN, C. W. (1973). Utilization of energy for growth and maintenance in continuous and batch cultures of micro-organisms. A reevaluation of the method for the determination of ATP production by measuring molar growth yields. *Biochimica et Biophysica Acta*, **301**, 53–70.

STOUTHAMER, A. H. & BETTENHAUSSEN, C. W. (1975). Determination of the efficiency of oxidative phosphorylation in continuous cultures of *Aerobacter aerogenes*. *Archives of Microbiology*, **102**, 187–92.

STRÖM, T., FERENCI, T. & QUAYLE, J. R. (1974). The carbon assimilation pathways of *Methylococcus capsulatus*, *Pseudomonas methanica* and *Methylosinus trichosporium* (OB3B) during growth on methane. *Biochemical Journal*, **144**, 465–76.

TEMPEST, D. W., HUNTER, J. R. & SYKES, J. (1965). Magnesium-limited growth of *Aerobacter aerogenes* in a chemostat. *Journal of General Microbiology*, **39**, 355–66.

VAN VERSEVELD, H. W. & STOUTHAMER, A. H. (1976). Oxidative phosphorylation in *Micrococcus denitrificans*. Calculation of the P/O ratio in growing cells. *Archives of Microbiology*, **107**, 241–7.

DE VRIES, W., KAPTEIJN, W. M. C., BEEK, E. G. VAN DER & STOUTHAMER, A. H. (1970). Molar growth yields and fermentation balances of *Lactobacillus casei* L3 in batch cultures and in continuous cultures. *Journal of General Microbiology*, **63**, 333–45.

DE VRIES, W. & STOUTHAMER, A. H. (1968). Fermentation of glucose, lactose, galactose, mannitol and xylose by bifidobacteria. *Journal of Bacteriology*, **96**, 472–8.

DE VRIES, W., WIJCK-KAPTEIJN, W. M. C. VAN & STOUTHAMER, A. H. (1973). Generation of ATP during cytochrome-linked anaerobic electron transport in propionic acid bacteria. *Journal of General Microbiology*, **76**, 31–41.

WATSON, T. G. (1970). Effect of sodium chloride on steady-state growth and metabolism of *Saccharomyces cerevisiae*. *Journal of General Microbiology*, **64**, 91–9.

WHITAKER, A. M. & ELSDEN, S. R. (1963). The relation between growth and oxygen consumption in microorganisms. *Journal of General Microbiology*, **31**, xxii.

WIMPENNY, J. W. T. & COLE, J. A. (1967). The regulation of metabolism in facultative bacteria. III. The effect of nitrate. *Biochimica et Biophysica Acta*, **148**, 233–42.

ZUMFT, W. G. & MORTENSON, L. E. (1975). The nitrogen-fixing complex of bacteria. *Biochimica et Biophysica Acta*, **416**, 1–52.

BACTERIAL CHEMOTAXIS AND SOME ENZYMES IN ENERGY METABOLISM

D. E. KOSHLAND, Jr

Department of Biochemistry, University of California,
Berkeley 94720, USA

CHEMOTAXIS AND ENERGY METABOLISM

Bacteria have a well developed sensory system which allows them to swim towards attractants, which are usually nutrients, and away from repellents, which are usually indicators of toxic conditions. Thus, they respond to environmental conditions in ways which enhance their survival. In fact, it is quite easy to select for bacteria with improved motility by generating conditions which make a high degree of motility advantageous to the organism. This bacterial need is probably two-fold: the need for cellular constituents and the need for energy. To insure that it obtains adequate energy the bacterial cell has developed a behavioral response on which it expends a small but reasonable fraction of its energy resources in order to migrate to more favorable conditions. The nature of this behavioral response and its relationships to some of the enzymes involved in energy metabolism will be the subject of this paper.

Chemotaxis in bacteria was discovered in the 1880s by the great biologists Engelmann (1881) and Pfeffer (1883). The phenomenon of chemotaxis is not only pervasive in bacteriological species (see reviews by Weibull, 1960; Ziegler, 1962) but occurs in all forms of life; for example in insects which are guided by pheromones, and the use by lymphocytes of chemicals to lead them to their prey. We shall make no attempts to cover the comparative biology of chemotaxis but will concentrate on bacterial chemotaxis because of its utility as a model sensory system. Adler, who has contributed so much to the modern development of this system, has written excellent reviews (1969, 1975) of the field, and some approaches to chemotaxis as a model sensory system have been reviewed by us (Koshland, 1974, 1976a, b).

RANDOM WALK

If one observes bacteria under the microscope, they appear to travel roughly in a straight line and then turn abruptly. Sometimes the bacteria appear to be tumbling head over tail for a brief period before swimming off in a new direction. This led early workers to deduce that some type of random walk was likely for bacteria (Weibull, 1960). These qualitative observations were placed on a quantitative basis by Berg & Brown (1972) who utilized an automatic tracking apparatus. The quantitative measurement of bacterial motion showed (1) that the length of the runs, i.e. the distance between tumbles, was Poissonian, and (2) that the angle of the turn averaged 62° with a Poissonian distribution about this average. When the bacteria were observed in a concentration gradient, the pattern changed so that the length of an average run was increased when the bacteria were travelling up a gradient of attractant. The length of the run on going down a gradient remained the same as in an isotropic solution.

TEMPORAL SENSING MECHANISM

How could an organism as small as a bacterium sense a gradient? Two logical alternatives seemed obvious, based on knowledge of other sensory systems: an instantaneous spatial sensing or a temporal sensing mechanism. In the first, the bacterium would compare the concentration of attractant at its 'head' with the concentration of attractant at its 'tail' in a given instant of time. In the second, the bacterium, swimming through space, would compare the concentration over its whole body at time t_2 with that at time t_1. The choice between these alternatives for bacteria was resolved by Macnab & Koshland (1972) who used a rapid mixing device to plunge the bacteria rapidly from one uniform concentration of attractant (C_i) into a second uniform concentration (C_f). Control experiments had established that the absolute concentration of attractant did not affect the motility pattern. This was called a temporal gradient experiment since it involved a change in sensory information over time while producing a new isotropic chemical distribution. The thorough mixing meant that the bacteria were observed in the absence of any instantaneous difference in concentration between their heads and tails. The results showed clearly that the bacteria sense over time, i.e. they tumbled more frequently when the concentration was decreased rapidly, less frequently when the concentration was increased. This means that they have some kind of 'memory' which allows them to compare the environment of their past with that of their present and to interpret this signal.

Berg & Brown (1972) studied bacteria in gradients and showed that the angle of turning did not change in gradients, whereas the frequency of tumbling did. These results and the 'memory' experiments are consis-

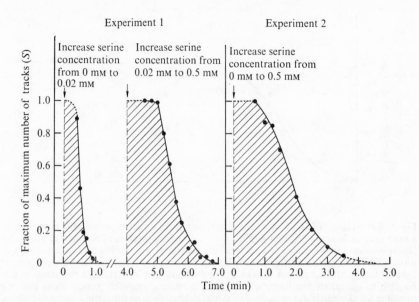

Fig. 1. Demonstration of additivity of stimuli. In experiment 1 bacteria are stimulated first with a 0→0.02 mM serine stimulus and then with a 0.02→0.5 mM stimulus. In each case after the sudden concentration change the bacteria are allowed to recover to normal non-gradient motility. In experiment 2, the bacteria are subjected to one large stimulus of 0→0.5 mM. The areas under the first two curves (the recovery times) are 0.5 min and 1.5 min respectively. The area under the third curve is 2.0 min.

tent and lead to a simple picture. Bacteria migrate by regulating tumbling frequency. They tumble less when going in favorable directions and more when going in unfavorable ones, thus biasing their migration towards the favorable directions.

RELATION BETWEEN IN-VIVO RESPONSE AND AN IN-VITRO RECEPTOR CONSTANT

The bacterial response described above has characteristics of a neuron in the sense that a chemical (attractant, repellent or neurotransmitter, respectively) induces a response in the cell which leads to an all-or-none event (a tumble or a depolarization, respectively). It seemed desirable, therefore, to quantitate the response in a manner similar to an action potential and this was done by the quantitative assay procedure (Spudich & Koshland, 1975).

This procedure, described in more detail in Fig. 1, involves subjecting the bacteria to a sudden temporal gradient of attractant or repellent. The sudden increase of attractant, for example, suppresses tumbling so

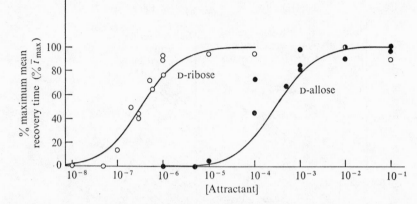

Fig. 2. A correlation of an in-vivo behavioral response of a bacterium with the properties of a pure protein *in vitro* (Spudich & Koshland, 1975). The method of Spudich & Koshland (1975) involves a rapid mixing of a bacterial suspension with a solution containing attractant (or repellent). At intervals after the mixing the bacteria are examined with a stroboscopic light for 0.8 s with four flashes of light. From the photograph obtained in this manner, it is possible to quantitate the bacteria which are swimming smoothly versus those that are tumbling for any particular time. By following the bacteria over time, the smooth swimming after an attractant stimulus is seen to decrease and the bacteria return to normal non-gradient motility. The average time (by integration under the curve) it takes to return to normal is defined as the recovery time. Points represent recovery times of bacterial tumbling after being subjected to stimuli of an attractant whose concentration is shown on the abscissa. Solid lines were calculated assuming previously determined dissociation constants (Aksamit & Koshland, 1974) for attractants to the purified ribose receptor (3.3×10^{-7} M for ribose and 3.0×10^{-4} M for allose).

that all of the bacteria swim smoothly. As time goes on, however, more and more tumbling occurs until the bacteria are all returned to their normal motility pattern in the absence of a gradient. The qualitative observations seen by an observer looking through a microscope can be replaced by a quantitative measurement by photographing the bacteria for standard intervals at various times after the stimulus. The number of bacteria swimming smoothly or tumbling can be counted objectively on such a photographic film and plotted on co-ordinate paper. The results of such an analysis of several stimuli are shown in Fig. 1. The recovery time, which is the average time it takes the bacterial population to return to normal behavior, has been shown to be calculated by the area under the curve (Spudich & Koshland, 1975).

By using this procedure for measuring tumbling frequency, we were able to make a precise quantitative correlation of the properties of the purified ribose receptor with the behavior of the intact organism. The ribose receptor is a periplasmic protein isolated by conventional procedures involving osmotic shock of *Salmonella typhimurium* (Aksamit &

Koshland, 1974). A similar periplasmic protein had been identified as the galactose receptor for *Escherichia coli* by Hazelbauer & Adler in 1971. The purified ribose receptor was found to be a protein of 30 000 molecular weight and to have a dissociation constant (K_d) for ribose of 3×10^{-7} M as determined by equilibrium dialysis (Aksamit & Koshland, 1974); the dissociation constant of the in-vivo receptor for ribose chemotaxis was found to be 3×10^{-7} M (Spudich & Koshland, 1975). Moreover allose, which has a dissociation constant of 3×10^{-4} M for the purified receptor, gave a behavioral response curve which corresponded to a value of 3×10^{-4} M (Fig. 2). Sugars such as arabinose which did not interact with the receptor did not cause chemotaxis. Thus the receptor showed the same specificity *in vivo* and *in vitro*, a finding of interest to the study of higher species in which the isolation of receptors is not as simple.

The quantitative studies (Spudich & Koshland, 1975) further showed that the stimulus is proportional to the change in receptor occupancy, $\Delta(RC)$, where R is the receptor and C is the chemoeffector. The change in receptor occupancy was calculated according to the mass law relationships where C_i represents the concentration of chemoeffector surrounding the bacterium initially and C_f represents the final concentration after the increase in chemoeffector generated in the mixing apparatus. Then:

$$\text{Initial receptor occupancy,} \qquad (RC)_i = \frac{(R)(C_i)}{K_d}. \qquad (1)$$

$$\text{Final receptor occupancy,} \qquad (RC)_f = \frac{(R)(C_f)}{K_d}. \qquad (2)$$

$$\text{Change in receptor occupancy,} \; \Delta(RC) = (RC)_f - (RC)_i. \qquad (3)$$

The recovery times from attractant stimuli were found to be additive algebraically. For example, Fig. 1 shows the results of comparing two small stimuli ($0 \rightarrow 0.02$ mM and $0.02 \rightarrow 0.5$ mM) to one overall stimulus ($0 \rightarrow 0.5$ mM). It is seen that the areas under the two smaller stimuli (their recovery times) equal the area under the large stimulus. Repellents were shown to act in the same way but the signs were inverse, i.e. the recovery times of increased repellents would subtract from the recovery times of increased attractants (Tsang, Macnab & Koshland, 1973). Specificity patterns indicate that there are receptors for repellents as well as attractants (Tso & Adler, 1974a). Moreover, the bacteria can add the signals from attractants and repellents in an algebraic way (Tsang *et al.*, 1973; Tso & Adler, 1974b), not unlike the combined effect of inhibitory and excitatory signals in higher species.

Table 1. *Response of* Salmonella typhimurium *and* Escherichia coli
to temporal gradients of mono- and divalent cations

	Recovery time (min)				
Chemical and gradient (mM)	Tumbling mutant of *S. typhimurium* (ST171)		Wild-type *E. coli* (K12)	ATPase mutant of *E. coli* (An120)	ATPase mutant of *E. coli* (SW46)
	Control	+DCCD[a]			
MgCl₂ (0→10)	1.03	< 0.2	1.2	< 0.2	< 0.2
CaCl₂ (0→10)	1.45	< 0.2	1.5	< 0.2	< 0.2
ZnCl₂ (0→5)	1.9	< 0.2	0.8	< 0.2	< 0.2
NaCl (0→100)	< 0.2	n.d.	< 0.2	n.d.	n.d.
KCl (0→100)	0.55	n.d.	0.6	n.d.	n.d.
L-serine (0→0.1)	1.2	1.3	4.2	4.1	4.4
Ribose (0→1)	0.8	0.7	1.0	0.5	0.55
(0→10)	0.8	0.8	1.0	0.8	0.8
Galactose (0→1)	< 0.2	< 0.2	1.1	0.6	0.5
(0→10)	< 0.2	< 0.2	1.1	1.2	1.2

Bacteria were grown in Vogel–Bonner citrate buffer (VBC) (supplemented with L-histidine (50 μg ml^{-1}) and thymine (30 μg ml^{-1}) for ST171, and with arginine (50 μg ml^{-1}) and thiamine-HCl (10^{-4} M) for An120) at 30 °C and harvested by centrifugation at early logarithmic growth (10⁸ cells per ml). The bacterial pellet was resuspended in VBC (pH 7.0) lacking magnesium or in 10 mM sodium phosphate buffer (pH 7.0) plus 10^{-4} M EDTA. All measurements were made at 30 °C.

[a] *N,N¹*-dicyclohexyl carbodiimide (DCCD) is a potent and highly selective inhibitor of Mg^{2+}, Ca^{2+}-ATPase activity in membranes.

Mg^{2+}, Ca^{2+}-ADENOSINE TRIPHOSPHATASE

These studies laid the groundwork for the examination of a most intriguing relationship between chemotaxis and a central enzyme in energy metabolism (Zukin & Koshland, 1976).

In Table 1 are shown the response times of a tumbling mutant, *S. typhimurium* ST171, subjected to temporal gradients of some representative cations and some known attractants. Suppression of tumbling is caused by increases in concentration of Mg^{2+}, Ca^{2+} and Zn^{2+}, and also of Mn^{2+} and Co^{2+} (not shown). Varying the anion for a given salt resulted in no change in the observed recovery time, a result which indicates that the bacteria are responding specifically to the divalent cation added. Weak responses were observed on transfer to high concentrations (100 mM) of K$^+$ but not Na$^+$. The absence of an effect in the case of Na$^+$ indicates that the positive results in the case of the divalent cations are not caused by a generalized ionic strength effect. Moreover, the normal swimming pattern observed after a pre-incubation in various Mg^{2+} concentrations in the range 0 to 10.0 mM shows that the bacteria respond to changes in divalent cation concentrations, rather than to

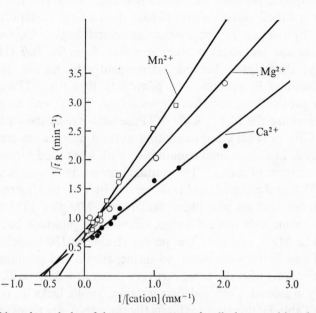

Fig. 3. Double reciprocal plot of dose–response curve for divalent metal ions in *S. typhimurium* chemotaxis. Data for recovery times (t_R) are obtained by integrating under curves such as those of Fig. 1. Data are obtained for a *che*T unco-ordinated mutant in VBC medium lacking magnesium at pH 7 and 30 °C.

absolute values. Wild-type *E. coli* K12 gave essentially the same pattern of responses as *S. typhimurium* (Table 1).

The suggestion of a saturation effect was examined by studying the recovery times of the intact bacteria over a concentration range of metal ion gradients and analyzing the results by double reciprocal plots. As can be seen in Fig. 3, the data fall on straight lines indicating a hyperbolic binding relationship between the metal and a receptor. The theoretical lines are calculated assuming the recovery time is proportional to the change in receptor occupancy, as had been previously shown for the attractants ribose and allose (Spudich & Koshland, 1975). Thus the dissociation constants for metal binding can be determined. The resulting values for K_d were 1.54 mM for Ca^{2+}, 1.7 mM for Mg^{2+}, 2.0 mM for Zn^{2+}, 2.7 mM for Mn^{2+} and 14 mM for Co^{2+}. The demonstration of distinct curves and a specificity pattern indicate that the metal ions operate through binding to a protein receptor molecule and are not acting as a membrane gradient, in which case the response would be simply a function of the difference between external and internal cation concentration.

The response maxima for different metals measured for the intact bacteria and their variation with pH are similar to those reported for the purified F_1 portion of the membrane-bound Mg^{2+}, Ca^{2+}-adenosine triphosphatase (ATPase) protein complex from *E. coli* (Hanson & Kennedy, 1973), and checked independently by us for the purified *S. typhimurium* F_1 protein. The possibility that the ATPase was the receptor protein was, therefore, investigated. The *E. coli uncA* mutant, An120, lacking the Mg^{2+}, Ca^{2+}-ATPase activity, as shown by Butlin, Cox & Gibson (1971), but exhibiting normal magnesium transport as shown by J. Lusk (personal communication), was tested for the divalent cation response (Table 1). The mutant, grown in Vogel–Bonner salts with EDTA and glycerol, and resuspended in VBC lacking magnesium, or in 10 mM sodium phosphate buffer, pH 7.0, plus 10^{-4} M EDTA, exhibits chemotaxis toward serine, ribose and galactose but does not respond to Mg^{2+} or Ca^{2+}. The parent strain, An180 (which has wild-type ATPase activity) responds to serine, ribose and galactose, and to divalent cations with the same recovery times as wild-type *E. coli* K12. Similarly, a second ATPase-deficient strain, SW46, lacks the response to cations (Table 1); this strain exhibits the same recovery times for cations, amino acids and sugars as does An120. The parent strain of SW46, *E. coli* 1100, does respond to all divalent cations tested with the same recovery times as K12.

These results present strong *prima facie* evidence that the Mg^{2+}, Ca^{2+}-ATPase, or a portion of it, is the chemotaxis receptor for the divalent metal ions. Considerable evidence from studies using antibodies to the soluble F_1 portion of the ATPase complex indicates that the F_1 component, which catalyzes the phosphohydrolase reaction and which binds the metals, is on the inside of the cytoplasmic membrane (Futai, 1974). Direct interaction between divalent cations such as Mg^{2+} and the ATPase protein could occur in either of two ways: (1) the cations interact with some small fraction of the F_1 which might be exposed on the outer surface of the membrane; or (2) there exists some mechanism by which the cations or the F_1 oligomer, or one of its constituents, can traverse the membrane. Two systems for the transport of Mg^{2+} have now been identified, both with K_m values of 15–50 μM (Nelson & Kennedy, 1972). Both of these systems would be saturated well below 1 mM Mg^{2+}. Recent experiments by J. Lusk, however, show that in the millimolar range, Mg^{2+} uptake occurs by passive diffusion (J. Lusk, personal communication). Such a mechanism could account for the observed Mg^{2+} taxis, with ATPase as the chemoreceptor.

The finding of the Mg^{2+}, Ca^{2+}-ATPase as a receptor for chemotaxis has many ramifications some of which have been discussed (Zukin & Koshland, 1976). One of the most important seems to be an emerging generalization about the dual uses of the same peptide chain. In this case one or more of the subunits of the ATPase serves as an energy generating source and also as a sensor in chemotaxis. In the case of the ribose binding protein described above, the peptide chain is the first receptor in the transport of ribose and also in the chemotactic response system for ribose. The utilization of a single polypeptide chain in several capacities would be very intriguing in relation to the economy and organization of the cell and it will be interesting to see how far this generalization can be extended to other systems.

ATP AND OXIDATIVE PHOSPHORYLATION

In order to exhibit chemotaxis bacteria must be motile, and both biological processes require energy. There are two principal sources of energy which play a role; the chemiosmotic energy of Mitchell generated in the membrane (Mitchell, 1961, 1963, 1966), and the central energy compound ATP. It does not follow that both taxis and motility require the same energy source and indeed studies have shown that they do not.

In the case of *E. coli*, Larsen, Adler, Gargus & Hogg (1974a) applied these tests to the problems of motility and chemotaxis. It was found that mutants blocked in the conversion of ATP to the intermediate of oxidative phosphorylation because of damage to the ATPase, were motile in the presence of oxygen, showing that their flagellar machinery was intact. Conversely, in wild-type cells carbonylcyanide *m*-chlorophenyl hydrazone (CCCP), which uncouples oxidative phosphorylation, can inhibit motility even in the presence of ATP. Finally arsenate, which reduces ATP levels in the bacterium but does not significantly reduce the level of the intermediate in oxidative phosphorylation, did not affect motility. Thus it is clear that the energy source for motility is the intermediate of oxidative phosphorylation and is not ATP.

On the other hand, ATP is required for chemotaxis (Larsen *et al.*, 1974a). The addition of arsenate to motile cells does not inhibit their motility but prevents their chemotaxis. This loss of chemotactic ability is coincident with a massive decrease in the amount of ATP present. Addition of lactate and oxygen to these ATP-deprived cells maintains their motility by providing the high-energy intermediate of oxidative phosphorylation, but is incapable of allowing them to sense a gradient. These results not only establish the energy bases for the two processes in bacteria, but confirm the conclusion reached from mutant studies that motility and chemotaxis are interrelated but separate processes which can be controlled by separate genes and by separate enzymes.

NATURE OF THE BACTERIAL SYSTEM

The above studies and others reviewed elsewhere (Koshland, 1976*a*, *b*), although far from complete, do give a rudimentary picture of the bacterial sensing system. On the surface of the bacterium are located 20 or more receptors which 'report' to the bacterium with regard to environmental conditions. Some of the receptors are designed to direct bacterial motion towards attractants, which are usually nutrients, and some are designed to direct bacteria away from toxic or crowded conditions. Most of the receptors are highly specific, interacting with only one or two physiological compounds, but a few are only moderately specific, e.g. one receptor reacts with a variety of fatty acids. The repellent receptors do not respond to all toxic conditions, nor is there an attractant receptor for all nutrients which are essential to the cell. This indicates that the receptors are monitoring the environment on a sampling basis rather than on a comprehensive basis. An excellent survey of the receptor specificities and number of receptors has been given by Adler in his recent review (Adler, 1975).

The receptors fall in no single class. Some are periplasmic proteins, e.g. the galactose receptor identified by Hazelbauer & Adler (1971) and the ribose receptor identified by us (Aksamit & Koshland, 1974). Another is one of the subunits of the phosphotransferase system which attracts glucose (Adler & Epstein, 1974), and thirdly there is the divalent metal ion receptor described above (Zukin & Koshland, 1976). The latter two are not periplasmic proteins and are not released by osmotic shock, but they are in the membrane. Attempts by us to solubilize the serine receptor have so far been unsuccessful but they indicate that it is a tenacious membrane-bound protein, difficult to solubilize, with many of the same characteristics as mammalian receptors. Thus the receptors identified so far show a wide range of protein types although all are identified with membrane components.

The receptor transmits information through a generalized transmission system which is composed of at least nine gene products in *S. typhimurium* (Aswad & Koshland, 1974; Koshland, Warrick, Taylor & Spudich, 1976) and four gene products in *E. coli* (Armstrong & Adler, 1969; Parkinson, 1974). This analytical system compares changes in concentration based on changes in the fraction of receptor occupancy. It

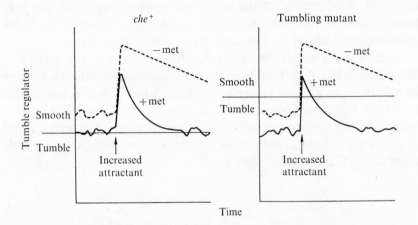

Fig. 4. A model for the regulation of tumbling frequency by gradients. When the tumble regulator concentration (wavy solid or dotted line) rises above the threshold value (horizontal solid line), tumbling is suppressed (smooth motility); when it falls below, tumbling (motility) is increased. In the absence of a gradient the regulator level fluctuates randomly near the threshold level, but a rapid temporal increase in attractant makes the tumble regulator concentration rise far above the threshold level, suppressing tumbling completely until the level returns to normal. The regulator is maintained at a steady-state level by a delicate balance between its synthesis and degradation. S-adenosyl methionine may be necessary for the degradation reaction. Thus, during methionine (met) starvation of the wild-type *S. typhimurium* (*che*+), the level of regulator is higher than normal so tumbling is suppressed. When a constantly tumbling mutant is perturbed by a temporal gradient it takes longer than normal to return to its steady-state level when methionine is absent. By assuming that the constantly tumbling mutant has an abnormally high threshold, one can easily explain its behavior in both the presence and absence of methionine.

is unaffected by an unchanging concentration of attractant or repellent. The signal generated by a gradient is transmitted to the flagella and results in suppression of tumbling when the bacteria move in a favorable direction (up an attractant gradient or down a repellent gradient), and an increase in tumbling when moving in an unfavorable direction (down an attractant gradient or up a repellent gradient). The quantitative contributions of these two effects, however, are not equal, the response in the positive direction being the more powerful. The system is able to integrate the responses from different signals, and to add them with proper algebraic signs, i.e. to integrate the effects of inhibition and activation.

The signal is ultimately transmitted to the flagella in a way which generates reversal of flagellar rotation (Berg, 1974; Larsen *et al.*, 1974*b*; Silverman & Simon, 1974). Reversal of flagellar rotation causes the flagella to fly apart in the case of the multi-flagellated bacteria, thereby generating a tumble (MacNab & Koshland, 1974). The frequency of the tumbling, then, represents the behavioral response by which the bacterium biases its random walk so that it can migrate towards favorable environmental situations and away from unfavorable ones.

A mechanism which can rationalize the behavior of the bacteria is illustrated in Fig. 4. A 'tumble regulator' whose concentration is varying in a Poissonian manner generates tumbling when it falls below certain threshold values and suppresses tumbling when it rises above them. Increasing receptor occupancy stimulates the formation of a compound X which we will use as the symbol for the parameter called 'the tumble regulator' (cf. the scheme below).

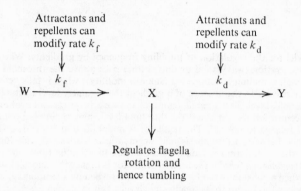

If this stimulus leads to an increased rate of formation of X relative to its rate of decomposition, X will accumulate as the bacteria go up a gradient of attractant and will therefore suppress tumbling. Its depletion when bacteria swim down a positive gradient leads to a decreased concentration of X and hence an increased probability of tumbling. A mutant which tumbles constantly has a level of tumble regulator below the threshold, a situation which could be produced either by an insufficient concentration of the tumble regulator or a change in threshold level. Conversely, a mutant which never tumbles produces tumble regulator above the threshold level.

In Fig. 4 this approach is applied to some situations which have been observed. In the wild-type *S. typhimurium che*[+] strain the tumble regulator fluctuates in a Poissonian manner in an isotropic environment. A sudden increase in attractant increases the level of X and suppresses tumbling for a period before returning to normal. In an auxotroph deprived of methionine, the level of tumble regulator is always above threshold level so tumbling is always inhibited (Aswad & Koshland, 1974).

In the constantly tumbling mutant, the tumble regulator is always below threshold level and the bacteria are constantly receiving signals to tumble. If this mutant is given a sudden increase in attractant artificially, the level of X rises above the threshold briefly (Macnab & Koshland, 1972; Spudich & Koshland, 1976). Travelling up a gradient could achieve this for the bacterium, but it never travels long enough in a straight line to straighten out its sensing system. (The analogy to human psychopathology is obvious.)

Thus, a very simple on–off switch can be utilized to develop a behavioral pattern. The regulator which controls this switch is geared into the metabolism of the cell by a sensory system which is driven by energy and controlled by sources of energy. The details of the operation of this regulator remain to be elucidated but the general pattern is beginning to emerge.

The author wishes to acknowledge support from the United States Public Health Service (AM9765) and the National Science Foundation (BMS 71-0133A03).

REFERENCES

ADLER, J. (1969). Chemoreceptors in bacteria. *Science*, **166**, 1588–97.

ADLER, J. (1975). Chemotaxis in bacteria. *Annual Review of Biochemistry*, **44**, 341–56.

ADLER, J. & EPSTEIN, W. (1974). Phosphotransferase-system enzymes as chemoreceptors for certain sugars in *E. coli* chemotaxis. *Proceedings of the National Academy of Sciences, USA*, **71**, 2895–9.

AKSAMIT, R. & KOSHLAND, D. E. JR (1974). Identification of the ribose binding protein as the receptor for ribose chemotaxis in *Salmonella typhimurium*. *Biochemistry*, **13**, 4473–8.

ARMSTRONG, J. B. & ADLER, J. (1969). Location of genes for motility and chemotaxis on the *E. coli* genetic map. *Journal of Bacteriology*, **97**, 156–61.

ASWAD, D. & KOSHLAND, D. E. JR (1974). Role of methionine in bacterial chemotaxis. *Journal of Bacteriology*, **118**, 640–5.

ASWAD, D. & KOSHLAND, D. E. JR (1975). Isolation, characterization, and complementation of *S. typhimurium* chemotaxis mutants. *Journal of Molecular Biology*, **97**, 225–35.

BERG, H. C. (1974). Dynamic properties of bacterial flagellar motors. *Nature, London*, **249**, 77–9.

BERG, H. C. & BROWN, D. A. (1972). Chemotaxis in *Escherichia coli* analysed by three dimensional tracking. *Nature, London*, **239**, 500–4.

BUTLIN, J. D., COX, G. B. & GIBSON, F. (1971). Oxidative phosphorylation in *Escherichia coli* K12. *Biochemical Journal*, **124**, 75–81.

ENGELMANN, T. W. (1881). Neue Methode zur Untersuchung der Sauerstoffausscheidung pflanzlicher und thierischer Organismen. *Pflügers Archiv für die gesamte Physiologie*, **25**, 285–92.

FUTAI, M. (1974). Orientation of membrane vesicles from *Escherichia coli* prepared by different procedures. *Journal of Membrane Biology*, **15**, 15–28.

HANSON, R. L. & KENNEDY, E. P. (1973). Energy-transducing adenosine triphosphatase from *E. coli*: purification, properties and inhibition by antibody. *Journal of Bacteriology*, **114**, 772–81.

HAZELBAUER, G. H. & ADLER, J. (1971). Role of the galactose binding protein in chemotaxis of *E. coli* toward galactose. *Nature New Biology*, **230**, 101–4.

KOSHLAND, D. E. JR (1974). Chemotaxis as a model for sensory systems. *FEBS Letters*, **40**, S3–S9.

KOSHLAND, D. E. JR (1976a). Bacterial chemotaxis as a simple model for a sensory system. *Trends in Biochemical Sciences*, **1**, 1–3.

KOSHLAND, D. E. JR (1976b). Sensory response in bacteria. *Advances in Neurochemistry*, in press.

KOSHLAND, D. E. JR, WARRICK, H., TAYLOR, B. & SPUDICH, J. (1976). The control of flagellar rotation in bacterial behavior. *Cold Spring Harbor Symposia on Quantitative Biology*, in press.

LARSEN, S. H., ADLER, J., GARGUS, J. J. & HOGG, R. W. (1974a). Chemomechanical coupling without ATP. The source of energy for motility and chemotaxis in bacteria. *Proceedings of the National Academy of Sciences, USA*, **71**, 1239–43.

LARSEN S. H., READER, R. W., KORT, E. N., TSO, W.-W. & ADLER, J. (1974b). Change in direction of flagellar rotation is the basis of the chemotactic response. *Nature, London*, **249**, 74–7.

MACNAB, R. M. & KOSHLAND, D. E. JR (1972). The gradient sensing mechanism in bacterial chemotaxis. *Proceedings of the National Academy of Sciences, USA*, **69**, 2509–12.

MACNAB, R. M. & KOSHLAND, D. E. JR (1974). Bacterial motility and chemotaxis: light-induced tumbling response and visualization of individual flagella. *Journal of Molecular Biology*, **84**, 399–406.

MITCHELL, P. (1961). Coupling of phosphorylation to electron and hydrogen transfer by a chemi-osmotic type of mechanism. *Nature, London*, **191**, 144–8.

MITCHELL, P. (1963). Molecule, group and electron translocation through natural membranes. *Biochemical Society Symposia*, **22**, 142–68.

MITCHELL, P. (1966). Chemiosmotic coupling in oxidative and photosynthetic phosphorylation. *Biological Reviews*, **41**, 445–502.

NELSON, D. L. & KENNEDY, E. P. (1972). Transport of magnesium by a repressible and nonrepressible system in *Escherichia coli*. *Proceedings of the National Academy of Sciences, USA*, **69**, 1091–3.

PARKINSON, J. S. (1974). Data processing by the chemotaxis machinery of *E. coli*. *Nature, London*, **252**, 317–19.

PFEFFER, W. (1883). Locomotorische Richtungsbewegungen durch chemische Reize. *Berichte der deutsche botanische Gesellschaft*, **1**, 524–33.

SILVERMAN, M. R. & SIMON, M. I. (1974). Flagellar rotation and the mechanism of bacterial motility. *Nature, London*, **249**, 73–4.

SPUDICH, J. & KOSHLAND, D. E. JR (1975). Quantitation of the sensory response in bacterial chemotaxis. *Proceedings of the National Academy of Sciences, USA*, **72**, 710–13.

TSANG, N., MACNAB, R. & KOSHLAND, D. E. JR (1973). Common mechanism for repellents and attractants in bacterial chemotaxis. *Science*, **181**, 60–3.

TSO, W.-W. & ADLER, J. (1974a). Negative chemotaxis in *Escherichia coli*. *Journal of Bacteriology*, **118**, 560–76.

TSO, W.-W. & ADLER, J. (1974b). 'Decision' making in bacteria: chemotactic response of *Escherichia coli* to conflicting stimuli. *Science*, **184**, 1292–4.

WEIBULL, C. (1960). Movement. In *The Bacteria*, ed. I. C. Gunsalus & R. Y. Stanier, vol. 1, pp. 153–205. New York & London: Academic Press.

ZIEGLER, H. (1962). Chemotaxis. In *Handbuch der Pflanzenphysiologie*, vol. 1711, pp. 484–532.

ZUKIN, R. S. & KOSHLAND, D. E. JR (1976). The Mg^{2+}, Ca^{2+}-ATPase as a receptor for divalent cations in bacterial sensing. *Science*, **193**, 405.

LIGHT ENERGY CONVERSION IN HALOBACTERIA

D. OESTERHELT, R. GOTTSCHLICH, R. HARTMANN, H. MICHEL AND G. WAGNER

Institute for Biochemistry, University of Würzburg, BRD-87 Würzburg, Röntgenring 11, W. Germany

INTRODUCTION

Halophilism is a good example of biological adaption to a special environment. Extremely halophilic bacteria grow in nearly saturated sodium chloride solutions and occur in nature wherever this growth condition is met, e.g. in open ponds where salt is produced by evaporation of sea water (for review see Larsen, 1967).

Cells of the genus *Halobacterium* are rod-shaped and usually flagellated at their polar ends. Halobacteria lack the typical components of bacterial cell walls, for instance muramic or diaminopimelic acids (Kushner *et al.*, 1964). A wall-like structure, consisting of glycoproteins, covers the cell surface and gives a certain rigidity to the cell. However, even weak mechanical forces will break the cells, and this allows the easy preparation of cell envelopes for the study of transport processes. Exposure of whole cells or cell envelopes to pure water results in lysis, with disintegration of the cell wall components and the cell membrane itself (Stoeckenius & Kunau, 1968).

Little is known about the metabolic pathways in Halobacteria, but no significant differences in their metabolic apparatus compared to non-halophilic bacteria have been found so far. Although most of the *Halobacterium* strains are aerobic, oxidising the amino acids in the medium, they encounter a considerably lower oxygen concentration than their non-halophilic counterparts. This is because the solubility of oxygen in 4 M sodium chloride solution is five times lower than in water. It is not known whether this condition could have induced the development of the novel bioenergetic system found so far only in Halobacteria. Light energy in these bacteria is used to drive phosphorylation in a process mediated not by chlorophyll, as in photosynthetic organisms, but by the retinal protein complex bacteriorhodopsin. The bacteria develop the capability for photophosphorylation under conditions where oxidative phosphorylation is slowed down by the lack of oxygen.

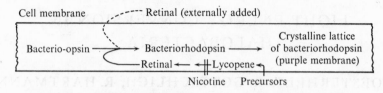

Fig. 1. Formation of bacteriorhodopsin in *Halobacterium halobium*.

The total system comprises bacteriorhodopsin as a light-driven proton pump and an ATP-synthase. This enzyme is found in all bioenergetic systems which equilibrate ATP levels with electrochemical ion gradients (for review see Oesterhelt, 1976).

FORMATION OF BACTERIORHODOPSIN AND THE 'PURPLE MEMBRANE'

Under conditions of limited oxygen supply, Halobacteria synthesise bacterio-opsin. This is inserted into the cell membrane where it combines with retinal to form bacteriorhodopsin. Analysis of the membrane fractions obtained by lysis of the cells in pure water followed by sucrose density centrifugation, reveals bacteriorhodopsin to be spread out among several of these fractions. Under conditions of continued bacteriorhodopsin formation, however, it crystallises out and forms large patches within the cell membrane which are called the 'purple membrane'.

The formation of bacteriorhodopsin in growing cells can be manipulated by the rate of aeration or by the addition of nicotine. This alkaloid blocks the cyclisation of lycopene to β-carotene and therefore prevents the formation of retinal from its precursor. The cells can still synthesise bacterio-opsin to some extent under these conditions (Sumper & Herrmann, 1976) and externally added retinal then induces bacteriorhodopsin formation (Sumper, Reitmeier & Oesterhelt, 1976). Thus, reconstitution of bacteriorhodopsin's bioenergetic function can be studied by the addition of retinal or retinal analogues to cells grown in the presence of nicotine (Oesterhelt & Christoffel, 1976).

The purple membrane patches formed by increasing concentrations of bacteriorhodopsin in the cell membrane can be visualised by electron microscopy of freeze-fractured cells (Blaurock & Stoeckenius, 1971). More than half of the cell membrane can consist of these patches. Lysis of the cells by water, as mentioned already, fragments the cell membrane and allows the isolation of large amounts of purple membrane. On a sucrose density gradient this membrane fraction bands at a buoyant density of 1.18 g cm^{-3}; the structural organisation of its protein and lipid has been studied (Blaurock & Stoeckenius, 1971; Blaurock, 1975; Henderson & Unwin, 1975).

MOLECULAR PARAMETERS OF BACTERIORHODOPSIN

Bacteriorhodopsin is the only protein of the purple membrane and comprises 75% of the membrane dry weight. It is associated with 25% lipid, corresponding to about 10 lipid molecules per bacteriorhodopsin molecule. The protein is arranged in a hexagonal crystalline array and spans the entire cross-section of the membrane. It consists of a single polypeptide chain with a molecular weight of 26000. Bacteriorhodopsin contains retinal bound to the protein in a ratio of 1:1. Upon association a chromophore is formed with an absorption maximum in the visible range around 560 nm, giving the membrane its purple colour. The molar extinction coefficient of the chromophore is 63000 (M^{-1} cm^{-1}). Upon illumination the bacteriorhodopsin chromophore, called 'purple complex', undergoes a photochemical reaction with a high quantum yield (0.8). A series of subsequent reactions leads to the regeneration of the purple complex thus establishing a photochemical cycle. Two retinal isomers have been found to be able to constitute the chromophore: 13-*cis* and all-*trans* retinal. In the dark both isomers are present; after short illumination only all-*trans* retinal is found. The reconversion to 13-*cis* retinal with a half time of 30 min at 35 °C in the dark, is rather slow compared to the cycling frequency of bacteriorhodopsin which is in the millisecond range. Although no indication for the participation of 13-*cis* retinal during the photochemical cycle in the light has been reported, this cannot be rigorously excluded at the present time.

The photophysics, photochemistry and function of bacteriorhodopsin can be studied at different levels of complexity: in intact cells, in cell vesicles free of cytoplasm (cell envelopes), in bacteriorhodopsin-containing liposomes, in purple membrane suspensions in water and organic solvent/water mixtures, and in lipid-free bacteriorhodopsin-containing detergent micelles (Happe & Overath, 1976). In the last few years an increasing number of laboratories have been involved in the research on bacteriorhodopsin (Boguslavsky *et al.*, 1975; Chance, Porte, Hess & Oesterhelt, 1975; Dencher & Wilms, 1975; Heyn, Bauer & Dencher, 1975; Kanner & Racker, 1975; Bogomolni, Baker, Lozier & Stoeckenius, 1976; Lanyi, Renthal & MacDonald, 1976; Sherman,

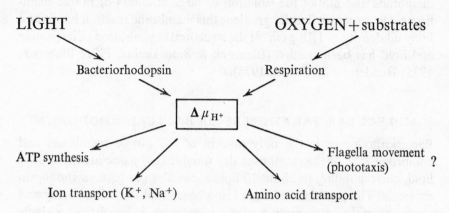

Fig. 2. Bioenergetic functions in Halobacteria.

Slifkin & Caplan, 1976; Yaguzhinsky, Boguslavsky, Volkov & Rakhmaninova, 1976). The discussion presented in this paper will be limited to the bioenergetic function of bacteriorhodopsin in intact cells.

FUNCTION OF BACTERIORHODOPSIN IN THE INTACT CELL

A general scheme of bioenergetic functions in Halobacteria is given in Fig. 2. Like other aerobic bacteria, Halobacteria create an electrochemical proton gradient across their cell membrane when they consume oxygen. This electrogenic proton extrusion is linked to the flow of electrons through their respiratory chain. Bacteriorhodopsin also creates an electrochemical proton gradient which, however, is not linked to any electron flow, i.e. bacteriorhodopsin acts purely as a light-driven proton pump. In aerated and illuminated cells both systems compete with each other. As a result, light inhibition of respiration is found. From quantitative measurements one can estimate that about 24 quanta of light energy will prevent the consumption of one molecule of oxygen, indicating a high efficiency of the light energy converting system in the cell (Krippahl & Oesterhelt, 1973).

Under anaerobic conditions bacteriorhodopsin-containing cells still carry out all the bioenergetic functions summarised in the lower part of Fig. 2. The electrochemical proton gradient will drive phosphorylation (i.e. ATP synthesis), potassium uptake and amino acid transport, as was recently shown by Hubbard, Rinehart & Baker (1976) and by MacDonald & Lanyi (1976). The gradient may even be the source of energy for movement of the flagella. A phototactic response allowing the

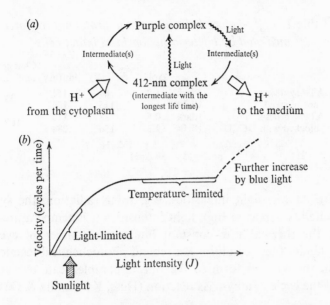

Fig. 3. Overall kinetics of the photochemical cycle in bacteriorhodopsin. The velocity (v) is proportional to the steady-state concentration of the 412-nm complex. Formation of the complex is proportional to the extinction coefficient ($\mathscr{E}$), quantum yield ($\mathscr{S}$) and light intensity (J); the decay is proportional to k_2. Thus $v = V_{max}/1 + (k_2/\mathscr{E} \times \mathscr{S})(1/J)$.

bacteria to stay in green light was recently described by Hildebrand & Dencher (1975). The action spectrum of this response was found to be identical to the chromophore absorption band of bacteriorhodopsin.

CREATION OF AN ELECTROCHEMICAL PROTON GRADIENT BY BACTERIORHODOPSIN

The efficiency of bacteriorhodopsin is intimately linked with its overall kinetics in the cell. Upon absorption of a quantum of light the purple complex passes through various intermediates and is then converted into the 412-nm complex with concomitant release of a proton which is finally found in the medium (Fig. 3a). At non-saturating light intensities the rate constant of the formation of the 412-nm form is dependent on light intensity, extinction coefficient and quantum yield. The decay of the 412-nm form occurs with a rate constant k_2, which has a value of about 100 s^{-1} at room temperature. A proton is taken up from the cytoplasm during the regeneration of the purple complex. On this basis the bacteriorhodopsin cycle can be expressed as a function of light intensity. The result is a simple saturation curve of the Michaelis–Menten type

Table 1. *Light-induced changes of the protonmotive force mediated by bacteriorhodopsin in intact cells*

		ΔpH	$\Delta\psi$(mV)	pmf(mV)	Change of pmf (mV)
ATP synthesis	{dark	1.0	87	151	35
not blocked	{light	1.2	113	186	
ATP synthesis	{dark	0.5	88	115	169
blocked with DCCD	{light	2.2	156	284	

$$\Delta\bar{\mu}_{\text{H}^+} = F\Delta\psi - 2.3\ RT\Delta\text{pH}$$
$$\text{pmf} = \Delta\psi - 59\ \Delta\text{pH}$$

(Fig. 3*b*). At low light intensities, e.g. solar radiation, the system is light-limited, whereas at high light intensities it is temperature-limited because the thermal rate constant limits the number of cycles per second. Under these conditions, a second photochemical reaction leading from the 412-nm form to the purple complex can accelerate the speed of the cycle in a by-pass reaction (Hess, Kuschmitz & Oesterhelt, 1976).

The photochemical cycle at its maximal velocity results in the electrogenic extrusion of about 100 protons s^{-1} per bacteriorhodopsin molecule, creating a membrane potential which is positive on the outside of the cell. Its size can be measured by a method first introduced by Skulachev some years ago (Bakeeva *et al.*, 1970). A cation like tetraphenylmethylphosphonium (TPMP$^+$), ^{14}C-labelled in its methyl group, will accumulate inside the cell until its diffusion potential equals the membrane potential. The cells can then be separated from the medium by quick centrifugation and the distribution of the labelled cation between supernatant and cell pellet measured. From its ratio of accumulation the membrane potential can be calculated.

Because the cell membrane is not fully impermeable to ions, these will respond to the membrane potential and in this way allow the protons to build up a pH gradient in addition to the membrane potential. Both potential and pH gradient contribute to the total electrochemical proton gradient known as protonmotive force (pmf). pH differences across the cell membrane are measured in a similar way to the membrane potential by using the distribution of a weak acid such as 5,5-dimethyl-2,4-oxazolidinedione (DMO) as an indicator.

Some of the experimental results using these methods are summarised in Table 1 (Michel & Oesterhelt, 1976). At pH 6, control cells show light-induced enhancement of their pH gradient by 0.2 units. The membrane potential increases at the same time from 87 mV to 113 mV,

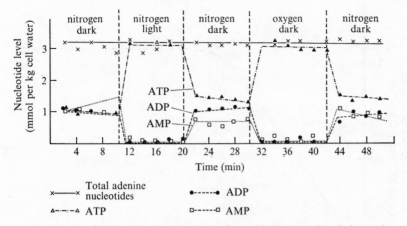

Fig. 4. Nucleotide levels in *H. halobium* during oxidative phosphorylation and photophosphorylation. (From Gottschlich 1976.)

so that the total change of the protonmotive force is 35 mV. The same experiment carried out with cells not capable of phosphorylation because their ATP-synthase is blocked with dicyclohexyl-carbodiimide (DCCD), shows an enhancement of the proton-motive force of the cells by 169 mV upon illumination. There is a good agreement between these results obtained from intact cells and those obtained by Renthal & Lanyi (1976) using vesicles and a fluorimetric method for the measurement of light-induced pmf changes. They found a light-induced increase in pmf of 172 mV if the vesicles contained 3 M potassium chloride and were suspended in 3 M sodium chloride.

PHOTOPHOSPHORYLATION AND OXIDATIVE PHOSPHORYLATION

The next question we have to deal with concerns the changes of adenine nucleotide levels upon illumination. Fig. 4 compares changes during photophosphorylation, as indicated by nitrogen and light, with changes during oxidative phosphorylation, as indicated by oxygen and dark. In both cases all adenine nucleotides are converted into ATP and, as expected, their sum remains constant. We have good indications that the electrochemical proton gradient and the phosphorylation potential of the cell satisfy the Gibbs free energy relationship. We therefore assume that only negligible ATP turnover occurs in starved cells under nitrogen by pathways other than ATP-synthase coupled to the proton gradient.

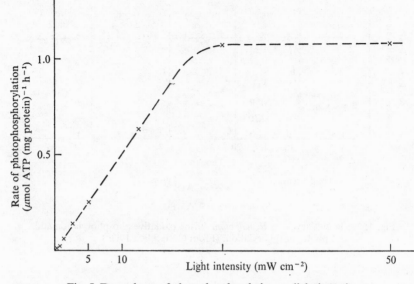

Fig. 5. Dependence of photophosphorylation on light intensity.

The ATP produced can be used for the accumulation of phosphorylated intermediates if suitable acceptors are added to illuminated cell suspensions. The entire phosphate pool of the cells, which is 20 times higher than the adenine nucleotide pool, is depleted if glycerol is present and corresponding amounts of glycerol-3-phosphate are then found (Gottschlich, 1976). From this experiment it can be predicted that the cell is able also to transfer the phosphorylation potential of ATP to other, as yet unidentified, acceptors in the cell. At this point it should be mentioned that very little is known about the metabolism of Halobacteria, and much work will have to be devoted to the question of whether Halobacteria exhibit general differences in their metabolic pathways as compared to other bacteria.

Considering ATP synthesis as induced by the proton-pumping function of bacteriorhodopsin, the time scale of the two processes has to be compared. A first, puzzling observation is that the onset of ATP synthesis occurs faster than any pH change observed in the medium using pH electrodes (Oesterhelt, 1974). It therefore seems likely that the membrane potential, which is accompanied by very small pH changes, rather than the pH gradient itself is the component of the protonmotive force which drives ATP synthesis. In fact we can calculate from the electrical properties of the cell membrane, the volume of the cell, the number of bacteriorhodopsin molecules per cell (approx. 10^6) and their turnover number (approx. 125 per second at 25 °C), that the time necessary to produce a membrane potential of 300 mV is (assuming maximal concentration of bacteriorhodopsin and saturating light intensity) less than 8 ms.

Although this might be a very rough estimate, the conclusion from such a calculation clearly is that at high light intensities bacteriorhodopsin can create a membrane potential within the first cycle which is large enough to drive photophosphorylation. At low light intensities the membrane potential, and therefore the rate of phosphorylation, is expected to be proportional to light intensity (Fig. 5). No significant lag period in the rate of phosphorylation is observed when light intensity is increased from zero. At high light intensities the rate of photophosphorylation becomes constant, very probably because ATP-synthase limits the rate. At this intensity bacteriorhodopsin still cycles at less than 10% of its maximal frequency.

The thylakoid membrane of chloroplasts has a higher cation permeability than the halobacterial cell membrane. Under continuous illumination (steady-state conditions) the membrane potential in chloroplasts is very low (Rottenberg, Grunwald & Avron, 1972) and the large pH gradient is the main component in the protonmotive force used for phosphorylation. Evidently that does not mean that the ATP-synthase in chloroplasts is not able to use the membrane potential for phosphorylation, as was shown by light flash experiments (Gräber & Witt, 1976).

Direct evidence for the membrane potential as a driving force in Halobacteria is given in Fig. 6. The rise of the ATP level upon illumination without inhibitor should be compared with the rise when DMO is added. No significant difference in ATP level is observed although the pH gradient is diminished at the concentration of DMO used in this experiment. TPMP+, which greatly diminishes the membrane potential at concentrations around 1 mM (Michel & Oesterhelt, 1976), decreases dramatically the rate of phosphorylation. The combination of both inhibitors completely abolishes photophosphorylation.

The special features of photophosphorylation in Halobacteria can be summarised as follows.

Photophosphorylation is:
(1) not affected by electron transport inhibitors;
(2) inhibited by uncoupling agents;
(3) inhibited after treatment of the ATP-synthase with DCCD or Phloretin;
(4) inhibited by lipophilic cations affecting the membrane potential.

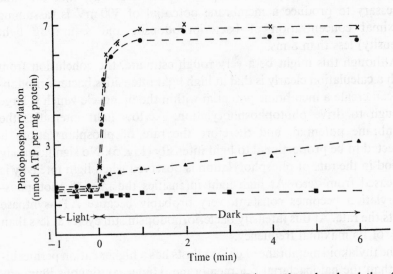

Fig. 6. Influence of the inhibitors DMO and TPMP$^+$ on photophosphorylation (light intensity 50 mW/cm^{-2}. ●, control; ×, with DMO; ▲, with TPMP$^+$; ■, with TPMP$^+$+ DMO.

In addition, phosphorylation is induced by sudden changes in external pH, and also by the addition of monactin. The fact that monactin induces phosphorylation is explained by the high internal potassium concentration of Halobacteria, which only slowly decreases by diffusion into the medium under control conditions. Increase of the K$^+$ permeability by addition of the ion carrier then induces a transient increase in membrane potential driving ATP synthesis.

PHOTOPHOSPHORYLATION AND PH CHANGES OF THE MEDIUM

A difficult problem in our understanding of bacteriorhodopsin function is the correlation of ATP synthesis with pH changes observed in the medium. In intact cells not only bacteriorhodopsin and ATP-synthase but also net consumption of protons during ATP synthesis, a postulated electrogenic sodium–proton exchange (Lanyi & MacDonald, 1976), and the permeability of the cell membrane to protons, potassium and sodium contribute to the changes with time of internal and external pH upon illumination, as well as to the steady state of pH values under continuous illumination. In addition, all these processes will change in

(a)

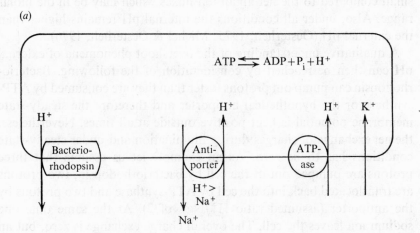

Rate: Bacteriorhodopsin > ATPase > Antiporter > K$^+$ > Passive H$^+$ permeation

(b)

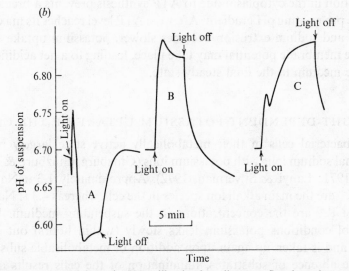

Fig. 7. Ion fluxes in Halobacteria. Upon illumination all proton-flux-dependent processes contribute to a new steady state of the external pH. Depending on the state of the cell, this may lead to a net acidification or net alkalinization of the medium and to overshoot phenomena. A, B and C are three different cell populations.

velocity after the start of illumination until a steady state of ion fluxes and ATP level is reached. This explains the rather complicated time course of the pH traces observed (Fig. 7b). The overall picture may be very different for cells in different states, e.g. with varying internal sodium and potassium concentrations before illumination. It should be mentioned that the changes in proton concentrations are always very small compared to the net alkali ion fluxes which may be in the molar range. Also, under all conditions the internal pH remains higher than the external pH (Oesterhelt, 1975; Michel & Oesterhelt, 1976).

A qualitative understanding of the overshoot phenomena of external pH can then be reached by consideration of the following. Bacteriorhodopsin can pump out protons faster than they are consumed by ATP-synthase or the hypothetical antiporter and therefore the steady-state membrane potential is kept positive outside at all times. Nevertheless, the net exchange of charges during illumination and under steady-state conditions has to be zero. As an example let us assume that three protons are pumped out of the cell by bacteriorhodopsin, two protons are translocated back into the cell by ATP-synthase and two protons by the antiporter (assumed ratio H^+/Na^+ of 2). At the same time one sodium ion leaves the cell. The overall charge exchange is zero, but an alkalinisation of the medium is observed. Additionally the proton consumption in the cytoplasm due to ATP synthesis prevents a breakdown of the pre-existing pH gradient. After the ATP level reaches its maximum value and sodium extrusion becomes slower, potassium uptake driven by the membrane potential may take place, leading to a net acidification of the medium in the final steady state.

LIGHT-DEPENDENT POTASSIUM UPTAKE OF THE CELL

Halobacterial cells in their metabolically active state largely replace internal sodium ions with potassium ions (Ginzburg, Ginzburg & Tosteson, 1971; Lanyi & Silverman, 1972). Approximately 1.3 M Na^+ and 3 M K^+ are the main alkali ion species in the cell, whereas 4.3 M Na^+ and 27 mM K^+ are the concentrations in the suspending medium. Under control conditions potassium leaks slowly (within hours) out of the cells, and is taken up again upon addition of metabolisable substrates. In the absence of substrates, illumination of the cells results also in potassium uptake, indicating that the electrochemical proton gradient itself, or the ATP from it, drives potassium uptake. In the latter case the rate of potassium uptake should depend on the ATP level of the cells. Both potassium uptake and ATP level are functions of light intensity and are compared in Fig. 8.

As can be seen from Fig. 8, the rate of potassium uptake of a cell suspension still increases when ATP has already reached its maximum level. Therefore the result of this experiment points towards a membrane-potential-induced potassium inflow, rather than the existence of a Na^+/K^+-stimulated ATPase in Halobacteria.

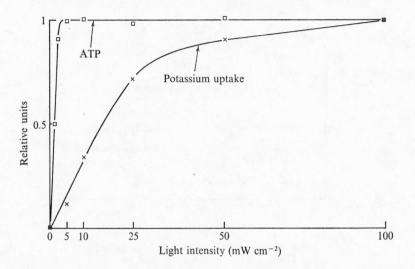

Fig. 8. ATP level and light-induced potassium uptake.

RECONSTITUTION OF BACTERIORHODOPSIN
FUNCTION IN INTACT CELLS

Retinal in the purple complex of bacteriorhodopsin is unable to react with hydroxylamine (though one of the intermediates of the photo-chemical cycle can produce retinal oxime by reaction with hydroxyl-amine). A chromophore-free membrane can therefore by prepared by illumination of purple membranes (or intact cells) in the presence of hydroxylamine (Oesterhelt & Schuhmann, 1974). After removal of this agent the chromophore can be reconstituted by addition of retinal. Because the treatment of cells with hydroxylamine could damage essen-tial cell functions, reconstitution experiments are best carried out with cells grown in the presence of nicotine. Under these conditions, as men-tioned above, only bacterio-opsin and no retinal is synthesised by the cells. As expected such cells do not show any light-induced pH changes or photophosphorylation capacity.

The addition of retinal not only reconstitutes the purple complex of bacteriorhodopsin, but also its proton-pumping function and photo-phosphorylation (Oesterhelt & Christoffel, 1976). This is also a very suitable system for studying the essential structural features of a retinal molecule for its ability to produce not only a chromophore with bacterio-opsin, but also a *functional* chromophore by restoring the bio-

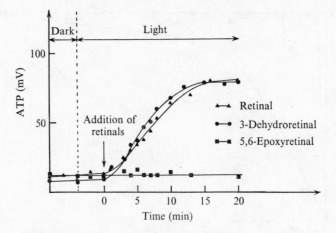

Fig. 9. Reconstitution of photophosphorylation.

energetic function of bacteriorhodopsin. As shown in Fig. 9, 3-dehydro-retinal (but not 5,6-epoxyretinal) is also able to reconstitute photo-phosphorylation.

OUTLOOK

Besides the bioenergetic aspects of bacteriorhodopsin discussed in this article, there are further aspects which make the purple membrane system a very attractive one for study. Membrane structure, differentiation and biogenesis can all be studied using this halobacterial system, which is probably the most simple model system known. The chemical mechanism of this proton-pumping bacteriorhodopsin, although completely unknown at present, has a high chance of being understood in the future. The similarity between rhodopsins and bacteriorhodopsin in some aspects of their chemistry and photochemistry introduces evolutionary considerations of retinal protein complexes both as light sensors and as light energy transducers.

REFERENCES

BAKEEVA, L. E., GRINIUS, L. L., JASAITIS, A. A., KULIENE, V. V., LEVITSKY, P. O., LIBERMAN, E. A., SEVERINA, I. I. & SKULACHEV, V. P. (1970). Conversion of biomembrane-produced energy into electric form. II. Intact mitochondria. *Biochimica et Biophysica Acta*, **216**, 13–21.

BLAUROCK, A. E. (1975). Bacteriorhodopsin: a trans-membrane pump containing α-helix. *Journal of Molecular Biology*, **93**, 139–58.

BLAUROCK, A. E. & STOECKENIUS, W. (1971). Rhodopsin-like protein from the purple membrane of *Halobacterium halobium. Nature New Biology*, **233**, 149–54.

BOGOMOLNI, R. A., BAKER, R. A., LOZIER, R. H. & STOECKENIUS, W. (1976). Light-driven proton translocations in *Halobacterium halobium. Biochimica et Biophysica Acta*, **440**, 68–88.

BOGUSLAVSKY, L. I., KONDRASHIN, A. A., KOZLOV, I. A., METELSKY, S. T., SKULACHEV, V. P. & VOLKOV, A. G. (1975). Charge transfer between water and octane phases by soluble mitochondrial ATPase (F_1), bacteriorhodopsin and respiratory chain enzymes. *FEBS Letters*, **50**, 223–6.

CHANCE, B., PORTE, M., HESS, B. & OESTERHELT, D. (1975). Low temperature kinetics of H^+ changes of bacterial rhodopsin. *Biophysical Journal*, **15**, 913–17.

DENCHER, N. & WILMS, M. (1975). Flash photometric experiments on the photochemical cycle of bacteriorhodopsin. *Biophysical Structure and Mechanism*, **1**, 259–71.

GINZBURG, M., GINZBURG, B. Z. & TOSTESON, D. C. (1971). The effects of anions on K^+ binding in a *Halobacterium* species. *Journal of Membrane Biology*, **6**, 259–68.

GOTTSCHLICH, R. (1976). Der Einfluss von Licht und Sauerstoff auf den Energiegehalt des Adenylatsystems sowie auf den Phosphatspiegel von *Halobacterium halobium*. Diploma, Würzburg.

GRÄBER, P. & WITT, H. T. (1976). Relations between the electrical potential, pH gradient, proton flux and phosphorylation in the photosynthetic membrane. *Biochimica et Biophysica Acta*, **423**, 141–63.

HAPPE, M. & OVERATH, P. (1976). Reconstitution of light-driven proton pump from bacteriorhodopsin depleted of purple membrane lipids. In *Tenth International Congress on Biochemistry, Hamburg*. Abstract No. 06-2-206.

HENDERSON, R. & UNWIN, P. N. T. (1975). Three-dimensional model of purple membrane obtained by electron microscopy. *Nature New Biology*, **257**, 28–32.

HESS, B., KUSCHMITZ, D. & OESTERHELT, D. (1976). Photochemical reaction of the 412 nm chromophore state of bacteriorhodopsin. In *Tenth International Congress on Biochemistry*, Hamburg. Abstract No. 06-2-206.

HEYN, M. P., BAUER, P.-J. & DENCHER, N. A. (1975). A natural CD label to probe the structure of the purple membrane from *Halobacterium halobium* by means of exciton coupling effects. *Biochemical and Biophysical Research Communications*, **67**, 897–903.

HILDEBRAND, E. & DENCHER, N. (1975). Two photosystems controlling behavioural responses of *Halobacterium halobium. Nature New Biology*, **257**, 46–8.

HUBBARD, J. S., RINEHART, C. A. & BAKER, R. A. (1976). Energy coupling in the active transport of amino acids by bacteriorhodopsin-containing cells of *Halobacterium halobium. Journal of Bacteriology*, **125**, 181–90.

KANNER, B. I. & RACKER, E. (1975). Light-dependent proton and rubidium translocation in membrane vesicles from *Halobacterium halobium*. *Biochemical and Biophysical Research Communications*, **64**, 1054–61.

KRIPPAHL, G. & OESTERHELT, D. (1973). Light inhibition of respiration in *Halobacterium halobium*. *FEBS Letters*, **36**, 72–6.

KUSHNER, D. J., BAYLEY, S. T., BORING, J., KATES, M. & GIBBONS, N. E. (1964). Morphological and chemical properties of cell envelopes of the extreme halophile, *Halobacterium cutirubrum*. *Canadian Journal of Microbiology*, **10**, 483–97.

LANYI, J. K. & MACDONALD, R. E. (1976). Light-dependent cation gradients and electrical potential in *Halobacterium halobium* cell. *Federation Proceedings*, in press.

LANYI, J. K., RENTHAL, R. & MACDONALD, R. E. (1976). Light-induced glutamate transport in *Halobacterium halobium* envelope vesicles. II. Evidence that the driving force is a light-dependent sodium gradient. *Biochemistry*, **15**, 1603–10.

LANYI, J. K. & SILVERMAN, M. P. (1972). The state of binding of intracellular K^+ in *Halobacterium cutirubrum*. *Canadian Journal of Microbiology*, **18**, 993–5.

LARSEN, H. (1967). Biochemical aspects of extreme halophilism. *Advances in Microbial Physiology*, **1**, 97–132.

MACDONALD, R. E. & LANYI, J. K. (1976). Light-activated amino acid transport in *Halobacterium halobium* envelope vesicles. *Federation Proceedings*, in press.

MICHEL, H. & OESTERHELT, D. (1976). Light-induced changes of the pH gradient and the membrane potential in *H. halobium*. *FEBS Letters*, **65**, 175–8.

OESTERHELT, D. (1974). Bacteriorhodopsin as a light driven proton pump. In *Membrane Proteins in Transport and Phosphorylation*, ed. G. S. Azzone, M. Klingenberg & N. Siliprandi, pp. 79–84. Amsterdam: Elsevier.

OESTERHELT, D. (1975). The purple membrane of *Halobacterium halobium*: a new system for light energy conversion. *Ciba Foundation Symposia*, **31**, 147–67.

OESTERHELT, D. (1976). Bacteriorhodopsin as an example of a light-driven proton pump. *Angewandte Chemie*, **15**, 17–24.

OESTERHELT, D. & CHRISTOFFEL, V. (1976). Reconstitution of a proton pump. *Biochemical Society Transactions*, **4**, 556–9.

OESTERHELT, D. & SCHUHMANN, L. (1974). Reconstitution of bacteriorhodopsin. *FEBS Letters*, **44**, 262–5.

RENTHAL, R. & LANYI, J. K. (1976). Light-induced membrane potential and pH gradient in *Halobacterium halobium* envelope vesicles. *Biochemistry*, **15**, 2136–43.

ROTTENBERG, H., GRUNWALD, T. & AVRON, M. (1972). Determination of ΔpH in chloroplasts. 1. Distribution of ^{14}C methylamine. *European Journal of Biochemistry*, **25**, 54–63.

SHERMAN, W. V., SLIFKIN, M. A. & CAPLAN, S. R. (1976). Kinetic studies of phototransients in bacteriorhodopsin. *Biochimica et Biophysica Acta*, **423**, 238–48.

STOECKENIUS, W. & KUNAU, W. H. (1968). Further characterization of particulate fractions from lysed cell envelopes of *Halobacterium halobium* and isolation of gas vacuole membranes. *Journal of Cell Biology*, **38**, 337–57.

SUMPER, M. & HERRMANN, G. (1976). Biogenesis of purple membrane: regulation of bacterio-opsin synthesis. *FEBS Letters*, in press.

SUMPER M., REITMEIER, H. & OESTERHELT, D. (1976). Biosynthesis of the purple membrane of *Halobacteria*. *Angewandte Chemie* (*International edn*), **15**, 187–94.

YAGUZHINSKY, L. S., BOGUSLAVSKY, L. I., VOLKOV, A. G. & RAKHMANINOVA, A. B. (1976). Synthesis of ATP coupled with action of membrane protonic pumps at the octane–water interface. *Nature New Biology*, **259**, 494–5.

COUPLING OF ENERGY WITH ELECTRON TRANSFER REACTIONS IN CHEMOLITHOTROPHIC BACTERIA

M. I. H. ALEEM

Thomas Hunt Morgan School of Biological Sciences,
University of Kentucky, Lexington, Kentucky, USA

INTRODUCTION

Chemolithotrophic bacteria are uniquely characterized by their ability to extract energy from the oxidation of inorganic compounds and synthesize cellular components solely from carbon dioxide. Thus they are able to grow in a completely inorganic mineral salt medium containing their specific electron donor, small quantities of phosphate, magnesium, iron, a nitrogen source such as ammonium sulfate, and carbon dioxide or bicarbonate. In most cases, molecular oxygen serves as the final electron acceptor but a few chemolithotrophs can respire on nitrate in place of oxygen under anaerobic conditions. When both oxygen and nitrate are present in the growth medium, oxygen is preferentially utilized as the terminal acceptor. Most of the chemolithotrophs are facultative, while a few are still considered as obligate chemolithoautotrophs because of their inability to grow in the absence of the oxidizable inorganic substrate and to synthesize cellular material completely from the organic carbon compounds. The latter may, however, be incorporated to a limited extent into the newly synthesized cells, but the bulk of the cellular components is derived from carbon dioxide which is assimilated via the ribulose-bisphosphate carboxylase cycle involving two key enzymes, namely phosphoribulokinase and ribulose-bisphosphate carboxylase. The presence of these enzymes characterizes and differentiates chemolithotrophs from the heterotrophic organisms.

The principal mode of energy generation in all of the chemolithotrophic bacteria is through oxidative phosphorylation. However, in some species of sulfur-oxidizing bacteria such as *Thiobacillus thioparus*, *T. denitrificans*, and *T. ferrooxidans*, substrate-level phosphorylation may contribute significantly to the overall energy production under chemolithotrophic growth on inorganic sulfur compounds. The endergonic carbon dioxide-based biosynthesis in the chemolithotrophs is tightly coupled to the exergonic inorganic-oxidation reactions responsible for the generation of energy through the electron transport chain, the efficiency of which appears to be under the direct control of carbon dioxide and its reductive assimilation. The latter process is driven at the expense of ATP and NADH, both of which are generated within the electron transport chain during the electron transfer from inorganic electron donor to the final electron acceptor, which may be molecular oxygen under aerobic conditions or, in some cases, nitrate under anaerobic conditions. With the exception of molecular hydrogen, the redox potentials of the inorganic substrates employed by the chemolithotrophs are much more positive than the redox potentials of the pyridine nucleotides, and thus the latter cannot be reduced directly because the generation of a strong reductant such as NADH cannot be achieved directly by the electrons of weak reductants such as nitrite, ferrous iron, ammonium or hydroxylamine, and reduced or partially oxidized inorganic sulfur compounds. It is the oxidation–reduction potentials of these substrates that evidently determine the number of ATP molecules produced per pair of electrons transferred to molecular oxygen; in all cases the reduction of NAD^+ by these electron donors is energy-dependent and involves reversal of the oxidative phosphorylation sequence either from the level of cytochromes or flavoprotein, depending upon the substrate utilized. Quite obviously the functions of the electron transport chain with respect to the forward and reverse electron flows are regulated by the phosphorylation state of the adenine nucleotides and the ratio of oxidized to reduced state of the pyridine nucleotides. The associated energy transduction systems are likewise controlled and regulated by the same factors. Thus the simplicity of the electron transport chain in terms of energy generation, and the complexity of its operation to shuffle electrons back and forth, have attracted the imaginations of many able investigators. As a result, a considerable progress has been made during the past two decades concerning the intimate details of the intermediary metabolism of chemolithotrophic organisms. Since a number of excellent reviews dealing with the chemolithotrophic energy metabolism have appeared in recent years (Gibbs & Schiff, 1960; Peck, 1968; Wallace & Nicholas, 1969; Trudinger, 1969; Aleem, 1970; Kelly, 1971; Suzuki, 1974; Schlegel, 1975), only the more recent work dealing with the nitrifying bacteria, sulfur-oxidizing bacteria and hydrogen-oxidizing bacteria will be discussed in this presentation.

NITRIFYING BACTERIA

Energy transduction in Nitrobacter species

The unique features of chemolithotrophic metabolism in *Nitrobacter* are the mechanisms involved in nitrite oxidation and the coupled generation of energy and reducing power, both of which are essential to drive the carbon dioxide-based cell biosynthesis. Each of these aspects will be considered separately.

Nitrite oxidation

The first observation concerning the reduction of cytochromes 551 and 589 in nitrite-treated whole cells of *Nitrobacter* was made by Lees & Simpson (1957). Butt & Lees (1958) subsequently proposed the following scheme for nitrite oxidation:

$$\text{cyt}_{551} \cdot \text{Fe}^{3+} + \text{NO}_2^- \rightarrow \text{cyt}_{551} \cdot \text{Fe}^{2+} + \text{'NO}_2\text{'}$$
$$\text{cyt}_{551} \cdot \text{Fe}^{2+} + \text{'NO}_2\text{'} + \tfrac{1}{2}\text{O}_2 \rightarrow \text{cyt}_{551} \cdot \text{Fe}^{3+} + \text{NO}_3^-$$

According to this scheme, nitrite oxidation is mediated by cytochrome c, and nitrate is formed by the incorporation of molecular oxygen into nitrite. Soon after, Aleem & Nason (1959) characterized 'nitrite oxidase' as the cytochrome-electron-transport particle which catalyzed electron transport from nitrite to molecular oxygen as follows:

$$\text{NO}_2^- \rightarrow \text{NO}_2^-\text{-cyt } c \text{ reductase} \rightarrow \text{cyt } c \rightarrow$$
$$\text{cyt } a_1 \rightarrow \text{cyt oxidase} \rightarrow \text{O}_2$$

Since the electrons from nitrite reduce the terminal acceptor to water during nitrite oxidation according to the above scheme, it would appear highly unlikely that nitrite is oxidized at the expense of molecular oxygen. Thus, by using ^{18}O-labelled oxygen in molecular oxygen and in water, Aleem, Hoch & Varner (1965) provided unequivocal evidence that nitrite was oxidized by whole cells or cell-free extracts at the expense of the oxygen atom from water and not molecular oxygen. These workers demonstrated, in addition, that nitrite could also be oxidized in the absence of molecular oxygen with added ferricyanide as the electron acceptor.

In an earlier study, Kiesow (1964) claimed that chlorate can replace molecular oxygen for NADH oxidation and coupled phosphorylation in the reaction

$$\text{NADH} + \tfrac{1}{2}\text{O}_2 + m\text{ADP}^{3-} + m\text{HPO}_4^{2-} + (m+1)\text{H}^+ \rightleftharpoons \text{NAD}^+ +$$
$$m\text{ATP}^{4-} + (m+1)\text{H}_2\text{O}$$

under anaerobic conditions, and that nitrite is oxidized by the products of this reaction with concomitant generation of NADH and nitrate as shown below:

$$NAD^+ + NO_2^- + nATP^{4-} + (n+1)H_2O \rightleftharpoons NADH + NO_3^- + \\ nADP^{3-} + nHPO_4^{2-} + (n+1)H^+.$$

The values of m and n are 3 and 2, respectively. According to Kiesow's scheme, nitrite oxidation and coupled pyridine nucleotide reduction consume two ATPs and oxidation of the reduced pyridine nucleotide by molecular oxygen generates three ATPs. Thus he concludes that the oxidation of nitrite does not generate energy but rather it is energy-dependent. He also concludes that the third oxygen atom in nitrate originates from water. He further states, 'The assimilation of energy from inorganic sources begins with the disintegration of water into protons and electrons. The splitting of water might be achieved with a variety of energy sources, including light as it acts upon a suitable photosensitive compound. We have already observed that light in the presence of FMN performs the reduction of DPN in *Nitrobacter* particles. These experiments will be described in a later publication.' Unfortunately no experimental data have so far been provided by Kiesow to substantiate these rather unusual claims.

Based on the ^{18}O incorporation from water into nitrite, Aleem (1970) proposed the following mechanism of nitrite oxidation:

$$NO_2^- + H_2^{18}O \rightarrow NO_2^- \cdot H_2^{18}O$$
$$NO_2^- \cdot H_2^{18}O + 2 \text{ cyt } a_1 \cdot Fe^{3+} \rightarrow N^{18}O_3 + 2 \text{ cyt } a_1 \cdot Fe^{2+} + 2H^+$$
$$2 \text{ cyt } a_1 \cdot Fe^{2+} + 2 \text{ cyt } aa_3 \cdot Fe^{3+} \rightarrow 2 \text{ cyt } a_1 \cdot Fe^{3+} + 2 \text{ cyt } aa_3 \cdot Fe^{2+}$$
$$2 \text{ cyt } aa_3 \cdot Fe^{2+} + 2H^+ + \tfrac{1}{2}O_2 \rightarrow 2 \text{ cyt } aa_3 \cdot Fe^{3+} + H_2O$$

Sum: $NO_2^- + \tfrac{1}{2}O_2 \rightarrow NO_3^-$

The above scheme indicates several important aspects of nitrite oxidation: namely, the hydration of the nitrite molecule prior to removal of electrons from nitrite and protons from water, cytochrome a_1 as the primary electron acceptor from nitrite, the incorporation into nitrite of the oxygen atom from water, the generation of the proton current, and finally transport of electrons and protons to molecular oxygen with the mediation of cytochrome oxidase components. Thus the protonmotive force generated by the nitrite-dependent proton-translocating aniso-tropic respiratory chain involving the cytochrome oxidase loop may be thought of as conserving energy in the terminal segment of the electron transport chain.

Involvements of cytochromes in nitrite oxidation

Investigations by Sewell, Aleem & Wilson (1972) have revealed that *Nitrobacter* contains at least five cytochromes. These are cytochrome c (midpoint potential at pH 7.0, $E_{m\,7.0} = 274$ mV), cytochrome a ($E_{m\,7.0} = 240$ mV), cytochrome a_3 ($E_{m\,7.0} = 400$ mV), and two a_1-type cytochromes ($E_{m\,7.0} = 352$ mV and $E_{m\,7.0} = 100$ mV). Neither of the a_1 cytochrome components binds azide or carbon monoxide, and the midpoint potentials of these components as well as cytochrome c showed no pH-dependence. When carbon monoxide binds with cytochrome a_3, the half-reduction potential of cytochrome a becomes more positive, from 240 mV to 270 mV. In *Nitrobacter* the cytochromes a and a_3 have a strong heme–heme interaction and are very similar in midpoint potentials, spectral properties and reactivities to those of cytochromes a and a_3 of rat-liver and pigeon-heart mitochondria. Essentially similar midpoint potentials of the five cytochromes were reported independently by Cobley (1973). However, Ingledew, Cobley & Chappell (1974) observed that, although the oxidation–reduction potentials of cytochromes c, a, and a_3 showed no pH-dependence, the high-potential component of cytochrome a_1 had a gradient of approximately 25 mV/pH unit and the lower-potential component a gradient of 50 mV/pH unit. In both cases the midpoint potentials became more positive as the pH decreased, indicating that protons facilitated the reduction of cytochrome a_1 components. An analogous situation was observed (D. L. Sewell and M. I. H. Aleem, unpublished data) with respect to the midpoint potential of the nitrite/nitrate couple in the presence of *Nitrobacter* cell-free extracts (Table 1). The E_m becomes more positive from pH 9.0 ($E_m = 328$ mV) to pH 6.8 ($E_m = 414$ mV). The observed optimal pH for nitrite oxidation is around pH 8.0. The E_m for nitrite and cytochrome a_1-high-potential component at this pH would be 360 mV and 327 mV respectively. It is thus apparent that the reduction of the high-potential a_1 component by nitrite poses no thermodynamic barrier. However, since the E_m of cytochrome c shows no pH-dependence, its reduction by nitrite or cytochrome a_1 would be endergonic.

Considerable evidence has accumulated during recent years that the prime site of entry of nitrite in the electron transport chain is cytochrome a_1 and that the reduction of cytochrome c by nitrite is energy-dependent (Kiesow, 1967; Aleem, 1967, 1968; Sewell & Aleem, 1969; Cobley, 1973; Ingledew *et al.*, 1974; Cobley, 1976*b*; W. J. Ingledew, personal communication). Although O'Kelley, Becker & Nason (1970) have claimed that the reduction of cytochrome c by nitrite occurs

Table 1. *The pH dependence of the midpoint potentials of the nitrite/nitrate couple and the cytochrome a_1 components*

The 20 000g supernatant fraction was dialyzed against two changes of 100-fold volumes containing 1 mM $MgCl_2$, 10 μM glutathione (reduced) and 50 mM of the appropriate buffer. The reaction mixture contained 7–8 mg of protein ml^{-1} and 100 μM KCN. Alternately the reaction mixture minus KCN was evacuated in Thunburg-type cuvettes. The absorbance change at 550–540 nm was titrated reductively (nitrite addition) and oxidatively (nitrate addition). The E_h of the reaction mixture was calculated from the reduction of cytochrome c. The pH was obtained by using 100 mM phosphate (pH 6.3–7.3) or 100 mM Tris(hydroxymethyl)-aminomethane–HCl (pH 7.3–9.0) as buffers. The calculations of the E_m of the cytochrome a_1 components are based on the observations of Ingledew, Cobley & Chappell (1974). The $E_{m\,7.0}$ values are used from the data of Sewell, Aleem & Wilson (1972). The E_m values for nitrite/nitrate are taken from unpublished data of D. L. Sewell and M. I. H. Aleem.

	Midpoint potential (mV)		
pH	NO_2^-/NO_3^-	High $E_m a_1$	Low $E_m a_1$
Phosphate 6.8	414	357	110
Phosphate 7.3	390	345	85
Tris–HCl 7.3	394	345	85
Tris–HCl 7.6	381	337	70
Tris–HCl 8.0	360	327	50
Tris–HCl 9.0	328	302	0

without any input of energy, their own data, along with their criticism, will be discussed later under the section 'Generation of reducing power'. In the light of the interesting observations by the above-mentioned investigators, the pathway of electron transfer may be presented according to the scheme in Fig. 1.

The essential features of this scheme are the same as reported earlier (Aleem, 1968, 1970). Some modifications have been introduced because of the vectorial arrangements of the electron carriers and the aniso-tropic nature of the electron transport chain (Cobley, 1976a; W. J. Ingledew, personal communications). According to this scheme, cyto-chrome c attains energization by the protonmotive force (Δp) generated by proton-translocation as the electrons from a hydrated nitrite molecule couple with cytochrome a_1, which in turn reduces cytochrome c; con-comitant with this is the incorporation of the oxygen atom from water into nitrite to yield the end-product, nitrate. The subsequent transfer of electrons from cytochrome c to oxygen, mediated by cytochrome a and a_3, is coupled to ATP generation catalyzed by the reversible action of the membrane-localized anisotropic ATPase.

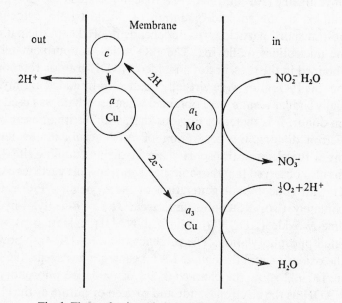

Fig. 1. The mechanism of nitrite oxidation in *Nitrobacter*.

Generation of energy

The first conclusive evidence for the occurrence of oxidative phosphorylation in chemolithotrophic bacteria came from the observed P/O ratios of 0.2, 0.1, 0.05 and 0.03 in *Nitrobacter* particles, concomitant with the respective oxidation of nitrite, succinate, NADH and ascorbate (Aleem & Nason, 1960); the phosphorylation product was ATP, as identified by column chromatography and radioautography. These results have since been confirmed by various investigators (Malavolta, Delwiche & Burge, 1960; Fischer & Laudelout, 1965; Kiesow, 1964; Aleem, 1968; O'Kelley *et al.*, 1970; Cobley & Chappell, 1974; Cobley, 1976*a*). Kiesow (1964) reported that NADH oxidation by *Nitrobacter* particles could yield P/O ratios of 3.0 and $P/2e^-$ ratios of 2.0, with oxygen and nitrate as the respective electron acceptors. Since reversal of the latter reaction involves nitrite oxidation by the pyridine nucleotide, Kiesow was led to the conclusion that the oxidation of nitrite in *Nitrobacter* is obligatorily coupled with NAD^+ reduction, which would require two ATPs, while the oxidation of NADH by molecular oxygen would yield three ATPs; thus he proposed that nitrite oxidation consumes energy. Since nitrite oxidation is the sole energy source for growth and biosyntheses in *Nitrobacter*, the status of oxidative phosphorylation

was reinvestigated (Aleem, 1968). It was demonstrated that ATP generation coupled to the specific oxidation of nitrite by the cytochrome-
electron-transport particles was independent of the participation of
pyridine nucleotides or flavins. The P/O ratios of approximately one
were obtained in the presence of an added NADH-trap or rotenone; in
addition, the P/O ratios were virtually unaffected by added antimycin A.
Essentially similar results were obtained when ascorbate was used as the
electron donor. It is therefore obvious that, with nitrite or ascorbate as
the electron donors, the generation of ATP occurs in the terminal
segment of the electron transport chain involving coupling site 3.

It was also observed that the same electron-transport particles oxidized
NADH with concomitant generation of ATP, yielding P/O ratios of
approximately two. Rotenone, antimycin A or 2n-heptyl-4-hydroxy-
quinoline-N-oxide (HOQNO) inhibited NADH oxidation as well as
associated phosphorylation. The presence of an NADH-trap prevented
both the oxidation and the coupled ATP generation (Aleem, 1968). The
data in Table 2 show the characteristics of oxidative phosphorylation
with NADH as the electron donor and oxygen or nitrate as the electron
acceptor. The P/NO$_3$$^-$ ratio is approximately two-thirds of the P/O ratio.
The range in the concentration of the inhibitors used shows approximately 50% inhibition of the oxygen consumed or nitrate reduced except
for cyanide. In this case, 0.033 mM cyanide caused 53% inhibition of
oxygen uptake and associated ATP generation, while 0.5 mM cyanide
caused only 13% inhibition of nitrate reduction and had no effect on
coupled phosphorylation. Both the aerobic and anaerobic NADH oxidation and coupled phosphorylation were inhibited by about 70% in the
presence of 1 mM azide (data not shown). Carbonylcyanide m-chloro-
phenylhydrazone (CCCP) was the more potent uncoupler of oxidative
phosphorylation as compared with 2,4-dinitrophenol (2,4-DNP). Almost complete uncoupling of oxidative phosphorylation occurred with
either oxygen or nitrate as the terminal acceptor in the presence of 5 μM
CCCP; the corresponding stimulation in the oxygen consumption or
nitrate reduction was 34% and 46%, respectively. In view of these
observations it is clear that the aerobic NADH oxidation and
associated ATP formation involves all of the three coupling sites and
the NADH–nitrate reductase system appears to involve sites 1 and 2.

It might be of interest to mention that attempts by D. L. Sewell and
M. I. H. Aleem to demonstrate acid-induced ATP formation with an
initial exposure of the otherwise phosphorylating *Nitrobacter* particles
to acid pH (2.0–3.8) followed by a rapid increase to pH 8.0 (Jagendorf &
Uribe, 1966; Cole & Aleem, 1973) have not so far been successful.

Table 2. *Phosphorylation coupled to the aerobic and anaerobic oxidation of NADH by Nitrobacter particles*

The reaction mixture (3.0 ml) contained 0.33 mM NADH, 1.0 mM ADP, 2.5 mM phosphate (pH 7.0), 3.0 mM $MgCl_2$, 47 mM Tris–HCl buffer (pH 7.0) and electron transport particles (1.5 mg protein). The reaction vessel was shaken aerobically or anaerobically (continuously flushed with nitrogen gas) at 30 °C and the experiment was terminated 5 min after the addition of NADH. The inhibitors were added 5 min prior to the addition of NADH. The P/O ratios or P/NO_3^- ratios were corrected for endogenous ATP produced, and oxygen or nitrate reduced.

Additions	No. of determinations	Final concn. (mM)	O_2 consumed (ng atom)	ATP formed (nmol)	NO_3^- reduced (nmol)	ATP formed (nmol)	P/O	P/NO_3^-
NADH	5	0.33	223	250	244	172	1.13	0.70
+amytal	4	0.50	136	103	164	100	0.73	0.59
+antimycin A	4	8[a]	76	39	100	56	0.48	0.59
+HOQNO	4	13[a]	164	107	147	62	0.65	0.41
+CN⁻	4	0.033	104	122	—	—	1.19	—
		0.50	—	—	212	172	—	0.82
+CCCP	4	0.005	299	3	356	6	0.01	0.01
+2,4-DNP	4	0.10	181	95	195	94	0.52	0.47

D. L. Sewell and M. I. H. Aleem (unpublished data).
[a] in μg mg⁻¹ protein

Generation of reducing power

The transfer of electrons from nitrite (with the mediation of cytochromes a_1 and c) to either molecular oxygen or to the pyridine nucleotides involves thermodynamic barriers because cytochrome c plays a central role in both the forward and the reverse electron transfer. The reduction of cytochrome a_1 by nitrite is thermodynamically feasible; however, an energy-equivalence of approximately 7 kcal (normally yielded by the hydrolysis of one ATP molecule) appears to be required for the reduction of cytochrome c by nitrite. The interesting observations by Ingledew & Chappell (1975) indicate that cytochrome c is energized in the presence of ATP and the midpoint potential at pH 7.0 is shifted from 274 mV to 360 mV, especially when the redox couple used is nitrite/nitrate. However, it would appear that the protonmotive force generated as a result of proton translocation brought about by the electron transfer from a hydrated nitrite molecule to cytochrome a_1 normally bridges the thermodynamic barrier involved in cytochrome c reduction. Thus, upon energization, cytochrome c can easily equilibrate with the redox state of cytochrome a_1, a phenomenon observed frequently in intact cells and highly coupled vesicular preparations. The oxygen-uptake with nitrite by the *Nitrobacter* particles can be completely inhibited by 0.5 mM 2,4-DNP, 0.05 mM thyroxin, or 0.05 mM dicumarol (Aleem & Nason, 1960); under these conditions the reduction of cytochrome a_1 is not markedly affected but cytochrome c reduction is very significantly suppressed. As a matter of fact, when the *Nitrobacter* particles respire on nitrite and become anaerobic, cytochrome c is oxidized but cytochrome a_1 stays reduced (Aleem, 1968). Cytochrome c can be reduced again by introduction of oxygen (which presumably causes ATP generation), or the reduced state of cytochrome c can be prolonged in the presence of added ATP (Kiesow, 1967). Added mammalian cytochrome c ($E_{m\,7.0}$ 250 mV) can couple with the *Nitrobacter* cytochrome c; however, the reduction of endogenous or exogenous cytochrome c by nitrite is ATP-dependent (Sewell & Aleem, 1969). The proposal by Suzuki (1974) that ATP requirement for cytochrome c reduction by nitrite is due to the efficient removal of nitrate from the cells or vesicles appears to be plausible; however, it cannot account for the observations that cytochrome c is oxidized under anaerobic conditions without ATP while cytochrome a_1 stays reduced. The latter in fact should also be oxidized by nitrate since it mediates electron transfer from cytochrome c to nitrate.

Once cytochrome c is reduced by nitrite, the electrons must gravitate rapidly towards oxygen to generate energy since the reduction of NAD^+ by ferrocytochrome c requires approximately 27 kcal mol^{-1}. The synthesis of ATP requires the transfer of two reducing equivalents over a potential span of 250–300 mV. The overall potential span between nitrite ($E_{m\,7.0} = 420$ mV) and molecular oxygen ($E_{m\,7.0} = 820$ mV) would suggest that cytochrome a_3 ($E_{m\,7.0} = 400$ mV) should be highly oxidized in the aerobic steady state. The terminal oxidase in *Nitrobacter* is cytochrome a_3 (Sewell *et al.*, 1972), and with a second-order rate constant of greater than 10^8 M^{-1} sec^{-1} and an oxygen concentration of 100 μM, a 1 μM concentration of cytochrome a_3 would consume oxygen at the rate of 10 μM O_2 min^{-1}, with a redox potential (E_h) approaching 700 mV. On thermodynamic considerations it is quite feasible to propose that the site 3 of ATP formation in *Nitrobacter* is located between cytochrome c and cytochrome a_3. In harmony with the proposal of Mitchell (1970), the proton translocation as a result of downhill electron transfer reaches an equilibrium with the proton-translocating ATPase since the latter, as well as the proton-translocating electron transport chain, is localized in the same 'nitrite oxidase' particles. At this point the carbon dioxide-based biosynthetic metabolism exerts pressure for the utilization of reducing power which is generated by the ATP-dependent reversal of electron transfer from ferrocytochrome c to NAD^+ with the mediation of the flavoprotein system (Aleem, Lees & Nicholas, 1963; Sewell & Aleem, 1969). Thus the ATP utilization by the NAD^+ reduction and the reductive carbon dioxide assimilation affect the equilibrium between the proton-translocating electron transport chain and the proton-translocating ATPase by lowering the $[ATP]/[ADP] \times [HPO_4^{2-}]$ ratio. The latter factor, in Mitchell's language (1970), 'results in the regeneration of ATP by the circulation of the proton current between the oxidoreduction loop and the ATPase system'. This concept explains elegantly the basis of the control mechanisms in the chemolithotrophs, especially when the proton-translocating respiratory chain is engaged partly in the energy-generating forward electron flow and partly in the energy-driven reverse electron flow.

The coupling of nitrite oxidation with NAD^+ reduction was observed by Kiesow (1963). Since the reduction of NAD^+ by nitrite is an endergonic reaction, nitrite oxidation must serve as the sole energy source. The proposal that nitrite oxidation does not result in the production of energy but rather it consumes energy (Kiesow, 1964, 1967) is inconsistent with the thermodynamics of the reaction; $\Delta F'$ for the aerobic oxidation of nitrite to nitrate is about -18 kcal mol^{-1}. Moreover, this

Table 3. *Effect of electron transport chain inhibitors on various oxidoreductase systems in* Nitrobacter *particles*

The aerobic oxidation of nitrite, ascorbate and NADH was measured polaro-graphically; 5 mM NO_2^- with 0.1 M Tris–HCl (pH 8.0); 5 mM ascorbate with 32 μM mammalian cytochrome c (Sigma horse-heart, type III) and 0.1 M Tris–HCl (pH 8.0); 1 mM NADH with 0.1 M Tris–HCl (pH 7.0); anaerobic oxidation was measured in the presence of 8 mM NO_3^-. In all cases 0.5 to 1.0 mg protein in 144 000g particles was used. Values are given in inhibition percent.

Inhibitor	Final concn (mM)	NO_2^-:O_2 oxidore-ductase	Ascorb-ate-cyt. c:O_2 oxidore-ductase	NADH:O_2 oxidore-ductase	NADH: NO_3^- oxidore-ductase	ATP-dependent NO_2^-: NAD$^+$ oxidore-ductase
Rotenone	0.01	10	18	80	90	100
Amytal	1.0	5	15	74	44	40
Antimycin A	5[b]	0	0	62	52	80
HOQNO	10[b]	0	0	60	30	85
Cyanide	0.05	100	100	75	0	50
Azide	0.10	70	28	23	0	—
CCCP	0.01	60	4	(21)[a]	(78)[a]	75
2,4-DNP	0.10	40	8	8	(9)[a]	100

[a] Percent stimulation.
[b] in μg mg^{-1} protein

reaction has been shown to yield approximately one mole of ATP per mole of nitrite oxidized and it does not involve any participation of pyridine nucleotides or flavins (Aleem, 1968). Since the $\Delta F'$ for the reduction of NAD$^+$ by nitrite and by ferrocytochrome c is 34 kcal mol^{-1} and 27 kcal mol^{-1}, respectively, the thermodynamic barrier cannot be bridged by two ATPs in either case. The reduction of NAD$^+$ by ferrocytochrome c would require at least three ATP equivalents, assuming that the ATP hydrolysis releases 10–12 kcal mol^{-1} in tightly coupled reactions in the cell. Quite obviously the generation of 3ATP/2e$^-$ transferred from NADH to oxygen occurs only under the conditions when oxidative phosphorylation proceeds at an efficiency of about 50–60%, as in the case of whole cells or intact mitochondria. Thus it is hard to envisage the balance of one net-gain of ATP when the NADH oxidase and the nitrite oxidation by the pyridine nucleotides are con-sidered to be tightly linked (Kiesow, 1964). Using cell-free preparations capable of catalyzing active phosphorylation with nitrite or NADH, Sewell & Aleem (1969) have observed the utilization of 4–6 ATP equivalents per equivalent of NAD$^+$ reduced by nitrite. There was a definite sequence in the transfer of electrons from nitrite to NAD$^+$. The initial reaction is the reduction of cytochrome a_1 by nitrite. The nitrite–

cytochrome a_1 reductase probably contains molybdenum and is inhibited by cyanide. The reduction of cytochrome c is energy-dependent and the reaction is linear with time in the presence of added ATP. The reversed electron flow from ferrocytochrome c to NAD^+ is mediated by the flavoproteins and is inhibited by rotenone and amytal. The reduction of added FMN by nitrite or ferrocytochrome c is ATP-dependent. The marked inhibition caused by antimycin A or HOQNO indicates the possible mediation of coupling site 2. CCCP and 2,4-DNP cause a strong inhibition of the energy-linked NAD^+ reduction. Oligomycin effectively blocks the energy transfer from ATP to drive electrons from nitrite to NAD^+. Because of the similar inhibition pattern of the NADH oxidase and the ATP-driven nitrite–NAD^+ reductase system, the latter appears to involve a reversal of the oxidative phosphorylation sequence (Table 3), which can be represented by the following scheme:

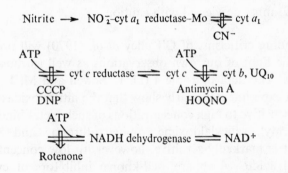

O'Kelley *et al.* (1970) reported that the reduction of cytochrome c catalyzed by the enzyme nitrite–cytochrome c reductase in *Nitrobacter* is energy-independent and claimed that the entry of nitrite into the electron transport chain occurs prior to or at cytochrome c. In addition, they criticized the data of Aleem (1968) and Sewell & Aleem (1969) with respect to the observed energy-linked reactions and the experimental techniques employed. The main points raised by O'Kelley *et al.* (1970) can be summarized as follows:

1. The failure of nitrite to reduce cytochrome c in the presence of 0.5 mM cyanide is not due to a block in energy generation but is explained by the extreme sensitivity of the nitrite–cytochrome c reductase to metal-binding agents.

2. The difference spectra obtained in the presence of added nitrite and 2,4-DNP exhibited no effect on the reduction of cytochromes a_1 and a_3 but markedly suppressed the reduction of cytochrome c; these spectra

are very similar to those obtained 5 min after the addition of nitrite in the absence of uncoupler.

3. Crude extracts were used to show ATP-dependent reversal of electron transfer from cytochrome a_1 to c, and the experiments were monitored solely by following rates of change in absorbance of cytochrome c and a_1.

4. Several important controls are lacking, in particular a reaction mixture with added ATP but without the substrate, nitrite. The latter control experiment is of particular importance in view of the apparent initial stimulation by ATP of cytochrome c reduction in the complete reaction mixture. Does a similar effect occur without the addition of nitrite?

5. The tendency to attribute the pronounced trough at 450 nm and 465 nm in the difference spectra of nitrite oxidase preparations to bleaching of a flavin component (Van Gool & Laudelout, 1966; Sewell & Aleem, 1969) must be viewed with caution.

The foregoing criticisms of O'Kelley et al. (1970) will now be considered in the light of our own observations as well as those by other investigators. The data in Table 4 (D. L. Sewell and M. I. H. Aleem, unpublished experiments) clearly show that the nitrite oxidase system is relatively insensitive to high concentrations of such metal-binding agents as α,α-dipyridyl, salicylaldoxime, o-phenanthroline and 8-hydroxyquinoline; it is markedly sensitive, however, to low concentrations of cyanide and azide, which are well-known inhibitors of cytochrome oxidase. An inhibition of about 50 % by 10 μM CCCP can hardly be attributed to its metal-binding effect. In addition, O'Kelley et al.'s own data (1970) clearly show that a 1 mM concentration of 2,4-DNP (which is 100-fold the concentration of CCCP) causes only 12 % inhibition of their 'nitrite–cytochrome c reductase'. The inhibition of the reduction by nitrite of the endogenous cytochromes caused by a low concentration (e.g. 80 μM) of 2,4-dibromophenol (Aleem, 1968) does not appear to be due to its metal-binding capacity. As a matter of fact, the data of O'Kelley et al. (1970) further show only about 20–30 % inhibition of their 'nitrite–cytochrome c reductase' by non-physiologically high concentrations of 5 mM each of o-phenanthroline and 8-hydroxyquinoline.

The fact that the difference spectra obtained in the presence of added nitrite and 80 μM 2,4-dibromophenol exhibit no effect on the reduction of cytochrome a_1 but markedly suppress the reduction of cytochrome c (Aleem, 1968) constitutes important evidence for the energy-linked coupling of nitrite with cytochrome c. This is in harmony with the

Table 4. *Effect of metal-binding agents on nitrite oxidase*
in Nitrobacter *particles*

Experimental conditions were similar to those in Table 3. Particles were incubated
with inhibitor for 5 min prior to nitrite addition.

Inhibitor	Final concentration		Inhibition (%)
	mM	μmol mg^{-1} protein	
8-Hydroxyquinoline	1.0	5.8	10
o-Phenanthroline	5.0	29.0	11
Salicylaldoxime	5.0	29.0	18
α,α-Dipyridyl	5.0	29.0	8
Cyanide	0.05	0.29	100
Azide	0.05	0.29	50
CCCP	0.01	0.058	53

observations of Kiesow (1967), who demonstrated that upon anaerobiosis cytochrome c is oxidized while cytochrome a_1 stays reduced in
nitrite-treated *Nitrobacter* particles. The redox state of cytochrome c
could be restored either by introduction of oxygen or by addition of
ATP; however, in the presence of added uncoupler, neither oxygen nor
ATP restored cytochrome c reduction by nitrite.

Unquestionably nitrite is the source of electrons for the ATP-
dependent reduction of cytochrome c, added flavin, or added NAD$^+$
(Sewell & Aleem, 1969). The reduction of these electron acceptors was
measured in a double-beam spectrophotometer and the difference
absorption was recorded in the absence and presence of added ATP or
nitrite. The details of these experiments have unfortunately been overlooked by O'Kelley *et al.* (1970). The data clearly show that the reduction of cytochrome c by nitrite was ATP-dependent. When the nitrite
was omitted from the reaction mixture, containing ATP, the reaction
levelled off quickly and very little reduction or oxidation of cytochrome
c and the associated ATP-dependent reduction of NAD$^+$ could be
observed. Finally the remark made by O'Kelley *et al.* (1970) that the
decrease in absorbance at 450 nm of crude cell-free preparations of
N. agilis may not be due to a flavin component is true only when no
flavin is added. However, we have measured the ATP-dependent reduction of *added* FMN. The validity of spectrophotometric techniques for
measuring the oxido-reduction of cytochromes and added flavin or
pyridine nucleotides can hardly be questioned. As mentioned earlier,
several investigators have observed that the reduction of cytochrome c
by nitrite is energy-dependent because of the strong inhibition caused by
the uncouplers both on nitrite oxidation as well as cytochrome c

reduction. The oxidation of cytochrome c reduced either by NADH or ascorbate is unaffected by the uncoupling agents.

In view of the foregoing discussion, an important question can be raised, 'Is there any nitrite–cytochrome c reductase?' The answer will be subject to speculations in view of the experimental data presented by O'Kelley *et al.* (1970). The reaction mixture used by these investigators contained 50 mM nitrite, 160 μM mammalian cytochrome c, and enzyme preparation. The reaction is linear only for 90 sec. The reaction levels off at higher enzyme concentrations. Assuming a mM extinction coefficient of mammalian cytochrome c as 19, the fully reduced 160 μM cytochrome c used in the assay for nitrite–cytochrome c reductase should exhibit an absorption of 3.04. The observed absorption of 0.09 per 3.0 ml reaction mixture corresponds to the reduction of only 5 μM cytochrome c. If the product of the reaction was nitrate, which these workers did not report, it can be in the order of 2.5 μM nitrate. This amount is far too low to cause any inhibition of the so-called 'nitrite–cytochrome c reductase' which, according to the data of O'Kelley *et al.*, is inhibited by about 58% at a 2000-fold higher (e.g. 5 mM) concentration of nitrate. Since the very rapid levelling off of the reaction after 90 sec (and even faster at higher enzyme concentrations) results in the reduction of only one-thirtieth of the added cytochrome c, the reaction can hardly be free from experimental artefact, especially when the measurements were made in a single-beam instrument using rather high concentrations of nitrite and exogenously added mammalian cytochrome c. Evidently further work is highly essential to establish the status of 'nitrite–cytochrome c reductase', which has been claimed by O'Kelley *et al.* (1970) to be non-energy-dependent.

Energy transduction in Nitrosomonas species

Among the chemolithotrophic organisms, the energy-transduction mechanisms are least understood in the *Nitrosomonas* species. The primary energy-yielding reaction for the growth of *Nitrosomonas* is the oxidation of ammonia to nitrite:

$$NH_3 + 1\tfrac{1}{2}O_2 \rightarrow NO_2^- + H_2O + H^+$$
$$\Delta F' = -65 \text{ kcal mol}^{-1}$$

Addition of ammonia or hydroxylamine to intact cells results in the reduction of cytochromes of the b, c, and a types (Aleem & Lees, 1963; Falcone, Shug & Nicholas, 1962). The cytochrome system in the cell-free extracts is reduced by hydroxylamine but not by ammonia. The soluble fraction of the cell-free extracts contains cytochrome o and cytochrome P450 (Rees & Nason, 1965) or cytochrome P460 (Erickson & Hooper, 1972). Both the cytochromes combine with carbon monoxide and their properties are similar to the mammalian or *Pseudomonas* cytochrome P450. The *Nitrosomonas* cytochrome P460 is suggested to be involved in the hydroxylation of ammonia to yield hydroxylamine and in the oxidation of the dehydrogenation product of hydroxylamine (e.g. NOH) to produce nitrite (A. B. Hooper, personal communications; Suzuki, 1974). However, the experimental evidence for the involvement of cytochrome P460 in hydroxylation or oxygenation reactions in *Nitrosomonas* is lacking and the actual mechanism of oxidation of ammonia to hydroxylamine is still obscure. The redox potential of the ammonia/hydroxylamine couple is so electropositive ($E_0' = 899$ mV) that none of the electron transport chain components can be directly reduced by ammonia (Aleem, 1970); and moreover, the reaction is slightly endergonic:

$$NH_3 + \tfrac{1}{2} O_2 \rightarrow NH_2OH$$
$$\Delta F' = 3.85 \text{ kcal mol}^{-1}$$

The only energy-yielding step is the oxidation of hydroxylamine to nitrite:

$$NH_2OH + O_2 \rightarrow NO_2^- + H_2O + H^+$$
$$\Delta F' = -68.89 \text{ kcal mol}^{-1}$$

Since the redox potential of the hydroxylamine/nitrite couple is 66 mV (Aleem, 1970), it can couple with the electron transport chain at the flavoprotein level. As a matter of fact, hydroxylamine oxidation has been observed to be sensitive to the flavoprotein inhibitors (Falcone *et al.*, 1962) and added FMN completely reversed the inhibition caused by atabrine (Aleem & Lees, 1963). On the basis of experimental observations with whole cells and cell-free preparations oxidizing ammonia or hydroxylamine, the following pathway of electron transport can be proposed:

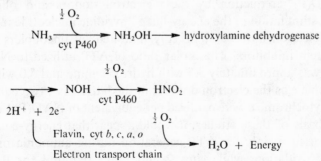

Since the redox potential of hydroxylamine is similar to that of the succinate/fumarate couple, at least two energy-coupling sites should be involved in the transport of electrons to molecular oxygen. The experimental evidence for this is lacking and the P/O ratios of 0.2 (Ramaiah & Nicholas, 1964) are too low to arrive at any meaningful conclusion. Interestingly enough, the H^+/O ratio of 2 (J. W. Drozd, personal communication) indicates that only one site is involved in proton translocation with either ammonia or hydroxylamine as the electron donor. Most probably the bulk of the electrons couple at the cytochrome c level and the ATP is generated only in the terminal portion of the electron transport chain.

Since *Nitrosomonas* is considered to be an obligate chemolithotroph, it must derive its reducing power from the oxidation of ammonia or hydroxylamine to achieve reductive carbon dioxide assimilation. The direct reduction of pyridine nucleotides by either ammonia or hydroxylamine is thermodynamically very unfavourable (Gibbs & Schiff, 1960):

$$NH_4^+ + OH^- + NAD(P)^+ \rightarrow NH_2OH + NAD(P)H + H^+$$
$$\Delta F' = 47 \text{ kcal mol}^{-1}$$

$$NH_2OH + H_2O + 2NAD(P)^+ \rightarrow NO_2^- + 2NAD(P)H + 3H^+$$
$$\Delta F' = 110 \text{ kcal mol}^{-1}$$

Since there is no ammonia dehydrogenase, and ammonia cannot couple with the electron transport chain, it is quite unlikely that pyridine nucleotides are reduced by the electrons from ammonia. However, the mechanism of pyridine nucleotide reduction by hydroxylamine would be analogous to the reduction of NAD^+ by succinate (Chance, 1961). Thus it was demonstrated (Aleem, 1966c) that cell-free preparations from *Nitrosomonas* catalyzed an active ATP-dependent $NAD(P)^+$ reduction by hydroxylamine, succinate or ferrocytochrome c. The aerobic oxidation of hydroxylamine or succinate also provided energy for $NAD(P)^+$ reduction by these electron donors and oligomycin caused a stimulation in the energy-linked pyridine nucleotide reduction. Reverse electron flow was markedly inhibited by uncouplers and the flavoprotein inhibitors. The molar ratio of ATP utilized to $NAD(P)^+$ reduced was approximately 3.5 with hydroxylamine and 5.0 with ferrocytochrome c as the electron donors; in the latter case, two equivalents of ferrocytochrome c were oxidized per equivalent of $NAD(P)^+$ reduced. On the basis of these studies, it would appear that energy-coupling at site 3 is used mainly for ATP generation during ammonia or hydroxylamine oxidation, while sites 2 and 1 are required for the ATP-driven reverse-electron flow to NAD^+. Undoubtedly further work is essential to establish the number of ATP molecules produced per pair of electrons transferred to oxygen during ammonia or hydroxylamine oxidation.

SULFUR-OXIDIZING BACTERIA

Energy transduction in Thiobacillus species

The biochemical reaction mechanisms involved in the oxidation of sulfur compounds have been discussed in an earlier article (Aleem, 1975). Although the Thiobacilli employ reduced or partially oxidized inorganic sulfur compounds such as sulfide, sulfur or thiosulfate as the energy-yielding substrates for growth, the pathways and the intermediates involved in the biological oxidation of these electron donors have not as yet been well established. The non-enzymic reactivities of the inorganic sulfur compounds as well as occurrence of different sulfur oxidation pathways in various Thiobacilli has been one of the major difficulties in understanding the biochemistry of sulfur oxidation. Although considerable progress has been made in the past few years, more extensive investigations are needed to elucidate sulfur-oxidation pathways and associated mechanisms involved in the generation of energy by oxidative phosphorylation as well as by substrate-level phosphorylation. It is also important to consider the thermodynamics of the generation of reducing power at the expense of the oxidation of various intermediates in the sulfur oxidation pathways in view of the fact that the oxidation of sulfide to sulfate involves the transfer of eight electrons.

In Thiobacilli there are three potential systems of sulfur oxidation. The first two are the energy-yielding reactions involving oxidation of inorganic sulfur compounds by molecular oxygen, and by oxyanions of nitrogen such as nitrate, nitrite, etc., under anaerobic conditions as represented by the following reactions:

$$\Delta F'$$
$$(\text{kcal mol}^{-1})$$

$H_2S + 2O_2 \rightarrow H_2SO_4$	-160
$S^0 + 1\frac{1}{2}O_2 + H_2O \rightarrow H_2SO_4$	-118
$S_2O_3^{2-} + 2O_2 + H_2O \rightarrow 2SO_4^{2-} + 2H^+$	-211
$SO_3^{2-} + \frac{1}{2}O_2 \rightarrow SO_4^{2-}$	-60
$5S_2O_3^{2-} + 8NO_3^- + H_2O \rightarrow 10SO_4^{2-} + 4N_2 + 2H^+$	-893

While all the Thiobacilli catalyze aerobic oxidation of inorganic sulfur compounds, the anaerobic oxidation with nitrate or nitrite as the final electron acceptor occurs only in species of *Thiobacillus denitrificans*.

The third sulfur oxidizing system is highly endergonic, but it is of great physiological significance in the Thiobacilli since it is involved in the generation of reducing power which is essential for reductive carbon dioxide assimilation for cell biosynthesis. The oxidation of sulfur compounds by pyridine nucleotides is exemplified by the following equations (Gibbs & Schiff, 1960):

$$H_2S + 3H_2O + 3NAD(P)^+ \rightarrow$$
$$SO_3^{2-} + 3NAD(P)H + 5H^+ \quad \Delta F' = 186 \text{ kcal mol}^{-1}$$
$$SO_3^{2-} + H_2O + NAD(P)^+ \rightarrow$$
$$SO_4^{2-} + NAD(P)H + H^+ \quad \Delta F' = 34 \text{ kcal mol}^{-1}$$

It is thus clear that the aerobic and the anaerobic sulfur oxidation involving molecular oxygen and nitrate, respectively, must provide all the energy for the generation of reduced pyridine nucleotides and the carbon dioxide-based metabolism of Thiobacilli under chemolithotrophic growth conditions. Some aspects of the energy transduction systems in these organisms shall now be considered.

Oxidative phosphorylation

Interesting studies by Kelly & Syrett (1963, 1966) demonstrating inhibition of carbon dioxide fixation by low concentrations of uncoupling agents without affecting the oxygen uptake during sulfide oxidation by cell suspensions of *Thiobacillus thioparus* led to two important conclusions: (*a*) occurrence of oxidative phosphorylation with an inorganic sulfur compound as the electron donor, and (*b*) the energy-linked generation of reduced pyridine nucleotides required for reductive carbon dioxide assimilation. The oxidation of sulfide, thiosulfate and sulfite is mediated by the cytochrome system (Trudinger, 1969; Aleem, 1965, Charles & Suzuki, 1966). Although the oxidation of sulfite was reported to accompany ATP formation in *T. novellus* (Charles & Suzuki, 1965) and *T. thioparus* (Davis & Johnson, 1967), the P/O ratios were only of the order of 0.1 and, moreover, 2,4-DNP caused only 30 % inhibition of phosphorylation. It is questionable therefore if the low level of ATP formation was due to oxidative phosphorylation. However, the ATP-dependent reversal of electron transfer reactions (Aleem, 1966*a*, *b*) and phosphorylation coupled to the oxidation of NADH, succinate and ascorbate catalyzed by *T. novellus* cell-free extracts (Cole & Aleem, 1970; 1973) established the operation of all of the three energy-coupling sites in the electron transport chain of this organism. The ATP formation coupled to thiosulfate oxidation, yielding P/O ratios of 0.8 in *T. neapolitanus* (Ross, Schoenhoff & Aleem, 1968) and 0.9 in *T. novellus*

(Cole & Aleem, 1970), was markedly sensitive to low concentrations of uncoupling agents and cytochrome oxidase inhibitors, but both the oxidation of thiosulfate and the coupled phosphorylation were not affected by either flavin inhibitors or site 2 inhibitors such as antimycin A or HOQNO. It was therefore concluded that only the terminal site of the electron transport chain was involved in energy generation since the electrons from thiosulfate entered at the cytochrome c level. Interestingly enough while the aerobic thiosulfate oxidation by *T. denitrificans* is unaffected by flavin inhibitors and HOQNO or antimycin A, the anaerobic thiosulfate oxidation coupled to nitrate reduction is severely inhibited by these electron transport chain inhibitors. The thiosulfate molecule appears to undergo cleavage under anaerobic conditions thus enabling the sulfide and the sulfite moieties to couple to the respiratory chain at the flavin level. Thus *T. denitrificans* regulates energy generation by a thiosulfate-cleavage enzyme under anaerobic conditions.

The mechanism of sulfide oxidation in Thiobacilli is not well understood, however, but its oxidation has been shown to be coupled to ATP formation yielding relatively high P/O ratios of 1.4 in *T. novellus* cell-free extracts (Cole & Aleem, 1973). The oxidation of sulfide and the associated phosphorylation is mediated by the conventional coupling sites although electrons from sulfide reduce added FMN ($E_0 = -200$ mV) but not NAD^+. It is questionable at present if energy coupling takes place at site 1 with sulfide as the electron donor since the reduction of NAD^+ is ATP-dependent (Saxena & Aleem, 1972).

Substrate-level phosphorylation

In those organisms such as *T. thioparus* where thiosulfate is cleaved to sulfite and disulfide, the major route of thiosulfate oxidation would

$$SSO_3^{2-} + RSH \rightarrow SO_3^{2-} + RSSH$$

appear to involve the subsequent oxidation of the sulfite and the sulfide (or sulfur) moieties; the latter being oxidized to sulfite:

$$H_2S + \tfrac{1}{2}O_2 \rightarrow S^0 + H_2O$$
$$S^0 + H_2O + O_2 \rightarrow SO_3^{2-} + 2H^+$$

The two sulfite molecules thus produced from thiosulfate can then be oxidized through the adenosine phosphosulfate (APS) pathway as reported by Peck & Fischer (1962):

$$2SO_3^{2-} + 2AMP \rightarrow 2APS + 4e^-$$
$$2APS + 2HPO_4^{2-} \rightarrow 2ADP + 2SO_4^{2-}$$
$$2ADP \rightarrow ATP + AMP$$

Thus each thiosulfate or two sulfite molecules oxidized through the above pathway result in the generation of one ATP molecule at the substrate-level. The four electrons released in the reaction sequence probably couple at the flavin-level and are subsequently transported to molecular oxygen through the cytochrome-electron-transport chain resulting in the generation of four ATP by oxidative phosphorylation:

$$4e^- + 4H^+ + O_2 \xrightarrow[\text{chain}]{\text{Electron transport}} 2H_2O$$

Thus the oxidation of a thiosulfate molecule should generate a total of five ATP molecules, which appears to be in harmony with the yield studies by Hempfling & Vishniac (1967). Thus 20 % of the phosphorylation should be uncoupler-insensitive in Thiobacilli that oxidize thiosulfate or sulfite through the APS pathway.

A variant to the ADP-sulfurylase reaction involved in the formation of ATP and sulfate from APS is the ATP–sulfurylase reaction:

$$S_2O_3^{2-} \xrightarrow{1\frac{1}{2}O_2} 2SO_3^{2-} \rightarrow 4e^- + 2APS \xrightarrow{2PP_i} 2ATP + 2SO_4^{2-}$$

This reaction should yield a total of six ATPs, two at the substrate-level and four by oxidative phosphorylation.

Reducing power

Since the coupling of thiosulfate with the electron transport chain in most of the Thiobacilli occurs at the cytochrome c-level, and of sulfide and sulfite at the flavin-level, these organisms have to generate reducing power in the form of NADH by an energy-driven reversal of electron transfer. The process is inhibited by uncouplers of oxidative phosphorylation, antimycin A or HOQNO and flavin inhibitors; thus the system appears to involve the reversal of the sequence of oxidative phosphorylation reactions. The energy requirements for NAD$^+$ reduction by thiosulfate are 2.5 ATP equivalents per equivalent of NAD$^+$ reduced in *T. neapolitanus*, but only one ATP equivalent in *T. novellus*. The reason for the lower energy requirement in the latter system is not apparent at present. However, several factors may be involved in thiosulfate metabolism such as generation of ATP at the substrate-level, splitting of the thiosulfate molecule to sulfite and disulfide thus permitting their entry at the flavin level, and/or involvement of a low-potential cytochrome as the acceptor of electrons for NAD$^+$ reduction. Quite obviously much more work is needed to understand the mechanism of energy-linked generation of reducing power in the sulfur-oxidizing bacteria.

HYDROGEN-OXIDIZING BACTERIA

The biochemical mechanisms involved in the generation and regulation of energy and reducing power in chemolithotrophically grown hydrogen bacteria are grossly different from each other with respect to their electron transport systems. Broadly speaking there are two types of hydrogenomonads: (1) those possessing a NAD-linked soluble hydrogen dehydrogenase as well as a membrane-bound hydrogenase that is not linked with the pyridine nucleotides, and (2) those possessing the membrane-bound hydrogenase only. Since the latter enzyme couples with the electron transport chain at the ubiquinone–cytochrome b complex, these organisms have to generate reducing power (e.g. NADH) through an energy-driven reverse electron flow.

Hydrogenomonas eutropha possesses both the soluble NAD^+-reducing and the particulate non-NAD^+-reducing hydrogenases (Atkinson & McFadden, 1954; Wittenberger & Repaske, 1961); and the *Hydrogenomonas* species are considered to use the particulate hydrogenase to generate ATP and the soluble to reduce the pyridine nucleotides, both of which are required to drive the carbon reduction cycle involved in cellular biosynthesis (Schlegel, 1975; Forrest & Walker, 1971). Indeed, coupling of phosphorylation in a NAD^+-independent hydrogen-oxidizing system has been reported in a *Hydrogenomonas* strain H-16 (Bongers, 1967).

We have compared the phosphorylation efficiencies and inhibitor responses of the electron transport chains in chemolithotrophically grown *Hydrogenomonas eutropha*, *Paracoccus* (previously *Micrococcus*) *denitrificans* and *Pseudomonas saccharophila* (Ishaque & Aleem, 1970; Knobloch, Ishaque & Aleem, 1971; Ishaque, Donawa & Aleem, 1971a, b; Donawa, Ishaque & Aleem, 1971; Ishaque, Donawa & Aleem, 1973). The data in Table 5 show that the oxidation of hydrogen, NADH, succinate and ascorbate was effectively coupled with ATP generation in *H. eutropha* and *P. denitrificans*, indicating the participation of all of the energy-conservation sites. These results are in harmony with the H^+/O ratios observed in *H. eutropha* (Beatrice & Chappell, 1974). In the case of *Ps. saccharophila*, however, the terminal energy-coupling site appears to be non-functional because of the lack of phosphorylation coupled to ascorbate oxidation. Another peculiarity of this system was that the P/O ratios with NADH were twice those obtained with hydrogen. In the presence of added NAD^+ and an NADH-trap consisting of pyruvate and lactate dehydrogenase (Table 6), the P/O ratios with hydrogen were virtually unaffected in *Ps. saccharophila* but decreased approximately 62 % in *H. eutropha and* 46 % in *P. denitrificans.*

Table 5. *Phosphorylation coupled to the oxidation of various electron donors by cell-free preparations from hydrogen-oxidizing bacteria (see discussion of results for details)*

	P/O ratios		
Substrate	Hydrogenomonas eutropha	Paracoccus denitrificans	Pseudomonas saccharophila
Hydrogen	1.60	1.57	0.66
NADH	1.80	0.95	1.22[b]
Succinate	0.91	1.00	0.15
Ascorbate	0.63	0.37[a]	0.00

[a] In the presence of 10 μM added mammalian cytochrome *c*.
[b] In the presence of generated NADH.

Table 6. *Effect of an NADH-trapping system on phosphorylation by cell-free preparations from hydrogen-oxidizing bacteria*

	P/O ratios		
Treatments	Hydrogenomonas eutropha	Paracoccus denitrificans	Pseudomonas saccharophila
Hydrogen alone	1.45	1.57	0.73
+NADH-trap	1.30	1.14	0.71
+NAD$^+$	1.60	1.25	0.64
+NAD$^+$+NADH-trap	0.55	0.84	0.63

In all cases, the phosphorylation was completely abolished by 10 μM CCCP thus establishing ATP formation by oxidative phosphorylation. The hydrogen-linked ATP generation in *H. eutropha* and *P. denitrificans* was significantly inhibited by 10 μM rotenone while the same rotenone concentration had no effect on hydrogen oxidation and coupled phosphorylation in *Ps. saccharophila*; however, rotenone blocked over 95 % of the ATP generation coupled to NADH oxidation in this organism. Thus the failure of rotenone to inhibit hydrogen-linked ATP generation but not the NADH-linked phosphorylation and the failure of ascorbate oxidation to generate ATP in *Ps. saccharophila* led to the conclusion that hydrogen oxidation and coupled ATP synthesis involved coupling site 2 only, while sites 1 and 2 were involved in energy generation with NADH as the electron donor. This is further strengthened by the fact that 0.8 μg antimycin A abolished phosphorylation with hydrogen or NADH in all of the organisms. In addition, oxidative phosphorylation was observed to be markedly sensitive to 0.1 mM cyanide in all cases.

Table 7. *Comparative aspects of oxidative phosphorylation linked to oxygen and nitrate reduction in autotrophically and heterotrophically grown* Pseudomonas saccharophila *(see discussion of results for details)*

	Autotrophic		Heterotrophic	
Substrate	P/O	P/NO$_3^-$	P/O	P/NO$_3^-$
Hydrogen	0.60	0.56	—	—
NADH	0.72	0.50	0.90	0.68
Succinate	0.17	0.11	0.51	0.32
Ascorbate	0.00	0.00	0.25	0.00

Nitrate respiration and coupled phosphorylation in Ps. saccharophila

Ps. *saccharophila* was adapted to grow autotrophically with hydrogen and heterotrophically with succinate as the electron donor and oxygen or nitrate as the electron acceptor. The P/O and P/NO$_3^-$ ratios with hydrogen were of the same order of magnitude; however, the P/NO$_3^-$ ratios with NADH and succinate were less than the corresponding P/O ratios obtained with cell-free extracts from autotrophically and heterotrophically grown cells (Table 7). Interestingly enough the aerobic ascorbate oxidation was coupled to ATP generation in heterotrophically grown organisms, however, no ATP formation occurred in the case of nitrate-linked ascorbate oxidation indicating that the terminal energy-coupling site is functional with oxygen but nonfunctional with nitrate as the final electron acceptor.

The phosphorylation associated with oxygen or nitrate reduction caused by electron flow from hydrogen or NADH was sensitive to HOQNO, cyanide, oligomycin and CCCP. While 100 μM rotenone had no effect on hydrogen-linked ATP generation in the presence of oxygen or nitrate, the heterotrophic phosphorylation associated with aerobic or anaerobic NADH oxidation was abolished by 0.2 μM rotenone. These observations indicate that hydrogen oxidation with nitrate in autotrophically grown organisms generates ATP at coupling site 2 only, while sites 1 and 2 are functional with NADH; all of the energy-coupling sites become functional in heterotrophically grown cells. In the latter case energy cannot be conserved at site 3 when nitrate serves as the terminal electron acceptor.

CONCLUDING REMARKS

It will be evident from this article that the electron transfer-linked energy-transduction reactions in chemolithotrophic bacteria play a vital role in their life processes. The latter are dependent, in most cases, on the supply of energy made available chiefly by oxidative phosphorylation and in some cases (e.g. a few *Thiobacillus* species) by substrate-level phosphorylation as well. The energy-driven cellular biosynthesis from carbon dioxide is the key reaction in chemolithotrophy which cannot proceed without reduced pyridine nucleotides which are generated by an energy-driven reversed electron transfer in nitrifying bacteria, sulfur-oxidizing bacteria, and those hydrogen-oxidizing bacteria possessing the particulate hydrogenase but lacking the soluble hydrogen dehydrogenase. It is quite obvious therefore that the energy-generating forward electron flow and the energy-consuming reverse electron flow are under the direct control of carbon dioxide. The initial reaction of course is the forward electron flow either towards oxygen or nitrate because unless energy is generated, carbon dioxide reduction for growth cannot be achieved. The overall regulatory process is not simple, however, since it is governed by several cellular parameters such as the inorganic electron donor to carbon dioxide, oxygen or nitrate ratios, the redox state of the NAD and NADP couples in cytoplasmic and membrane compartments, the phosphate potential i.e. $[ATP]/[ADP][HPO_4^{2-}]$, and the energy charge i.e. $[ATP] + \frac{1}{2}[ADP]/[ATP] + [ADP] + [AMP]$. It should thus be important to direct further studies relating to the control by carbon dioxide of the oxidation of inorganic energy-yielding substrates and its relationship to the cellular energy charge and the redox state of the pyridine nucleotide couples.

Although considerable progress has been made, there remain many unsolved problems dealing with the energetics of chemolithotrophic processes. The coupling of nitrite with the *Nitrobacter* electron transport chain and the mechanism of nitrite oxidation is still far from being completely understood. Although the scheme in Fig. 1 is tentative, if the hydroxyl radical of water is incorporated into nitrite, the uncharged HNO_3 would pass through the membrane without additional input of energy. Furthermore, the oxidation–reduction of molybdenum by nitrite has not been well established. Likewise, virtually nothing is known about the mechanism of ammonia oxidation and the role of P460, if any, in the reaction sequence catalyzed by *Nitrosomonas*.

It has still to be established why different *Thiobacillus* species involve different sulfur-oxidation pathways and to what extent are the contributions of oxidative and substrate-level phosphorylations in these species. Finally, it is not known why electron transport chains differ in different hydrogenomonads. In *H. eutropha* and *P. denitrificans*, 70 % and 50 % of the electrons from hydrogen flow through the pyridine nucleotide, while there is a complete bypass in *Ps. saccharophila*. Intriguing is the observation of the loss of energy-coupling site 3 in *Ps. saccharophila* under autotrophic growth, and in *P. denitrificans* under heterotrophic growth. Obviously much more work remains to be done to characterize the electron transport systems induced by growth conditions in many of the chemolithotrophs.

The author is greatly indebted to Dr B. A. Haddock and Dr W. John Ingledew for very stimulating discussions and for providing me with unpublished manuscripts on *Nitrobacter*. This work was supported in part by grants from the National Science Foundation and Biomedical Research Grant Support to the University of Kentucky.

REFERENCES

ALEEM, M. I. H. (1965). Thiosulfate oxidation and electron transport in *Thiobacillus novellus*. *Journal of Bacteriology*, **90**, 95–101.

ALEEM, M. I. H. (1966a). Generation of reducing power in chemosynthesis. III. Energy-linked reduction of pyridine nucleotide in *Thiobacillus novellus*. *Journal of Bacteriology*, **91**, 729–36.

ALEEM, M. I. H. (1966b). Generation of reducing power in chemosynthesis. IV. Energy-linked reduction of pyridine nucleotides by succinate in *Thiobacillus novellus*. *Biochimica et Biophysica Acta*, **128**, 1–12.

ALEEM, M. I. H. (1966c). Generation of reducing power in chemosynthesis. II. Energy-linked reduction of pyridine nucleotides in the chemoautotroph *Nitrosomonas europaea*. *Biochimica et Biophysica Acta*, **113**, 216–24.

ALEEM, M. I. H. (1967). Energy conversions in the chemoautotroph *Nitrobacter agilis*. *Bacteriological Proceedings*, **67**, 112.

ALEEM, M. I. H. (1968). Mechanism of oxidative phophorylation in the chemoautotroph *Nitrobacter agilis*. *Biochimica et Biophysica Acta*, **162**, 338–47.

ALEEM, M. I. H. (1970). Oxidation of inorganic nitrogen compounds. *Annual Review of Plant Physiology*, **21**, 67–90.

ALEEM, M. I. H. (1975). Biochemical reaction mechanisms in sulfur oxidation by chemosynthetic bacteria. *Plant and Soil*, **43**, 587–607.

ALEEM, M. I. H., HOCH, G. E. & VARNER, J. E. (1965). Water as the source of oxidizing and reducing power in bacterial chemosynthesis. *Proceedings of the National Academy of Sciences, USA*, **54**, 869–73.

ALEEM, M. I. H. & LEES, H. (1963). Autotrophic enzyme systems. I. Electron transport systems concerned with hydroxylamine oxidation in *Nitrosolmonas*. *Canadian Journal of Biochemistry and Physiology*, **41**, 763–78

ALEEM, M. I. H., LEES, H. & NICHOLAS, D. J. D. (1963). Adenosine triphosphate-dependent reduction of nicotinamide adenine dinucleotide by ferro-cytochrome *c* in chemoautotrophic bacteria. *Nature, London*, **200**, 759–61.

ALEEM, M. I. H. & NASON, A. (1959). Nitrite oxidase, a particulate cytochrome electron transport system from *Nitrobacter*. *Biochemical and Biophysica-Research Communications*, **1**, 323–7.

ALEEM, M. I. H. and NASON, A. (1960). Phosphorylation coupled to nitrite oxidation by particles from the chemoautotroph, *Nitrobacter agilis*. *Proceedings of the National Academy of Sciences, USA*, **46**, 763–9.

ATKINSON, D. E. & MCFADDEN, B. A. (1954). The biochemistry of *Hydrogenomonas*. I. The hydrogenase of *Hydrogenomonas facilis* in cell-free preparations. *Journal of Biological Chemistry*, **211**, 885–93.

BEATRICE, M. C. & CHAPPELL, J. B. (1974). Respiration-driven proton translocation in *Hydrogenomonas eutropha*. *Biochemical Society Transactions*, **2**, 151–3.

BONGERS, L. (1967). Phosphorylation in hydrogen bacteria. *Journal of Bacteriology*, **93**, 1615–23.

BUTT, W. D. & LEES, H. (1958). Cytochromes of *Nitrobacter*. *Nature, London*, **182**, 732–3.

CHANCE, B. (1961). The interaction of energy and electron transfer reactions in mitochondria. II. General properties of adenosine triphosphate-linked oxidation of cytochrome and reduction of pyridine nucleotides. *Journal of Biological Chemistry*, **236**, 1544–54.

CHARLES, A. M. & SUZUKI, I. (1965). Sulfite oxidase of a facultative autotroph, *Thiobacillus novellus*. *Biochemical and Biophysical Research Communications*, **19**, 686–90.

CHARLES, A. M. & SUZUKI, I. (1966). Mechanism of thiosulfate oxidation by *Thiobacillus novellus*. *Biochimca et Biophysica Acta*, **128**, 510–21.

COBLEY, J. G. (1973). The enigma of energy conservation in *Nitrobacter*. Ph.D. Thesis, University of Bristol.

COBLEY, J. G. (1976*a*). Energy-conserving reactions in phosphorylating electron-transport particles from *Nitrobacter winogradskyi*. *Biochemical Journal*, **156**, 481–92.

COBLEY, J. G. (1976*b*). Reduction of cytochromes by nitrite in electron-transport particles from *Nitrobacter winogradskyi*. *Biochemical Journal*, **156**, 493–9.

COBLEY, J. G. & CHAPPELL, J. B. (1974). Nitrite oxidation and the energized state in *Nitrobacter winogradskyi*. *Biochemical Society Transactions*, **2**, 146–9.

COLE, J. S. & ALEEM, M. I. H. (1970). Oxidative phosphorylation in *Thiobacillus novellus*. *Biochemical and Biophysical Research Communications*, **38**, 736–43.

COLE, J. S. & ALEEM, M. I. H. (1973). Electron-transport-linked compared with proton-induced ATP generation in *Thiobacillus novellus*. *Proceedings of the National Acedemy of Sciences, USA*, **70**, 3571–5.

DAVIS, E. A. & JOHNSON, E. J. (1967). Phosphorylation coupled to the oxidation of sulfite and 2-mercaptoethanol in extracts of *Thiobacillus thioparus*. *Canadian Journal of Microbiology*, **13**, 873–84.

DONAWA, A, ISHAQUE, M. & ALEEM, M. I. H. (1971). Energy conversion in autotrophically grown *Pseudomonas saccharophila*. *European Journal of Biochemistry*, **21**, 293–300.

ERICKSON, R. H. & HOOPER, A. B. (1972). Preliminary characterization of a variant CO-binding heme protein from *Nitrosomonas*. *Biochimica et Biophysica Acta*, **275**, 231–44.

FALCONE, A. B., SHUG, A. L. & NICHOLAS, D. J. D. (1962). Oxidation of hydroxylamine by particles from *Nitrosomonas*. *Biochemical and Biophysical Research Communications*, 9, 126–31.

FISCHER, I. & LAUDELOUT, H. (1965). Oxidative phosphorylation in *Nitrobacter winogradskyi*. *Biochimica et Biophysica Acta*, 110, 259–64.

FORREST, W. W. & WALKER, D. J. (1971). The generation and utilization of energy during growth. *Advances in Microbial Physiology*, 5, 213–74.

GIBBS, M. & SCHIFF, J. A. (1960). Chemosynthesis: the energy relations of chemoautotrophic organisms. In *Plant Physiology*, ed. F. C. Steward, pp. 279–319. New York: Academic Press.

HEMPFLING, W. P. & VISHNIAC, W. (1967). Yield coefficients of *Thiobacillus neapolitanus* in continuous culture. *Journal of Bacteriology*, 93, 874–8.

INGLEDEW, W. J. & CHAPPELL, J. B. (1975). ATP-induced changed in the midpoint potentials of *Nitrobacter* cytochromes: an indication of anisotropy. *Federation Proceedings*, 34, 488.

INGLEDEW, W. J., COBLEY, J. G. & CHAPPELL, J. B. (1974). Cytochromes of the *Nitrobacter* respiratory chain. *Biochemical Society Transactions*, 2, 149–51.

ISHAQUE, M. & ALEEM, M. I. H. (1970). Energy coupling in *Hydrogenomonas eutropha*. *Biochimica et Biophysica Acta*, 223, 389–97.

ISHAQUE, M., DONAWA, A. & ALEEM, M. I. H. (1971a). Electron transport and coupled energy generation in *Pseudomonas saccharophila*. *Canadian Journal of Biochemistry*, 49, 1175–82.

ISHAQUE, M., DONAWA, A. & ALEEM, M. I. H. (1971b) Oxidative phosphorylation in *Pseudomonas saccharophila* under autotrophic and heterotrophic growth conditions. *Biochemical and Biophysical Research Communications*, 44, 245–51.

ISHAQUE, M., DONAWA, A. & ALEEM, M. I. H. (1973). Energy coupling mechanisms under aerobic and anaerobic conditions in autotrophically grown *Pseudomonas saccharophila*. *Archives of Biochemistry and Biophysics*, 159, 570–9.

JAGENDORF, A. & URIBE, E. (1966). ATP formation caused by acid–base transition of spinach chloroblasts. *Proceedings of National Academy of Science USA*, 55, 170–7.

KELLY, D. P. (1971). Autotrophy: concepts of lithotrophic bacteria and their organic metabolism. *Annual Review of Microbiology*, 25, 177–210.

KELLY, D. P. & SYRETT, P. J. (1963). Effect of 2,4-dinitrophenol on carbon dioxide fixation by a *Thiobacillus*. *Nature, London*, 197, 1087–9.

KELLY, D. P. & SYRETT, P. J. (1966). Energy coupling during sulphur compound oxidation of *Thiobacillus* sp. strain C. *Journal of General Microbiology*, 43, 109–18.

KIESOW, L. (1963). Uber die Reduktion von Diphosphopyridinonucleotid bei der Chemosynthese. *Biochemische Zeitschrift*, 338, 400–6.

KIESOW, L. (1964). On the assimilation of energy from inorganic sources in autotrophic forms of life. *Proceedings of the National Academy of Sciences, USA*, 52, 980–8.

KIESOW, L. (1967). Energy-linked reactions in chemoautotrophic organisms. *Current Topics in Bioenergetics*, 2, 195–233.

KNOBLOCH, K., ISHAQUE, M. & ALEEM, M. I. H. (1971). Oxidative phosphorylation in autotrophically grown *Micrococcus denitrificans*. *Archiv für Microbiologie*, **76**, 114–25.

LEES, H. & SIMPSON, J. R. (1957). The biochemistry of the nitrifying organisms. 5. Nitrite oxidation by *Nitrobacter*. *Biochemical Journal*, **65**, 297–305.

MALAVOLTA, E., DELWICHE, C. C. & BURGE, W. D. (1960). Carbon dioxide fixation and phosphorylation by *Nitrobacter agilis*. *Biochemical and Biophysical Research Communications*, **2**, 445–9.

MITCHELL, P. (1970). Membranes of cells and organelles: Morphology, transport and metabolism. In *Organization and Control in Prokaryotic and Eukaryotic Cells*, ed. H. P. Charles & B. C. J. G. Knight, Society for General Microbiology Symposium 20, pp. 121–66, London: Cambridge University Press.

O'KELLEY, J. C., BECKER, G. E. & NASON, A. (1970). Characterization of the particulate nitrite oxidase and its component activities from the chemoautotroph *Nitrobacter agilis*. *Biochimica et Biophysica Acta*, **205**, 409–25.

PECK, H. D. (1968). Energy-coupling mechanisms in chemolithotrophic bacteria. *Annual Review of Microbiology*, **22**, 489–518.

PECK, H. D. & FISCHER, I. (1962). The oxidation of thiosulfate and phosphorylation in extracts of *Thiobacillus thioparus*. *Journal of Biological Chemistry*, **237**, 190–7.

RAMAIAH, A. & NICHOLAS, D. J. D. (1964). The synthesis of ATP and the incorporation of ^{32}P by cell-free preparations from *Nitrosomonas europaea*. *Biochimica et Biophysica Acta*, **86**, 459–65.

REES, M. & NASON, A. (1965). A P450-like cytochrome and a soluble terminal oxidase identified as cytochrome *o* from *Nitrosomonas europaea*. *Biochemical and Biophysical Research Communications*, **21**, 248–56.

ROSS, A. J., SCHOENHOFF, R. L. & ALEEM, M. I. H. (1968). Electron transport and coupled phosphorylation in the chemoautotroph *Thiobacillus neapolitanus*. *Biochemical and Biophysical Research Communications*, **32**, 301–6.

SAXENA, J. & ALEEM, M. I. H. (1972). Generation of reducing power in chemosynthesis. VII. Mechanisms of pyridine nucleotide reduction by thiosulfate in the chemoautotroph *Thiobacillus neapolitanus*. *Archiv für Mikrobiologie*, **84**, 317–26.

SCHLEGEL, H. G. (1975). Mechanisms of chemoautotrophy. In *Marine Ecology*, vol. II, ed. O. Kein, pp. 9–60. London: Wiley.

SEWELL, D. L. & ALEEM, M. I. H. (1969). Generation of reducing power in chemosynthesis. V. The mechanism of pyridine nucleotide reduction by nitrite in the chemoautotroph *Nitrobacter agilis*. *Biochimica et Biophysica Acta*, **172**, 467–75.

SEWELL, D. L., ALEEM, M. I. H. & WILSON, D. F. (1972). The oxidation–reduction potentials and rates of oxidation of the cytochromes of *Nitrobacter agilis*. *Archives of Biochemistry and Biophysics*, **153**, 312–19.

SUZUKI, I. (1974). Mechanisms of inorganic oxidation and energy coupling. *Annual Review of Microbiology*, **28**, 85–101.

TRUDINGER, P. A. (1969). Assimilatory and dissimilatory metabolism of inorganic sulphur compounds by micro-organisms. *Advances in Microbial Physiology*, **3**, 111–58.

VAN GOOL, A. & LAUDELOUT, H. (1966). The mechanism of nitrite oxidation by *Nitrobacter winogradskyi*. *Biochimica et Biophysica Acta*, **113**, 41–50.

WALLACE, W. & NICHOLAS, D. J. D. (1969). The biochemistry of nitrifying micro-organisms. *Biological Reviews*, **44**, 359–91.

WITTENBERGER, C. L. & REPASKE, R. (1961). Studies on hydrogen oxidation in cell-free extracts of *Hydrogenomonas eutropha*. *Biochimica et Biophysica Acta*, **47**, 541–52.

EPILOGUE: FROM ENERGETIC ABSTRACTION TO BIOCHEMICAL MECHANISM

PETER MITCHELL

Glynn Research Laboratories, Bodmin, Cornwall, PL30 4AU, UK

ABSTRACT

It has been said that nothing is new except arrangement. That is an apt description of the aspect of biology that energetic methods are competent to explore. The processes of new growth and of regeneration are energetic in that they involve the arrangement or rearrangement of electrons and atomic nuclei in various states of assembly in living organisms.

The free-energy function of thermodynamics has the form $\int\mathbf{F}d\mathbf{x}$, the scalar product of two vectors, $\mathbf{F}$ representing the force on a given species of particle, the $\mathbf{x}$ representing the co-linear displacement of the particle along a pathway from one domain (or one state of bonding) to another. Free energy must be expended in all actual thermodynamic reactions because a given particle, accelerated along the pathway $\mathbf{x}$ by the force $\mathbf{F}$, is frequently reflected back by collision with other particles, so that the kinetic energy of the particle, associated with its velocity in the $\mathbf{x}$ direction, is transformed to kinetic energy (called heat) associated with randomly directed (formally scalar) motion. The latter is conventionally expressed by TS, the product of the scalars temperature and entropy. It follows that efficient (potentially reversible) catalysis in a thermodynamic reaction system depends on the free-energy profile along the reaction pathway having a gentle downward slope, relatively free of humps or hollows that tend to reflect the reaction back and transform into heat the kinetic energy of the system associated with its (vectorial or higher order tensorial) velocity down the reaction pathway.

This simple energetic principle of molecular mechanics provides valuable insights into the general biochemical specifications, not only of individual enzymes, but also of the much more complex energy-transducing chemical and chemiosmotic reaction systems discussed in this volume. Most importantly, it explains the fundamental interrelationship between energy transfer and specific catalytic function.

INTRODUCTION

As this volume has shown, microbial energetics is a microcosm that reflects practically every aspect of the vast field of endeavour, called bioenergetics, in which energetic methods are applied to biological problems. In this epilogue, it is therefore hard to comment broadly on the past and future development of microbial energetics, as I have been requested to do, without developing a rather general and inevitably theoretical theme. I hope, however, that those of my readers who are not very partial to flights of theoretical fancy may find some solace in the fact that, as implied by the title of this contribution, my object is to indicate in an elementary way how the abstractions of energy in classical thermodynamics and in mechanics may be made to shed light on the details of biochemical reaction mechanisms as effectively and realistically as possible.

The general theme of vectorial chemistry or chemicalicity pursued here is a further development of the concepts of group translocation, chemiosmotic reactions and vectorial metabolism, which I have discussed in earlier papers (Mitchell, 1962, 1963, 1967a, 1968, 1970a, b, 1976a). The main background of knowledge and ideas, from which these conceptual developments have evolved, has been reviewed previously (Robertson, 1968; Mitchell, 1970a, 1972, 1973a). Some more recent or additional works that have contributed to or that are related to the view of chemicoconductive channels and circuits developed in the present paper are as follows: Lipmann, 1946; Fay, 1965; Wyman, 1965, 1975; Hawkins, 1966; Hill, 1968, 1976; Liebhafsky & Cairns, 1968; Dickerson, 1969; Oster, Perelson & Katchalsky, 1971; Citri, 1973; Lumry, 1974; Delaage, 1975; Jencks, 1975; Perelson, 1975; Hill & Simmons, 1976; Koshland, 1976; Lagarde, 1976.

It should, perhaps, be emphasised that I make no claims to originality in the present paper, the purpose of which is illustrative rather than innovative.

MOLECULAR MECHANICS

Energy as a function representing action or the potentiality for action

There has been a tendency for biochemists, following classical thermodynamics, to treat energy as if it were a substance. Quantities of energy are represented by scalar numbers, and the principle of energy conservation enables the migration of energy from one place to another to be treated like the migration of an indestructible substance. This particular abstraction of the concept of energy is obviously simplifying and useful in some respects (Guggenheim, 1933, 1949). But it has become counter-productive in that it isolates classical chemical thermodynamics from the general field of mechanics to which studies of the mechanisms and energetics of the processes of chemical transformation and synthesis naturally belong (Dickerson, 1969; Fay, 1965). It has thereby tended to perpetuate the conceptual rift between biochemistry and general biology by concealing the essentially directional attributes of energetic processes in biochemistry – so making it more difficult to describe the vitality of biological phenomena in biochemical terms (Mitchell, 1962, 1963).

As conceived in mechanics, energy is not a material, but it is a quantitative function that can be used to interrelate changes of the relative positions and motions of interacting material particles, such as the electrons and atomic nuclei of substrates and the active centres of enzymes involved in biochemical reactions, or such as the body, the locomotory parts and the environment of a mobile organism. Thus, the mechanical (rather than the classical thermodynamic) concept of energy helps to link our comprehension of biochemistry with that of general biology, because energy represents action or the potentiality for action. By action we mean the motion of material things relative to one another, as in nutrient uptake, organic synthesis, growth, morphogenesis, repair, reproduction, locomotion – properties that are especially characteristic of living things.

According to Newtonian principles, when two particles i and j come temporarily into contact, their interaction may be conceived in terms of a short-range force directed along the line joining their centres (Fig. 1). If the magnitude of this force is a single-valued function only of the distance between the centres of the particles, the action of particle j on particle i during the coupled process known as collision is conveniently given by the component $\mathbf{f}_j$ of this force exerted by particle j on particle i along the coupled or collisional pathway of i multiplied by the co-linear displacement $d\mathbf{r}_i$ of particle i integrated over this pathway extending, let us say, from a position α to a position ω. This function,

$$\int_\alpha^\omega \mathbf{f}_j d\mathbf{r}_i = w = \Delta U_i \qquad (1)$$

is described as the work w done on i by j or as the energy ΔU_i accepted by i from j. The beauty of this work function or energy function is that, according to Newtonian principles, the force $\mathbf{f}_j$ exerted on i by j must be matched by an equal and opposite force $\mathbf{f}_i$ exerted on j by i, and as the

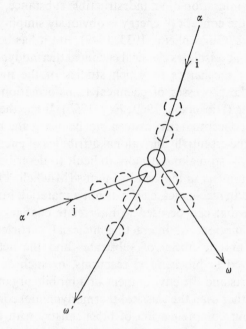

Fig. 1. Diagram representing collision and transfer of energy between two chemical particles i and j that follow pathways $\alpha \to \omega$ and $\alpha' \to \omega'$ respectively.

coupled pathways of i and j are equal in length along the line of the i pathway for purely geometric reasons, it follows that

$$\int_\alpha^\omega \mathbf{f}_j d\mathbf{r}_i = -\int_{\alpha'}^{\omega'} \mathbf{f}_i d\mathbf{r}_j \qquad (2)$$

or

$$\Delta U_i = -\Delta U_j \qquad (3)$$

In other words, the energy gain by i is equal to the energy loss by j, and energy behaves as if it is conserved in the interaction between i and j. The general notion of the conservation of non-radiative energy can be derived by the extension of this simple principle to any number of particles.

My object in beginning in this rather elementary way is to emphasise the fact that, in the non-radiative type of energy transfer with which we are mainly concerned in the coupling mechanisms of metabolism and biological transport, the energy function ΔU_i represents an action or a potentiality for action corresponding to the transport of the component i against an opposing force.

In the present context, it is obviously important whether the energy

ΔU_i donated to i by j may remain associated with i so that it is subsequently recoverable from i, or whether this energy may be dispersed amongst other components so that it ceases to be available. This depends on the relationships of the particle i to a system of components, generally called a thermodynamic system, with which i may interact. In particular, it depends upon the action on i of forces other than $\mathbf{f}_j$ that may originate in this thermodynamic i system, and on the extent of the space to which the particle i may be confined by the chemical reactivity and by the topology of this system. Perhaps I should interject that this, of course, is a matter that has been explored and developed by the natural accidental processes of biochemical evolution for many millions of years. Our conscious exploration of the natural achievements has, by comparison, only just begun.

The classical thermodynamic concept of the chemical potential μ_i, and the derived concept of the total or electrochemical potential $\bar{\mu}_i$ of the species of particle i (which may be a molecule, ion, electron or chemical group) represents the tendency of i to escape from a given phase or domain against a force that tends to retain it. Thus, as illustrated in Fig. 2(a), the total chemical potential difference $\Delta\bar{\mu}_i$, defined as $\bar{\mu}_i{}^\alpha - \bar{\mu}_i{}^\omega$, represents the work or energy available per molecule or particle of i as it escapes reversibly against an externally originating equilibrium poising force $\overset{E}{\mathbf{f}_i}$, equal to $-\bar{\mathbf{f}}_i$, along a pathway $\mathbf{r}$ from the region α of higher potential to the region ω of lower potential, or

$$\Delta\bar{\mu}_i = \int_\alpha^\omega \bar{\mathbf{f}}_i \mathrm{d}\mathbf{r}. \tag{4}$$

Alternatively, we can describe the total escaping force exerted per particle of i at any point along the pathway of escape between α and ω by the relationship

$$\bar{\mathbf{f}}_i = -\frac{\mathrm{d}\bar{\mu}_i}{\mathrm{d}\mathbf{r}}. \tag{5}$$

According to Newtonian principles, we can recognise two distinct types of force that contribute to the total (statistical) mechanical force $\bar{\mathbf{f}}_i$. One is the time-independent static or field-effect type of force, which includes hydrostatic pressure, electrical, primary chemical bonding, and secondary bonding forces – associated with the instantaneous position of i in the system. The other is the time-dependent accelerative type of force, which, when given as a time-average, corresponds to the essentially thermodynamic escaping or repulsive force – associated with the thermal vibratory momentum and collisional acceleration of particles of i in the system through an interval of time, and dependent on the degree of confinement or concentration of i. We can therefore describe the (statistical mechanical) time-average total force $\bar{f}_i$ acting on each particle of i as the sum of a set of forces, such as pressure $\overset{p}{f_i}$, electrical $\overset{\psi}{f_i}$, chemical $\overset{x}{f_i}$ and thermal $\overset{\theta}{f_i}$, as follows:

$$\bar{f}_i = \overset{p}{f_i} + \overset{\psi}{f_i} + \overset{x}{f_i} + \overset{\theta}{f_i}. \tag{6}$$

The significance of this equation may perhaps be more obvious when we write it in the form,

$$\bar{f}_i = -v_i\frac{dp}{dr} - z_i\frac{Fd\psi}{Ndr} - \frac{d\overset{x}{\mu_i}}{dr} - kT\frac{d(\ln c_i)}{dr}. \tag{7}$$

Here, the terms on the right correspond respectively to those of equation (6); p and ψ mean the pressure and electric potential, v_i, z_i, $\overset{x}{\mu_i}$ and c_i mean the partial molecular volume, electric charge, purely chemical potential and degree of confinement of i in the form subject to the forces given by equation (7), and F, N, k and T stand for the Faraday constant, the Avogadro number, the Boltzmann constant and the absolute temperature respectively. The degree of confinement, concentration or probability c_i may, for example, be more explicitly represented as a ratio of occupied to unoccupied sites c, or

$$c_i = n_{ic}/n_c. \tag{8}$$

The integrated form of equations (6) or (7), which may look more familiar, is

$$\Delta\bar{\mu}_i = \overset{p}{\Delta\mu_i} + \overset{\psi}{\Delta\mu_i} + \overset{x}{\Delta\mu_i} + \overset{\theta}{\Delta\mu_i}. \tag{9}$$

This equation may alternatively be written

$$\Delta\bar{\mu}_i = v_i\Delta p + z_i\frac{F}{N}\Delta\psi + \overset{x}{\Delta\mu_i} + kT\Delta\ln c_i. \tag{10}$$

The reader should note that the chemical potentials are given here as energy per molecule or particle, and not in the usual units of energy per mole. Also, the deltas written Δ represent potential differences in space, and correspond to potentials in domain α (corresponding to initial states) minus those in domain ω. They have the same magnitude but the opposite sign to the usual thermodynamic energy changes of i in time per molecule of i transferred, which are written Δ.

When a process involves the linked transfer of groups of particles of one or more species i, the total change of chemical potential is the sum of the individual chemical potential changes, and is conventionally given in terms of the Gibbs free-energy change ΔG. In other words,

$$-\Delta G = \Sigma_i \nu_i \Delta \bar{\mu}_i, \tag{11}$$

where Σ_i denotes the sum of terms in i, and ν_i is the number of particles of each species i travelling from α to ω in the linked reaction process. Thus, for example, an electronic redox potential difference of 0.25 electronvolt (eV) per electron transferred from a reductant couple β that donates 2 electrons per molecule to an oxidant couple α gives a Gibbs free-energy increase of -0.5 eV per reductant molecule (equivalent to 11531 cal mol^{-1} or 48.246 kJ mol^{-1}). It is rather convenient to measure biochemical energy changes in electronvolts per particle or molecule because the so-called phosphate potential difference between the ATP/ADP couple and the P_i/H_2O couple, which is generally indicative of the practical magnitude of intramolecular biochemical energy transfers, happens to be around 0.5 eV per phosphoryl group transferred.

Energy storage, conduction, transformation, and dissipation

Before going on to discuss the significance of the several contributions to the total (statistical mechanical) force $\bar{f}_i$ to which each particle of species i is subject, according to Fig. 2(a), between the α and ω domains, we have to consider what circumstances enable the energy represented by $\Delta \bar{\mu}_i$ to be retained or stored by the system, or how this energy may be conducted from one place to another, transformed from one chemical or physical form to another, or dissipated as heat.

In Fig. 2(a) the total equilibrium escaping force $\bar{f}_i$ on each particle of i is represented as being balanced by an equal and opposite equilibrium restraining force $\overset{E}{f}_i$ of external origin, which prevents the net escape of i from α to ω. There are two obvious fixed origins of this restraining force in biochemical systems.

Retention of chemical energy by latching of chemical bonds

When i represents an electron or a chemical group, and the domains α and ω represent donor and acceptor molecules (couples) to which the electron or group is chemically bound, the restraining force corresponds to the latching property of the chemical bonds in the donor and/or acceptor compounds. Thus, the escape of the group i from the donor couple α to the acceptor couple ω can only occur rapidly through a pathway provided by the catalytic centre region of an enzyme or enzyme

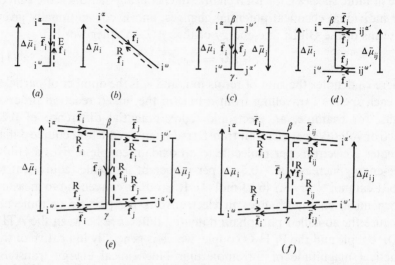

Fig. 2. Diagrams of relationships between total energy changes $\Delta\bar{\mu}$ and total forces acting on particle species i, j or ij along given pathways, such as $\alpha \rightarrow \omega$ or $\alpha' \rightarrow \omega'$. The arrows on the pathways represent the directions of the forces; and the arrows at input and output show the positive sense of the potential difference. Broken lines represent restraining $\overset{E}{f}$ or frictional $\overset{R}{f}$ forces. Further explanations are in the text. (a) Conservation of i energy. (b) Conduction or transmission of i energy. (c) Equilibrium transformation of i energy to j energy by symfer or symport of i with j. (d) Equilibrium transformation of i energy to ij energy by j unifer or uniport (equivalent to i/ij antifer or antiport). (e) and (f) as (c) and (d) respectively, but under net transfer conditions. Note that the formal sign of the frictional forces $\overset{R}{f}$ is reckoned from $\alpha \rightarrow \omega$ or from $\alpha' \rightarrow \omega'$, consistent with that of the total equilibrium forces $\bar{f}$ for a given component although arrows indicate the actual direction of the forces for the forward reaction.

system of appropriate specificity that can effectively lift the latch on the chemical bonds. In this case, the donor and acceptor couples represented by α and ω may be present in the same aqueous phase; and it is, of course, the fundamental latching property of certain types of chemical bond that enables the energy of environmental or intermediary metabolic group-donating and group-accepting couples to be locked in until the specific group is released through the catalytic centre of the appropriate enzyme or catalytic carrier system (Mitchell, 1972; Citri, 1973; Jencks, 1975; Koshland, 1976).

Retention of osmotic energy by the osmotic barrier function of membranes

When i is a hydrated solute, such as a hydrated H^+ ion, and the domains α and ω represent aqueous solutions separated by a topoogically closed lipid membrane of low permeability to the solute, the restraining force is attributable to the hydrophobic osmotic barrier phase of the lipid membrane which reflects back the momentum of the diffusing solute particles. Thus, the escape of the solute i from the donor phase α and its appearance in the acceptor phase ω can only occur rapidly through a pathway provided by the catalytic centre region of an enzyme or carrier system of appropriate specificity and conformational mobility, plugged through the membrane (Hamilton, 1975; Harold, 1976; Mitchell, 1967*a*, 1970*a*, *b*).

These chemical and osmotic types of restraining force on i, which are fixed in origin, enable the energy of i to be stored. However, the transfer of the chemical or osmotic energy of i, or its manifestation as action, depends on the net escape of the particles of i along the $\alpha \rightarrow \omega$ pathway represented by Fig. 2. There are two main kinds of circumstance, and corresponding types of system, that may determine the effect of the escape of i.

Energy conduction or transmission and energy connection

When, as indicated in Fig. 2(b), the externally originating equilibrium restraining force is zero, the equilibrium escaping force will accelerate the particles of i along the $\alpha \rightarrow \omega$ pathway until their net velocity rises to a value at which the force of frictional resistance $\overset{R}{f_i}$, due to collisions of i along the pathway is equal and opposite to the escaping force $\bar{f}_i$, or

$$\bar{f}_i + \overset{R}{f_i} = 0. \tag{12}$$

In that case, the energy $\Delta\bar{\mu}_i$ per particle of i will be manifested as the conduction or transmission of i from α to ω only against the frictional resistance of the system; and an amount of random thermal motion, or heat $\Delta\overset{R}{Q_i}$, equivalent to $\Delta\bar{\mu}_i$ per particle of i transported, will be dissipated under isothermal conditions. Thus,

$$\Delta G_i + \Delta\overset{R}{Q_i} = 0. \tag{13}$$

Many solute uniporters and chemical group transferases can be recognised as catalysts of this class of simple energy connection and conduction process. Perhaps it would be appropriate to call the enzymes of this class uniferases by analogy with uniporters.

Energy coupling and energy transformation

When, as indicated in Fig. 2(c), the total equilibrium escaping force $\bar{f}_i$ on i is balanced by an equal and opposite force $\bar{f}_j$ originating from a mobile particle j, and part of the pathway (β to γ) between α and ω can be traversed only by i and j together in their paired or coupled configuration, and not by i or j separately, it follows that, when i escapes from α to ω under equilibrium conditions,

$$\int_{\alpha}^{\omega} \mathbf{\bar{f}}_i dr_i = \int_{\alpha'}^{\omega'} \mathbf{\bar{f}}_j dr_j, \tag{14}$$

so that the free energy lost by i will be transferred reversibly to j, or

$$\Delta G_i + \Delta G_j = 0. \tag{15}$$

An alternative arrangement of coupling between the flows of i, j and ij is illustrated in Fig. 2(d). In this case, since j equilibrates between β and γ

$$\int_{\alpha}^{\omega} \mathbf{\bar{f}}_i dr_i = \int_{\alpha'}^{\omega'} \mathbf{f}_{ij} dr_{ij} \tag{16}$$

and

$$\Delta G_i + \Delta G_{ij} = 0. \tag{17}$$

This coupling arrangement is formally equivalent to i/ij antifer or antiport.

Our models of the fundamental process of reversible energetic coupling illustrated by Fig. 2(c) or 2(d) evidently represent a development from the unique kinematic event (Fig. 1) with which we began our analysis – in which particle i happened to collide with particle j – to a process in which an analogous interaction and energy transfer between the species of particle i and j occur repeatedly and predictably along a pathway that is prescribed by the molecular conformation and limited mechanical articulations within a specific catalytic system. The energetic cost of the specification of the coupled process by the channelling catalytic system is zero (equations (15) and (17)) as long as the interaction between i and j occurs at equilibrium. By an extension of the principle of coupling by equal and opposite forces between pairs of particles, it is also evident that, under equilibrium conditions, the energy is transferred according to equation (15) or (17) whether interaction between i and j occurs directly or whether it occurs indirectly via intermediaries including components of the channelling catalytic system – provided, of course, that these intermediaries are confined to an energetically closed cycling or reciprocating system.

As indicated in Fig. 2(e), to obtain a finite rate of net transport of ij, and thus of energy transfer in our model, a net force $\mathbf{\bar{f}}_{ij}$ is required to accelerate ij along the $\beta \rightarrow \gamma$ pathway, and likewise, net forces $\mathbf{\bar{f}}_i$ and $\mathbf{\bar{f}}_j$

are required along the $\alpha \to \beta$ and $\gamma \to \omega$ pathways and along the $\omega' \to \beta$ and $\gamma \to \alpha'$ pathways respectively. Thus,

$$\mathbf{\bar{f}}_{ij} = \mathbf{\bar{f}}_i - \mathbf{\bar{f}}_j > 0, \tag{18}$$

$$\mathbf{\bar{f}}_i > 0, \tag{19}$$

$$-\mathbf{\bar{f}}_j > 0. \tag{20}$$

In this case, the rate of energy transfer will settle at a value for which the frictional resistances $\overset{R}{\mathbf{f}_i}$, $\overset{R}{\mathbf{f}_j}$ and $\overset{R}{\mathbf{f}_{ij}}$ are equal and opposite to the driving forces through the conducting and catalytic pathway, so that

$$\mathbf{\bar{f}}_i - \mathbf{\bar{f}}_j + \mathbf{\bar{f}}_{ij} + \overset{R}{\mathbf{f}_i} - \overset{R}{\mathbf{f}_j} + \overset{R}{\mathbf{f}_{ij}} = 0. \tag{21}$$

A similar relationship is likewise obtained between the forces in the system of Fig. 2(f), but with purely conventional differences of sign. It follows by integration, as before, that

$$\Delta G_i + \Delta G_j + \Delta G_{ij} + \Delta \overset{R}{Q_i} + \Delta \overset{R}{Q_j} + \Delta \overset{R}{Q_{ij}} = 0, \tag{22}$$

showing that part of the free energy, represented by the terms in $\overset{R}{Q}$, is used to transport i, j and ij through the system against the respective frictional forces $\overset{R}{\mathbf{f}_i}$, $\overset{R}{\mathbf{f}_j}$ and $\overset{R}{\mathbf{f}_{ij}}$; and this appears as heat in the transfer of energy from i to j.

We can obviously generalise this model to describe a system that transfers energy between groups of particles of several different species by introducing the appropriate additional terms and stoichiometric coefficients in the equations. For example, in a coupled energy-transforming process involving the conduction of the species $i_m j_n$, containing m particles of i and n particles of j, and in the absence of side reactions, the free-energy changes would be given by

$$m\Delta G_i + n\Delta G_j + \Delta G_{i_m j_n} + m\Delta \overset{R}{Q_i} + n\Delta \overset{R}{Q_j} + \Delta \overset{R}{Q_{i_m j_n}} = 0. \tag{23}$$

The catalysts of this class of energy transformation by coupled conduction include: solute symporters and antiporters that catalyse osmotic energy transformation; certain bifunctional enzymes (which we might call symferases or antiferases) such as the phosphorylating glyceraldehyde-phosphate dehydrogenase and succinyl–coenzyme A synthetase that catalyse chemical energy transformation; and osmo-enzymes and osmoenzyme systems, such as protonmotive redox and photoredox chains and ionmotive ATPases that catalyse chemiosmotic energy transformation by chemical group-solute symport or group/solute antiport processes.

Interrelationships between energy-conducting and energy-transforming systems

It is particularly noteworthy that, while the main function of the non-radiative biochemical energy transferring or coupling system that we have been describing is generally to transform the free energy of i to that of j or ij, this type of system is also essentially a chemicoconductive or transport machine. If it is a chemical energy transforming system, such as the phosphorylating glyceraldehyde-phosphate dehydrogenase, which converts electronic energy to phosphoryl energy, it links the motion of chemical groups. If it is an osmotic energy transforming system, such as the β-galactoside–proton symporter, it links the motion of solutes. If it is a chemiosmotic energy transforming system, such as the protonmotive ATPase that converts chemical phosphoryl energy to osmotic protonic energy, it links the motion of chemical groups to that of solutes. It is the interlinked cyclic or reciprocating motion of the biochemical energy-transforming mechanism, like a miniature engine, that inevitably dissipates some of the energy of i in the process of transfer to j, as shown in equation (22) or (23).

In the special case that the species j of Fig. 2(e) and equation (22) or (23) moves through a closed cycle, ΔG_j will tend to zero at equilibrium and j will act as a carrier of i, so that the system will function only as a conductor of i. This is illustrated in Fig. 3(a), where i is shown diffusing as the carrier complex Ci_m, containing m particles of i. The conduction of hydrophilic solutes in the aqueous media, or of hydrophobic solutes in the lipid membrane media, generally occurs by simple diffusion. But there are exceptions. Proton conduction, for example, is greatly facilitated by water-soluble or lipid-soluble acid–base buffers which act as cycling proton carriers in the aqueous media or lipid media respectively, as in Fig. 3(a). The conduction of chemical groups in the aqueous or lipid media, unlike that of hydrophilic solutes in the aqueous media, cannot generally occur by simple diffusion of the chemical groups themselves but requires the circulation of appropriate carrier couples. In this case, however, owing to the latching property of the chemical bonds, the channelling of group conduction from one carrier couple to another in the aqueous media is dependent on group-transfer enzymes that act as specific latch-lifting connectors, as indicated in Fig. 3(b). For example, in classical respiratory chain systems, the NAD-linked substrate dehydrogenases act as connectors in the electron-conducting pathway from

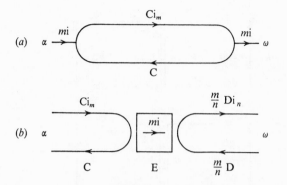

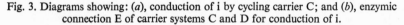

Fig. 3. Diagrams showing: (a), conduction of i by cycling carrier C; and (b), enzymic connection E of carrier systems C and D for conduction of i.

the substrates to the protonmotive NADH oxidase system. Likewise, the phosphoglycerate phosphokinase acts as a connector in the phosphoryl-conducting pathway from the phosphorylating glyceraldehyde-phosphate dehydrogenase to the ATP/ADP couple. In the relatively restricted non-aqueous media of enzyme and catalytic-carrier complexes in lipid membranes the conduction of chemical groups between one couple and another is evidently facilitated by intrinsic binding and conformational properties corresponding to the latch-lifting properties of the active centre catalytic channel regions of individual enzymes (Mitchell, 1963, 1972).

We can usefully develop the fundamental coupling principle of Fig. 2(c) by means of the chemical circuit or network notation of Fig. 3, the general principle of which is that the lines show the connected flows of the particles, the arrows with symbols denote the (microscopic) net forces, directions and stoichiometries of the flows, and the equilibrium chemicomotive potentials $\Delta\bar{\mu}_i$ and $\Delta\bar{\mu}_j$ are shown at the (purely formal) energy input (α,ω) and energy output (α',ω') respectively. Fig. 4(a) represents energy conversion by direct coupling between m particles of i and n particles of j. Figs. 4(b) and (c) represent the same overall process, but involving the cyclic motion of the carrier component C, either as a carrier of i and j together (as in osmotically coupled symport) or as a carrier of i and j alternately (as in osmotically coupled antiport). Fig. 4(d) represents a rearrangement of the flow system of Fig. 4(a), which is formally equivalent to $mi/i_m j_n$ antifer or antiport, represented here as being mediated by an enzyme E.

Fig. 5 illustrates how this notation may be used to represent classical

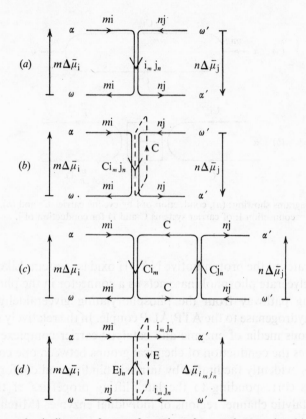

Fig. 4. Various circuit diagrams for transformation of energy between i and j: (a), direct $i_m j_n$ symfer or symport; (b), $i_m j_n$ symfer or symport mediated by carrier C; (c), i_m / j_n antifer or antiport mediated by carrier C. In (d), the system shows transformation of energy from i to ij by n_j unifer or uniport (formally $mi/i_m j_n$ antifer or antiport) mediated by enzyme E.

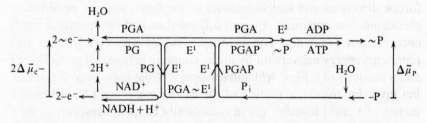

Fig. 5. Circuit diagram of substrate-level phosphorylation catalysed by glyceraldehyde-phosphate dehydrogenase, E^1, and phosphoglycerate kinase, E^2, transforming electronic energy $\sim e^-$ to phosphoryl energy $\sim P$. Abbreviations: PGA, 3-phosphoglycerate; PG, 3-phosphoglyceraldehyde; PGAP, 1,3-diphosphoglycerate. The H_2O and H^+ at the input, and the H_2O at the output, are included to show how the circuit is completed.

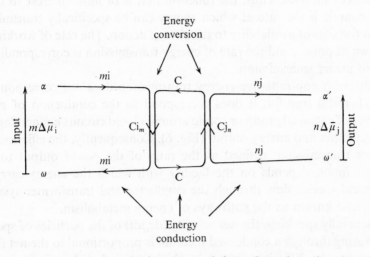

Fig. 6. Illustration of net conduction or transmission and transformation of power
in a system corresponding to Fig. 4(c).

substrate-level phosphorylation, catalysed by glyceraldehyde-phosphate dehydrogenase (E^1) and phosphoglycerate kinase (E^2), as a chemico-conductive circuit system, transforming electronic energy, denoted by $\sim e^-$ to phosphoryl energy, denoted by $\sim P$.

The specific transmission of energy and power

The conduction and transformation aspects of the net transmission of energy are summarised together in Fig. 6 for a system corresponding to that of Fig. 4(c), using the vertical scale to represent the total potential difference across the system per chemical equivalent. This diagram emphasises how closely interrelated are the transfers of material and of the function that we call energy. It also illustrates, however, that if energy-producing and energy-consuming systems are connected respectively to the input and output, there will be a net transmission of energy from input to output, but there will be no corresponding *net* transport of material, because i, C and j move in closed circuits.

As indicated earlier, the energy function represents the potentiality for the various types of biochemical, physiological and behavioural actio that we recognise as the vitality of living organisms. For that reason, the energy transmission systems that conduct environmentally available energy to the required sites of action in living organisms have clearly played a major part in natural selection. But it is axiomatic in competitive situations that it is not so much action as the rapidity of action that influences survival. Thus, the function that is of most interest to us in this context is the rate at which energy can be specifically transmitted from the sites of availability to the sites of action. The rate of working is known as power, and the rate of energy transmission is correspondingly called power transmission.

Although non-radiative energy transmission does not correspond to *net* chemical transfer, it does correspond to the conduction of given species of chemical group or solute around closed circuits connecting the energy input and energy output (Fig. 6). Consequently, the efficiency of power transmission, defined as the ratio of the power output to the power input, depends on the facility with which the energy-carrying chemical species flow through the conductor and transformer systems otherwise known as the pathways of energy metabolism.

Generally speaking, the net velocity (dr_i/dt) of the particles of species i diffusing through a condensed medium is proportional to the net force acting on the diffusible particles, which is equal and opposite to the force of frictional resistance (Mitchell, 1967a, 1970a). Thus,

$$\bar{f}_i = -\overset{R}{f}_i = f_i \frac{dr_i}{dt}, \tag{24}$$

where f_i represents the constant of proportionality, or frictional coefficient, corresponding to the frictional force at unit velocity. The rate (dn_i/dt) at which i will pass along the pathway, described as the frequency with which i particles pass a given point, or as the quantity of i transported in a given time, is the product of the velocity of i and the concentration n_i/r of i along the pathway. Therefore, from equation (24),

$$\frac{dn_i}{dt} = \frac{1}{f_i} : \frac{n_i}{r} \cdot \bar{f}_i \tag{25}$$

It follows from equations (6) and (7) that

$$\frac{dn_i}{dt} = \frac{1}{f_i} \cdot \frac{n_i}{r} (\overset{\phi}{f}_i + \overset{\theta}{f}_i) \tag{26}$$

or

$$\frac{dn_i}{dt} = \frac{1}{f_i} \cdot \frac{n_i}{r} \left[\overset{\phi}{f}_i - kT \frac{d(\ln c_i)}{dt} \right], \tag{27}$$

where $\overset{\phi}{f}_i$ stands for the sum of the static or field-effect forces corresponding to the pressure, electric and chemical forces of equation (6).

We may alternatively write equation (25) as

$$\frac{dn_i}{dt} = -\frac{1}{f_i} \cdot \frac{n_i}{r} \cdot \frac{d\bar{\mu}_i}{dr}. \tag{28}$$

For a given total potential difference $\Delta\bar{\mu}_i$ across the circuit (Fig. 6), the power transmission P_i will be given by

$$P_i = \Delta\bar{\mu}_i \frac{dn_i}{dt} \tag{29}$$

or

$$P_i = -\Delta\bar{\mu}_i \cdot \frac{1}{f_i} \cdot \frac{n_i}{r} \cdot \frac{d\bar{\mu}_i}{dr}. \tag{30}$$

Thus, for a given gradient of total potential $d\bar{\mu}_i/dr$, and for a corresponding expenditure of energy or power per unit length of the transmission pathway, the power transmitted will be proportional to the potential difference $\Delta\bar{\mu}_i$ across the circuit, inversely proportional to the frictional coefficient f_i along the power transmission pathway, and proportional to the concentration n_i/r of the energy-carrying particles, where the quantity n_i may be compared with n_{ic} of equation (8). The improvement in the efficiency of power transmission in artificial electric transmission systems by the use of high potential differences and low resistance conductors that contain a high concentration of mobile electrons is, of course, well known.

Equation (30) shows not only the conditions for efficient power transmission along the required pathways, but also the conditions for power loss through side reactions, which effectively act as short-circuit or shunt resistances. To minimise the loss of power through side reactions (such as spontaneous hydrolysis of energy-rich anhydrides or ion or substrate leakage through membranes), the conditions needed are the converse of those needed to maximise power transmission through the required channels. Using the electrical analogy: efficient power transmission depends on the series conductances along the transmission channels being as high as possible (i.e. high n_i/rf_i), and the side reaction shunt conductances between the transmission channels being as low as possible (i.e. low n_i/rf_i).

When we take account of the side reactions in a system such as that of Fig. 6, the stoichiometric coefficients m and n for the $i_m j_n$ complex are not the same as those of the input and output systems because some of the i current in the input circuit and some of the j current in the output circuit is short-circuited. There may, in addition, be some slip between the flow of i and j in the energy-transforming system that ideally couples the flow of m particles of i to n particles of j. Thus, the power input P^α and the power output P^ω are related to the corresponding chemical currents $(dn_i/dt)^\alpha$ and $(dn_j/dt)^\omega$, to the corresponding chemical potential differences $\Delta\bar\mu_i^\alpha$ and $\Delta\bar\mu_j^\omega$, and to the total heat production (or energy dissipation) $d\bar Q/dt$ by equations such as

$$P_i^\alpha = \Delta\bar\mu_i^\alpha(dn_i/dt)^\alpha, \tag{31}$$

$$P^\omega = \Delta\bar\mu_j^\omega(dn_j/dt)^\omega, \tag{32}$$

$$P^\alpha = P^\omega + d\bar Q/dt. \tag{33}$$

The total heat production $d\bar Q/dt$ arises from the sum of the series resistances (represented in equation (23)) and the short-circuiting shunt resistances that we are now considering. It follows that the stoichiometric flow ratio between input and output is given by

$$(dn_j/dt)^\alpha/(dn_i/dt)^\omega = \frac{n}{m}\left(1-\frac{\delta m}{m}\right)\left(1-\frac{\delta n}{n}\right), \tag{34}$$

where $\delta m/m$ and $\delta n/n$ represent the proportion of the i current and j current respectively lost in the short-circuiting side reactions. Thus, the efficiency of power transmission P^ω/P^α can be given as

$$P^\omega/P^\alpha = n'\Delta\bar\mu_j^\omega/m'\Delta\bar\mu_i^\alpha, \tag{35}$$

where

$$\frac{n'}{m'} = \frac{n}{m}\left(1-\frac{\delta m}{m}\right)\left(1-\frac{\delta n}{n}\right). \tag{36}$$

It is instructive to note that the series resistances along the power conduction channels are associated with gradients of the chemical potentials and thus with diminutions of the chemical potential differences $\Delta\bar\mu_i$ and $\Delta\bar\mu_j$ from input to output, whereas the side reaction shunt resistances are associated with gradients of the chemical currents and thus with diminutions of the current-flow stoichiometry from input to output. The energy-transforming system coupling the flow of m particles of i to n particles of j ideally gives an output potential that is m/n times its input and an output current (of j) that is n/m times its input (of i), thus acting quite analogously to a direct-current electrical transformer.

The high number or concentration of energy-carrying particles required to facilitate efficient biological power transmission is importantly related to the different forces contributing to the total potential (or Gibbs free energy) profile of i, or of carriers of i, along the conduction pathway or channel. Equations (9) and (10), describing the components of the total potential of i, can be given as

$$\Delta\bar\mu_i = \Delta\mu_i^\phi + kT\Delta\ln c_i, \tag{37}$$

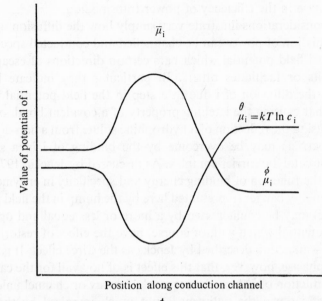

Position along conduction channel

Fig. 7. Diagram of field potential profile $\overset{\phi}{\mu_i}$ of a component i along a conduction channel, showing relationship to the thermodynamic potential $\overset{\theta}{\mu_i}$ and to the total potential $\bar{\mu}_i$ at equilibrium.

where $\overset{\phi}{\Delta\mu_i}$ represents the field potential difference along the conducting pathway due to the sum of the static or field-effect forces $\overset{\phi}{\mathbf{f}_i}$. This equation shows that when i is at equilibrium along the conduction channel

$$\overset{\phi}{\Delta\mu_i} + kT\Delta\ln c_i = 0. \qquad (38)$$

Therefore, as illustrated in Fig. 7, where there are humps or steps in the profile of the field potential $\overset{\phi}{\mu_i}$ along the conduction channel, there will be equal hollows in the profile of $kT\ln c_i$; and thus there will be corresponding hollows in the profile of c_i. A hump or step in the field potential $\overset{\phi}{\mu_i}$ of $+0.6$ eV per particle of i (equal to about 14 000 cal mol^{-1} or about 60 kJ mol^{-1}) produces a hollow or drop in c_i that corresponds to a dilution factor of about 10^{10} at 25 °C if the channel was not initially near saturation with i; so that, other factors being unchanged, the conduction of i and the transmission of power would be virtually barred. Even a hump in the field potential of only 0.06 eV or 60 electronmillivolts produces a dilution of i by a factor of ten when the channel is not near saturation with i; and, at constant f_i, this would cause a ten-fold local decrease in the efficiency of power transmission.

These considerations illustrate very simply how the diffusion of i and the transmission of power can be channelled and conducted specifically by the local field potential which bars certain directions of escape of i but permits or facilitates others. In particular they indicate how a barrier to the diffusion of i due to a step in the field potential profile (such as that causing the latching property of a covalent bond, or such as that causing the reflection of a hydrophilic solute from a non-aqueous osmotic barrier) may be overcome by the binding of i in a specific enzymic or catalytic carrier complex. As discussed by Jencks (1975) in a review of the function of binding energy and specificity in enzyme catalysis, the energy barrier (represented here by the hump in the field potential profile) may be counteracted by a more or less equal and opposite binding potential which will, of course, have the effect of restoring the level of c_i – an action described by Jencks as the Circe effect. It is essential to emphasise, however, that this effect is of no avail for the catalysis of the conduction of i along the required pathway or channel unless the mobility of i down the pathway is not much impaired by the tight complexation of i. This conduction requirement in enzymes or catalytic carriers presumably implies that the specific catalytic process depends importantly, not only on substrate-specific binding, but also on the specific translocation of i, prescribed by structural features that may include specifically articulated conformational mobility in the close-packed polypeptide and/or lipid system of which the i complex is a part and through which i is conducted (Mitchell, 1963, 1970a).

The essential catalytic conductive properties of enzymes and catalytic carriers outlined here apply both to what I have described as connective unifer or uniport systems, such as the phosphoglycerate kinase or such as valinomycin, that have a purely conductive function, and to the transformative symfer or symport systems, such as the phosphorylating glyceraldehyde-phosphate dehydrogenase, or such as protonmotive redox loop systems, that work by catalysing the coupled conduction of pairs of chemical species i and j. In the energy transformation systems, it is noteworthy that the forces of interaction that couple the conduction of i (usually through intermediaries C) to the conduction of j, are generally transmitted, at least in part, indirectly via the electrical and chemical forces that give rise to the binding energy in the catalytic channel complexes. Thus, even though, as is usually the case, there are direct chemical interactions between i and j, or between i and C and between C and j, part of the energy transferred from i to j is generally transmitted indirectly through the forces and displacements of the surrounding catalytic complex. In other words, enzymically catalysed directly coupled chemical reactions are generally partly mediated by what has been loosely called conformational energy transfer (Lumry, 1974). As I have pointed out before (Mitchell, 1963), conformational energy transfer may be expected to play at least as important a part in osmotically coupled and chemiosmotically coupled reactions as in chemically coupled reactions.

Returning to the fundamental importance of the concentration of the energy-carrying particles in biological power transmission systems, it should not be overlooked that when the energy-carrying particle i is conducted by a circulating carrier C, the rate of conduction of i in the steady state depends both on the concentration of the Ci complex or compound, and on the concentration of C, because the rate of forward conduction of Ci is equal to the rate of return conduction of C. Assuming that the frictional coefficients (f_{Ci} and f_C) for the loaded and unloaded carrier are the same, the maximum steady conduction rate will be achieved when the concentrations of loaded and unloaded carrier are equal. It is this essentially *kinetic* circumstance that explains the well known generalisation that biochemical carriers are usually found to be operating not far from their so-called mid-point potentials (Mitchell, 1967a, 1970a). Frictional coefficients depend upon local viscous effects and on the effective size of the thermally mobile complexes with which the diffusion of i is associated. The range of values of frictional coefficients is not therefore very wide in biological media and in the catalytic channels that have been evolved in enzymes and catalytic carrier systems.

We can conclude that efficient power transmission in biochemical systems has depended on the evolution of specific conduction mechanisms in which the energy-carrying species are maintained at a relatively high and relatively uniform concentration along the floor of conduction channels, and in which the profile of the field potential $\overset{\phi}{\mu_i}$ corresponding to the floor of the channels shows a gentle downward slope in the direction of power transmission, relatively free of humps and hollows, and following approximately the slope of the profile of the total potential $\bar{\mu}_i$ required to transmit the power from the input to the output systems (Mitchell, 1970a).

*The time-dependent kinetic basis of the concepts
of thermodynamic equilibrium and reversibility*

The efficient and (virtually) reversible transmission of chemicals and energy through specifically conducting catalytic channels, as discussed here, depends on the rates of diffusional equilibration along the floors of the channels being fast compared with the rates of diffusional equilibration by side reactions over the rims of these channels. Thus, the statement that a biochemical energy-transferring system operates reversibly (or virtually so) implies that the time-scale used must be long compared with the relaxation times for diffusion along the catalytic channels, but short compared with the relaxation times for the side reactions. As we have seen, this implies that the working substances must be specifically channelled at relatively high concentration or probability through continuous circuits between the input and output.

These elementary considerations illustrate, in a very general way, the useful application of thermodynamic methods to pathways, as well as to initial and final states. They reveal the fallacy in the criticism by Williams (1974) that 'Chemiosmotic theories have been used to obscure the distinction between thermodynamic relationships (which are time-independent) and pathways (which are time-dependent).' They also show that the attempt by Weber (1974) to demonstrate the impossibility of chemiosmotic coupling, as I have conceived it, fails because his view of equilibration in coupled metabolic systems is misconceived. His criticism was based on the remarkable generalisation that 'Every chemical compound generated in metabolism is the result of a reaction that runs towards thermodynamic equilibrium, with complete independence of every other reaction that occurs at the same time.' His more recent criticism (Weber, 1975) that 'We thus conclude that the *stationary electric field energy* $\Delta\psi_{\alpha\beta}$ associated with a concentration gradient of ions does not represent energy that can be usefully converted into chemical work' is even more astonishing. If this were correct, the transformation of chemical to electrical energy by the interaction of chemical and electrostatic forces at the electrode–electrolyte interfaces of fuel cells (Liebhafsky & Cairns, 1968) would be impossible; and even the measurement of a pH by the usual electrometric glass electrode technique, which depends on the balancing of the osmotic and electrostatic forces to which the H^+ ions are subject at the surfaces of the glass membrane, would be impossible. In contrast, the reader may enjoy consulting a marvellously precocious paper by Fritz Lipmann (1946) on metabolic process patterns, which foreshadows the present developments. In that paper he quoted the following relevant passage from Bergson (1944) on creative evolution: 'There is more in a transition than a series of states or possible cuts, more in a movement than a sequence of positions or possible stops.'

Interconversion between the thermodynamic or concentration potential $\overset{\theta}{\mu_i}$ and the field potentials $\overset{p}{\mu_i}$, $\overset{\psi}{\mu_i}$ and $\overset{x}{\mu_i}$

When a chemical particle of species i (which may be a molecule, ion, electron or chemical group) is in a system where it equilibrates between domains α and ω, so that its total potential difference between these two domains is zero, it follows (equation (9)) that

$$\overset{p}{\Delta\mu_i}+\overset{\psi}{\Delta\mu_i}+\overset{x}{\Delta\mu_i}+\overset{\theta}{\Delta\mu_i} = 0, \qquad (39)$$

where Δ stands for the value in domain α minus that in domain ω. Thus, the distribution of the energy of i between the different types of potential depends on the environment in the α and ω domains. If, for example, the domain α represents the binding site for i in a specific binding protein and ω represents an aqueous medium, and if the pressure and electric potential are uniform, the decrease in the purely chemical potential represented by a negative value of $\overset{x}{\Delta\mu_i}$, corresponding to the chemical binding energy of i in domain α, will appear as an equal increase in the thermodynamic potential of i represented by $\overset{\theta}{\Delta\mu_i}$ or by $kT\Delta\ln c_i$ (equations (9) and (10)). Thus, a specific binding protein may transform the chemical binding energy of i into thermodynamic energy, and (kinetically) facilitate the subsequent reaction and transfer of i in accordance with equation (28) by increasing its local concentration or probability (Mitchell, 1967a). Similar transformations are obviously possible between all four types of energetic potential represented in equation (39), but the transformations involving $\Delta\mu_i$ or $kT\Delta\ln c_i$ are especially significant in the general context of power transmission through the specific (catalytic) channelling of chemical currents, because, as I have endeavoured to explain, the effective conductance factor n_i/rf_i, in which n_i can be compared with n_{ic} of equation (8), plays a dominant part in delineating the topology of the network of power transmission channels that have been evolved in biological systems generally.

Particularly simple examples of the kinetic effects of the interconversion of chemical or electrical potential and thermodynamic or concentration potential are provided in the transmission of power by proticity in aqueous media. The equilibrium concentration of free (unhydrated) protons in aqueous media is practically negligible – less than 10^{-100} M – but the avid chemical binding of protons by water, mainly as the trihydrate of H_3O^+, brings the concentration to some 10^{-7} M in neutral aqueous solutions (Bell, 1959). Further, the binding of protons by diffusible acid–base buffers of small molecular weight, such as phosphate, carboxylic acids and amino acids, and by relatively immobile proteins and other rapidly proton-exchanging substances, such as are present in bacterial cell walls and in the periplasm, brings the total concentration of protons that are effectively mobile into a range of some 10^{-3} to 10^{-1} M. The high proton conductance of the aqueous media on either side of coupling membranes, essential for efficient power transmission by proticity, is thus attributable to the relatively high concentration of rapidly exchangeable protons in various states of chemical binding and not to the practically negligible concentration of free protons (Mitchell, 1967b). Essentially the same general principles relating the efficiency of chemical transport and power transmission to the concentration of the energy-carrying chemical species are, of course, applicable to the well known chemical-group carriers of classical biochemistry, such as the ATP/ADP couple and the P_i/H_2O couple, although the behaviour is complicated in such systems by the latching property of the covalent bonds in the carrier compounds and the consequent non-dissociability of the carrier chemical groups except in the catalytic channels of (connective or transformative) enzyme molecules of appropriate latch-lifting specificity.

Proton wells and proton traps

The originally hypothetical concept of the proton well (Mitchell, 1968) represents a proton-conducting crevice passing down into the dielectric of a coupling membrane from the electrically positive side, as illustrated in Fig. 8. Thus, according to equation (39), at uniform pressure (i.e. $\overset{p}{\Delta}\mu_{H^+} = 0$), the electric potential drop $\Delta\psi$ down the proton well in the dielectric is transformed to an equivalent increase of the thermodynamic and chemical potentials represented by $\overset{\theta}{\Delta}\mu_{H^+}$ and $\overset{\chi}{\Delta}\mu_{H^+}$, which are conventionally described together in terms of the chemical activity a_{H^+} or the pH by the expressions $kT\Delta\ln a_{H^+}$ or $-2.3\,kT\Delta$pH. The latter is equivalent to $-59\,\Delta$pH at 25 °C when the energy unit is the millielectronvolt.

There is now fairly good evidence that the F_0 component of the protonmotive ATPase (Racker, 1970) acts as a proton well, with the active site of F_1 at the bottom of it (Mitchell & Moyle, 1974; Mitchell,

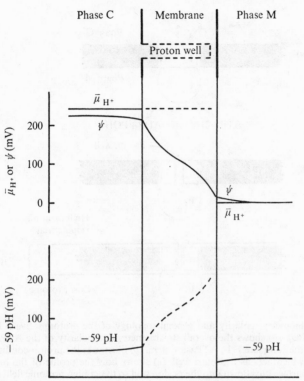

Fig. 8. Transformation of $\Delta\psi$ to $-59\Delta pH$ by a proton well in a lipid membrane (from Mitchell, 1968). The diagram shows the profiles of the total protonic potential $\bar{\mu}_{H^+}$ and its electric and pH components ψ and -59 pH respectively in the bulk of the aqueous media and membrane (continuous lines) and in the proton well (broken lines). The value of ψ in the proton well is assumed to be the same as in the bulk of the membrane. The total protonic potential difference and its bulk phase pH component are arbitrarily given as 250 mV and 25 mV respectively.

1976a), as represented crudely in Fig. 9. It is thought that, in this way, virtually the whole of the protonmotive potential difference $\Delta\bar{\mu}_{H^+}$ across the coupling membrane appears as an equivalent pH difference $-59\Delta pH$ across the active site of the enzyme. A total protonic potential of some 250 to 300 mV would thus be equivalent to a pH difference of some 4.5 units, so that if the F_1 side of the membrane were at about pH 8, the bottom of the proton well would be at about pH 3.5. The possibility that proton wells may function in systems other than the protonmotive ATPase, and the possibility that analogous wells for other ions might play a similar part in ionmotive chemiosmotic systems generally is obviously worth bearing in mind (Mitchell, 1976b).

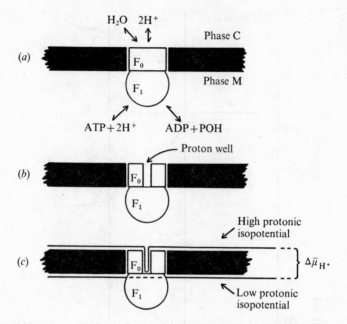

Fig. 9. Stoichiometry, polarity and general topology of the protonmotive ATPase (from Mitchell, 1976a). (a) shows the overall stoichiometry and polarity of the ATPase reaction catalysed by the complete F_0F_1 ATPase system. (b) illustrates the proton-conducting region through F_0, which acts as a proton well. (c) shows how the profile of the high protonic isopotential surface dips down into the proton well so that a high protonic field is effectively created in the active-centre region of F_1, where the profiles of the low and high protonic isopotential surfaces are close together. The protonic isopotential profile is shown as a broken line inside F_1, because the protonic potential will only be effectively determinate along the (unknown) route of access of the P_i and adenine nucleotides between phase M and the active centre in F_1.

The hypothetical concept of the proton trap is functionally the converse of the proton well. As indicated in Fig. 10, the chemical trapping of protons by specifically protonatable groups on either surface of a coupling membrane could convert part of a pH difference between the media on either side of the membrane into a local electric potential difference across the dielectric of the membrane. The principle is that when the pH values of the media on either side of the membrane are near the pK values of the proton-trapping groups on the membrane surfaces, the pH buffering power in the immediate vicinity of these groups may be much higher than in the bulk of the aqueous media: hence the pH profile and corresponding ψ profile represented by Fig. 10.

The fact that the total protonic potential difference $\Delta\bar{\mu}_{H^+}$ across the cytochrome oxidase complex influences the electronic spin state of the a

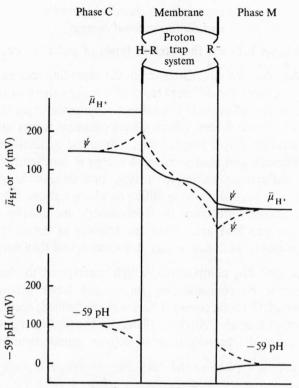

Fig. 10. Transformation of $-59\Delta pH$ to $\Delta\psi$ by a proton trap system. The diagram represents a component plugged through the membrane, having an acidic group RH exposed on either surface of the membrane. The conventions are as for Fig. 8, except that the broken lines represent the profiles of the ψ and $-59pH$ components of the total potential along a line normal to the membrane, through the centre of the proton trap system; and the bulk phase pH component of the total protonic potential difference is arbitrarily given as 100 mV. The diagram does not pretend to give a quantitative explanation for the extent of transformation of $-59\Delta pH$ to $\Delta\psi$, represented here as 100%.

cytochromes irrespective of whether the potential is applied mainly as an electric potential difference $\Delta\psi$ or as a corresponding pH difference $-59\Delta pH$, together with other facts described by Wikström and co-workers (Wikström, 1975; Wikström, Harmon, Ingledew & Chance, 1976), might, I suggest, imply that a proton trap mechanism is involved in the transformation of $-59\Delta pH$ to $\Delta\psi$ in the cytochrome c oxidase system; but the proton traps on either side of the membrane would themselves have to be situated in wells reaching some distance below the outer surfaces of the cytochrome oxidase complex – i.e. in the cytochrome a and a_3 polypeptide complexes respectively.

Interconversion of chemical, osmotic
and conformational energy

The distinctions between the various types of potential energy difference $\overset{p}{\Delta\mu_i}$, $\overset{\psi}{\Delta\mu_i}$, $\overset{x}{\Delta\mu_i}$ and $\overset{\theta}{\Delta\mu_i}$, discussed in the preceding section, are really distinctions between the different types of force or origins of the force to which the species of particle i is subject – hydrostatic pressure forces, (long-range) electric forces, (short-range) chemical forces and average thermal vibration forces respectively. The usual distinctions between chemical, osmotic and conformational energy in biochemistry are likewise really distinctions relating to the types or origins of the corresponding forces, but according to a different classification. Chemical (or metabolic) energy differences in biochemistry are usually defined as energy differences that arise from the transfer of covalently bonded chemical groups from donor to acceptor species, and they may therefore include $\overset{x}{\Delta\mu_i}$ and $\overset{\theta}{\Delta\mu_i}$ components, which correspond to chemical and thermodynamic or concentration forces, the latter generally being relatively small. Osmotic energy differences are defined, according to the chemiosmotic rationale (Mitchell, 1970a), as the total energy differences that arise from the translocation of solutes across membranes from donor to acceptor phases, and they may therefore include $\overset{p}{\Delta\mu_i}$, $\overset{\psi}{\Delta\mu_i}$, $\overset{x}{\Delta\mu_i}$ and $\overset{\theta}{\Delta\mu_i}$ components. In this case, $\overset{\psi}{\Delta\mu_i}$ and $\overset{\theta}{\Delta\mu_i}$ are generally the major components; and the total osmotic potential difference of i is related to the total force $\mathbf{f}_i$ tending to push i through the membrane from phase α to phase ω according to equation (4). Conformational energy differences are generally understood to refer to the free energy changes of relatively large molecules or complexes that arise from numerous stress–strain relationships within the molecules or complexes. A conformational energy change may be defined as the sum of the products of the forces and displacements of all the constituents of a molecule or complex X as it changes reversibly from conformation α to conformation ω. Thus, a purely conformational energy change described by $\Delta\bar{\mu}_X$ may involve all the different types of force acting in the complex X, such as are usually included (equation (6)) in the total potential change $\Delta\bar{\mu}_i$ of a component i of small molecular weight in moving from a domain α to a domain ω. Indeed the protein molecules or complexes which are customarily described as undergoing changes of conformational energy are essentially multi-component systems in which the components are physically interlinked and packed together in such a way that they may undergo certain specifically articulated movements, much as in a macroscopic engine. However, conformational energy changes are usually regarded as being largely determined by changes of electrical interaction and secondary bonding and packing of chemically unreactive hydrophobic polypeptide (or lipid) chains and hydrophilic anionic and cationic groups present in the protein or complex X, rather than by changes of primary or covalent bonding (Citri, 1973; Lumry, 1974; Jencks, 1975; Weber, 1975).

Changes of conformational state, which can be partly described by the (conformational) energy function, are especially interesting because they are certainly involved in the catalysis of osmotic reactions and chemiosmotic reactions across membranes, where the specific conformational changes of components of the coupling membrane are required to facilitate the translocation of solutes from the aqueous phase on one side to that on the other (Mitchell, 1963, 1970a, 1976a). Changes of conformational state also appear to be involved (to some extent) in enzyme-catalysed reactions generally, where chemical groups have to be transferred from chemical donor to acceptor groups via the catalytic channels in the condensed fabric of specific enzyme molecules (Citri, 1973; Jencks, 1975).

The mechanism of K^+/H^+ antiport catalysed by nigericin at low concentration illustrates the general principle of the interconversion of osmotic and conformational energy, as indicated in Fig. 11. Nigericin is a hydrophobic polyether having a carboxylic functional group (Pressman, 1968; Markin, Sokolov, Boguslavsky & Jaguzhinsky, 1975). In the deprotonated state, the functional group is anionic and hydrophilic, and the anionic nigericin–membrane complex X^- therefore takes up two alternative conformational states X_α^- or X_ω^-, in which the nigericin is anchored in the aqueous–membrane interface with the (unloaded) hydrophilic ionophore group either in the α or in the ω aqueous phase (Fig. 11a). But when the carboxylate functional group is either protonated or forms a potassium (or other alkali metal) salt, its hydrophilic properties are lost or masked, and the nigericin–membrane complex takes up conformational states in which the (loaded) hydrophobic functional group of the nigericin may pass into and through the lipid phase of the membrane. Therefore, as indicated in Fig. 11(b), if the H^+ ion activity is higher in aqueous phase α than ω and there are no alkali metal ions present, the anionic nigericin–membrane complexes will take up a conformational distribution $(n_{X^-}^\alpha/n_{X^-}^\omega < 1)$ in which the greater part of the anionic nigericin is anchored at the interface on the ω side. Moreover, at equilibrium, as the work required to cause this asymmetric

412   P. MITCHELL

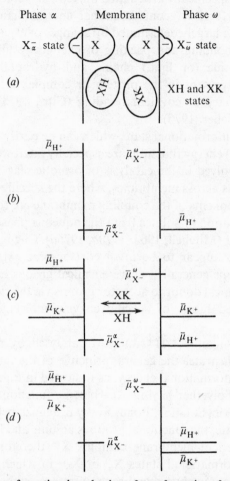

Fig. 11. Exclusively conformational mechanism of transformation of osmotic energy of H^+ to that of K^+ by nigericin. Diagram (a) illustrates the alternative conformational states of the nigericin anion, represented as X_α^- and X_ω^-, at the α and ω surfaces of the membrane respectively, compared with buried conformations for the protonated form XH and the potassium salt XK. The other diagrams indicate: (b), the potentials of the α and ω states ($\bar\mu_{X^-}^\alpha$ and $\bar\mu_{X^-}^\omega$ respectively) of the nigericin at a given potential difference of H^+; (c), shifts of potentials and migrations of XK and XH when K^+ is added at an equal concentration in phases α and ω; (d), the final equilibrium state with both H^+ and K^+ present.

conformational distribution of the anionic nigericin–membrane complexes is done by the transfer of one H^+ ion from aqueous phase α to aqueous phase ω per nigericin molecule in the complex, the conformational free energy difference $\Delta\bar\mu_{X^-}$ between the α and ω states of the anionic nigericin–membrane complexes must be given by

$$\Delta\bar\mu_{X^-} + \Delta\bar\mu_{H^+} = 0, \qquad (40)$$

where

$$\Delta\bar{\mu}_{X^-} = \Delta\mu_{X^-}^{\phi} + kT\Delta\ln(n_{X^-}^{\alpha}/n_{X^-}^{\omega}). \qquad (41)$$

Substituting K^+ for H^+, a similar argument leads to the equilibrium condition

$$\Delta\bar{\mu}_{X^-} + \Delta\bar{\mu}_{K^+} = 0. \qquad (42)$$

If, therefore, K^+ ions are introduced when $\Delta\bar{\mu}_{H^+} > 0$ (Fig. 11c), the conformational energy of the asymmetrically distributed anionic nigericin–membrane complexes can be used to pump the K^+ ions from ω to α until $\Delta\bar{\mu}_{K^+}$ and $\Delta\bar{\mu}_{H^+}$ are equal (Fig. 11d). This is a very simple but clear-cut example of the transformation of one form of osmotic energy (that of $\Delta\bar{\mu}_{H^+}$) to another (e.g. that of $\Delta\bar{\mu}_{K^+}$) exclusively via conformational energy ($\Delta\bar{\mu}_{X^-}$). In other words, direct interaction forces between the H^+ and K^+ ions are not involved in the coupling process, but the osmotic work is transmitted between the H^+ and K^+ ions exclusively through the forces on and the conformational displacements of the nigericin–membrane complexes. This exclusively conformational type of coupling mechanism may perhaps be a versatile and widely used mechanism for osmotic antiport reactions (Fig. 4c).

It is noteworthy that the criterion that we have used to distinguish between the directly coupled type of osmotic or chemical energy transformation reaction, in which conformational energy transfer contributes only partially, and the indirectly coupled type of energy transformation reaction, in which energy transfer occurs through an exclusively conformational energetic state and not through direct interaction between energy-donating and energy-accepting solutes or chemical groups (including enzyme-bound intermediaries), is the exclusion of direct interaction in the latter (Mitchell, 1976a, b). In other words, in the indirect exclusively conformational type of energy-transfer mechanism, there is no space–time overlap between the catalytic channels for the energy-donating and energy-accepting species. This means that in the (alternating) antiport type of osmotic energy transforming system (Fig. 4c) the catalytic channels for the energy-donating (e.g. H^+) and energy-accepting (e.g. K^+) species may overlap spatially through the mobile carrier centre (e.g. the functional group of nigericin) which moves alternately to either surface of the membrane, but they may nevertheless not overlap in time or, therefore, in space–time. Thus, the indirectly or conformationally coupled osmotic antiport type of system is in a rather special category mechanistically. By contrast, the osmotic symport or chemical symfer or antifer types of energy transformation reactions that are concerted (Fig. 4a, b and d) could not, by definition, be exclusively conformationally coupled unless the catalytic channels or active centres for the energy-donating and energy-accepting species were spatially separate – so that energy transfer would have to occur through forces and displacements transmitted through the polypeptide system intervening between the separate energy-donating and energy-accepting centres in the effectively duplex catalytic complex.

It has been recognised for many years that certain chemiosmotic reactions might proceed by the exclusively conformational type of coupling mechanism (Mitchell, 1963), especially in cases where the solute undergoing translocation could not form covalent intermediates and could not therefore be directly involved in the vectorial metabolic processes of group translocation. However, in chemical energy transformation by enzymes, biochemical experience has shown that in all cases so far studied the energy-transferring substrates interact directly with each other or via intermediary functional groups in the enzyme active centre, and only part of the energy is transferred by conformational interactions through the surrounding enzymic structure (Citri, 1973; Lumry, 1974; Jencks, 1975). For example, in *Escherichia coli*, the succinyl coenzyme A synthetase reaction proceeds via an enzyme-bound succinyl phosphate intermediate (Pearson & Bridger, 1975), as indicated in Fig. 12.

Classical protonmotive redox loops and the protonmotive ubiquinone cycle (Crofts, Crowther & Tierney, 1975; Gutman, Beinert & Singer, 1975; Mitchell, 1976a, b; several papers in this volume) represent chemiosmotic reaction mechanisms that are directly coupled by H^+–e^- symport or H uniport via hydrogen carriers in the specific osmoenzyme systems plugged through the coupling membrane. On the other hand, the protonmotive NAD(P) transhydrogenase (Mitchell, 1972; Moyle & Mitchell, 1973) and the protonmotive bacteriorhodopsin system (Oesterhelt, this volume) appear to be more complex mechanistically, despite their apparent chemical simplicity; and they might involve exclusively conformational chemiosmotic mechanisms. However, various degrees of indirectness of chemiosmotic mechanisms are conceivable, short of the exclusively conformational type of mechanism. In the case of the cation-translocating ATPases, for example, electrovalent interaction between the cations and the anionic phosphate groups could conceivably play a part in the chemiosmotic reaction mechanism (Mitchell, 1963, 1973b), even though the specificity of the binding and translocation of the cations would clearly have to be determined by complexation with other groups in the ATPase complexes. At all events, one of the most useful questions for generating experimental approaches

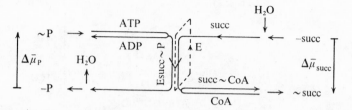

Fig. 12. Flow diagram of succinyl coenzyme A synthetase energy-transformation reaction, indicating participation of the enzyme-bound succinyl phosphate intermediary Esucc~P. Other conventions are as in Fig. 5.

to chemiosmotic reaction mechanisms generally is: how direct are the connections between the forces and displacements of the particles involved in the chemical reaction and the forces and displacements of the particles involved in the osmotic reaction?

The mechanism of the protonmotive ATPase is, perhaps, the most interesting and controversial of all the chemiosmotic reaction mechanisms now being studied. There are no known chemical intermediaries, and it is possible to formulate a particularly simple direct mechanism (Mitchell & Moyle, 1974; Mitchell, 1976a) that is essentially dependent on O^{2-} uniport (formally $2H^+/H_2O$ antiport) through the F_1 component of the ATPase. The arrangement of the ATPase in the membrane is represented in Fig. 9, and the proposed direct mechanism is indicated in Fig. 13 for the process of protonmotivated reversal of ATP hydrolysis. In this diagram, the time dimension is crudely represented by elongating F_1 and F_0 in the vertical dimension. The main feature of this scheme is that the entry of $2H^+$ through the proton-conducting region of F_0 (from bottom right) on the C-facing side of the active centre of F_1 is coupled to the translocation of the O^{2-} group of $ADPO^- + MgPO_4^-$ from the M-facing side of the active centre of F_1 (from bottom left), because the $2H^+$ ions attack the $MgPO_4^-$ from the C side, while the $ADPO^-$ attacks the $MgPO_4^- - MgPO_3^+$ transitional complex in the active centre from the M side as H_2O is withdrawn on the C side (lower and upper centre). The process is completed by the translocation of MgATP from the active centre on the M side (to top left). Two of the negative oxygens of PO_4^{3-} in the active centre of F_1 are shown as being electrically neutralised, and this may be due to Mg^{2+} as shown here. However, other groups in the active centre, not shown in Fig. 13, will presumably be involved in binding the P_i, ADP and ATP; and these, as well as or rather than the Mg^{2+} ions, may neutralise the charges on the PO_4^{3-} in the active centre.

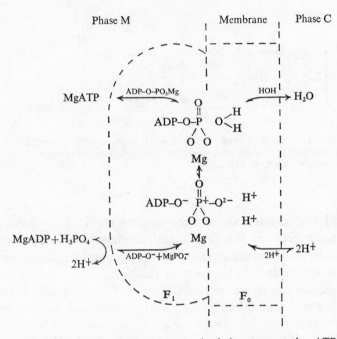

Fig. 13. Mechanism of protonmotivated reversal of the protonmotive ATPase (from Mitchell, 1976a). This diagram represents a possible O^{2-}-group translocation mechanism. The forward direction of the arrows (indicated by the double barbs) corresponds to proton-motivated ADP phosphorylation. Further details are given in the text.

Incidentally, the species shown as $MgP^+O_3-O^{2-}$ in Fig. 13 (lower centre) would obviously not exist in the unprotonated state, but it is represented in this way to identify the transfer of O^{2-} from $MgPO_4^-$ to $2H^+$ in the overall reaction.

The overall energetic consideration in this putative reversible proton-motive ATPase mechanism, as in any reversible mechanism, is that, close to equilibrium, every particle or complex undergoing net transfer in the virtually reversible process of the catalysed reaction must be subject to virtually no *net* force. The application of this principle to the O^{2-} group (or its equivalent) in the F_1-catalysed reaction of Fig. 13 is represented more explicitly by the flow diagram in Fig. 14. Here it is indicated that the O^{2-} group (or its equivalent) may equilibrate between the inorganic phosphate group in the $EADP-O^-$, $MgPO_4^-$ complex that is at a low effective water potential (i.e. that of the $(ADP+P_i)/ATP$ couple) and the $2H^+$ that is at a high water potential (i.e. that of the aqueous media), because there is a corresponding difference of the H^+ potential across the catalytic centre of the enzyme. Thus, $\Delta\bar{\mu}_{O^{2-}}$ is zero when $\Delta\bar{\mu}_{H_2O} = 2\Delta\bar{\mu}_{H^+}$. For example, a $\Delta\bar{\mu}_{H_2O}$ value (or phosphate potential) of 500 mV

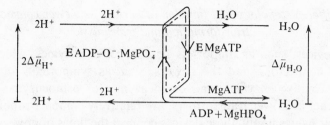

Fig. 14. Flow diagram of F_1-catalysed O^{2-}-group translocation reaction, indicating the role of enzyme-bound intermediaries between $EADP-O^-,MgPO_4^-$ and $EMgATP$ in the cyclic catalytic and energetic energy-transformation process. Further details are given in the text.

would be equivalent to a $\Delta\bar{\mu}_{H^+}$ value of 250 mV; so that if the M side of the catalytic centre of F_1 (Fig. 13) were at about pH 7, the other side (at the bottom of the proton well in F_0) would be at about pH 3. It should not, however, be overlooked that, according to orthodox principles of enzymology (Pauling, 1956; Jencks, 1975), the specific diffusibility of the O^{2-} group (or its equivalent) in the catalytic site of F_1 would be expected to depend on tight enzymic complexation of the O^{2-}-containing transitional intermediates, thus raising the effective concentration of the O^{2-} group and unlatching the O atom from covalent bonding with the phosphorus centre of inorganic phosphate. These tightly bound transitional complexes would, of course, be intermediate between the complexes indicated as $EADP-O^-,MgPO_4^-$ and $EMgATP$ in Fig. 14; and they may be expected to include singly protonated intermediaries involved in dissociative or associative mechanisms, which can be broadly represented as $EADP-OMgPO_3$, OH^- or $EADP-O-(MgPO_3^-)-OH$ respectively. The labilisation of the P–O bond of P_i in such enzyme-bound intermediaries, and the consequent tendency for them to exchange OH^- with H_2O of the medium, would lead one to expect a rather facile exchange between the O of the P_i in these complexes and the O of the aqueous medium. This tendency is supported by the excellent experimental work on exchange reactions by Boyer and co-workers (Boyer, Smith, Rosing & Kayalar, 1975); although Boyer himself prefers a quite different interpretation, which he regards as favouring an exclusively conformational reaction mechanism (Boyer, 1975). Slater has suggested a conformational ATPase mechanism essentially similar to that described by Boyer (Harris & Slater, 1975).

Different topology and function of devices for power connection,
transformation and transmission

The devices, such as uniporters or connective enzymes, which act as power connectors, and the devices, such as symporters, antiporters, transformative enzymes, and osmoenzymes or osmoenzyme systems, which act as power transformers, are relatively compact topologically. As their names imply, these devices connect the flows of power-carrying particles of given species, or transform the flow of one species of power-carrying particle, in the chemical or osmotic mode, to the flow of another species of power-carrying particle, in the same or in the other mode. The function of these connective and transformative devices generally involves conformational energy transfer as an essential component of the catalytic and power transfer and transformation mechanism, as discussed above. But it is important to note that conformational energy transfer operates only by direct contact through condensed complexes such as the polypeptide systems of proteins, as exemplified by haemoglobin (Edelstein, 1975); and it is therefore a relatively localised phenomenon.

Unlike the localised, topologically compact power connection and transformation devices, the power transmission devices are required by their function to be topologically extensive. These power transmission systems fall into two main categories: chemical and osmotic.

The chemical transmission of power occurs through aqueous intracellular media by the diffusion of pairs of group-carrying couples operating at different group potentials. Input and output have to occur through substrate-specific connective or transformative enzymes; and the distance over which power can be transmitted within microorganisms is of the same order as the extent of their intracellular aqueous media. Chemical power is also transmitted through extracellular aqueous and gaseous media by the diffusion of environmental redox couples operating at different potentials; and in the present state of evolution, the primary inputs are enzyme systems in photosynthetic organisms, and the outputs are enzyme systems in non-photosynthetic organisms. Chemical power may thus be transmitted over comparatively great distances.

The osmotic transmission of power occurs by the diffusion of certain solute species, particularly protons, at different potentials on either side of topologically closed coupling membrane systems. Thus, the distance over which power can be transmitted by proticity is limited by the extent of the coupling membrane system; and the input and output devices are osmoenzyme systems or porters that have to be plugged through the coupling membrane. In this context, the protonmotive ATPase may be regarded as a particularly important device in the power distribution system of many micro-organisms, because it acts as the transformer between the plasma membrane system or mitochondrial cristae membrane system that transmits power by proticity, and the cytoplasmic or mitochondrial matrix system that transmits power mainly by hydricity or phosphorylicity (Fig. 14).

CONCLUSION

The experimental work described in this volume has shown, and this epilogue has sought to emphasise that – despite a popular belief to the contrary – thermodynamic energetic methods are useful, not only for describing initial and final states, but also for describing intermediary states, and thus for investigating biochemical mechanisms.

The abstract function that we call energy, in considerations of non-radiative energy transfer, is a means of saying something about the relative positions and relative motions of the particles involved in the mechanics of biochemical reactions. But, as energy changes represent the *scalar* products of force and displacement *vectors*, they contain only a part of the information that we need to describe the vectorial space–time details of the biochemical, physiological and general biological mechanisms in which we are interested as biologists. Therefore, it is desirable to recognise the abstract energy function for what it is, and to try to combine the knowledge of energetics with other related information so as to work backwards from the scalar energy function to the vectorial forces and flows to which the chemical particles are actually subject in biological systems.

The homogeneous treatment of chemical and osmotic reactions used in this epilogue may perhaps help to reinforce our conclusion that, in the last analysis, all biological chemical reactions and osmotic reactions represent specific types of chemical-group and solute diffusion down potential gradients via a network of catalytically and structurally defined channels. Thus, I hope that my readers may be encouraged to pursue the method of the present, rather limited analysis further, not only in treating the minutiae of specific chemical, osmotic and chemiosmotic mechanisms, but also in treating broader physiological mechanisms where, as I have remarked before (Mitchell, 1970b), the flame-like dynamic morphological properties of living organisms are produced by a kind of continuous filtration of environmental chemical components through spatially organised catalytic systems that make up the organism as a whole.

I would like to thank Dr Jennifer Moyle for helpful discussion during the preparation of this paper, and Mrs Stephanie Key and Mr Robert Harper for help in preparing the manuscript. I am also indebted to Glynn Research Ltd for general financial support.

REFERENCES

BELL, R. P. (1959). *The Proton in Chemistry*. London: Methuen.

BERGSON, H. L. (1944). *Creative Evolution*. New York: Greenwood.

BOYER, P. D. (1975). A model for conformational coupling of membrane potential and proton translocation to ATP synthesis and to active transport. *FEBS Letters*, **58**, 1–6.

BOYER, P. D., SMITH, D. J., ROSING, J. & KAYALAR, C. (1975). Bound nucleotides and conformational changes in oxidative and photophosphorylation. In *Electron Transfer Chains and Oxidative Phosphorylation*, ed. E. Quagliariello, S. Papa, F. Palmieri, E. C. Slater & N. Siliprandi, pp. 361–72. Amsterdam: North-Holland Publishing Company.

CITRI, N. (1973). Conformational adaptability in enzymes. *Advances in Enzymology*, **37**, 397–648.

CROFTS, A. R., CROWTHER, D. & TIERNEY, G. V. (1975). Electrogenic electron transport in photosynthetic bacteria. In *Electron Transfer Chains and Oxidative Phosphorylation*, ed. E. Quagliariello, S. Papa, F. Palmieri, E. C. Slater & N. Siliprandi, pp. 233–41. Amsterdam: North-Holland Publishing Company.

DELAAGE, M. A. (1975). 'Carrier' theory and thermodynamics of irreversible processes. *Biochimica et Biophysica Acta*, **394**, 493–503.

DICKERSON, R. E. (1969). *Molecular Thermodynamics*. New York: W. A. Benjamin Inc.

EDELSTEIN, S. J. (1975). Cooperative interactions of haemoglobin. *Annual Review of Biochemistry*, **44**, 209–32.

FAY, J. A. (1965). *Molecular Thermodynamics*. Reading, Mass.: Addison-Wesley Publishing Co., Inc.

GUGGENHEIM, E. A. (1933). *Modern Thermodynamics by the Methods of Willard Gibbs*. London: Methuen.

GUGGENHEIM, E. A. (1949). *Thermodynamics*. Amsterdam: North-Holland Publishing Company.

GUTMAN, M., BEINERT, H. & SINGER, T. P. (1975). Coupling site I in relation to the Fe–S centers of NADH dehydrogenase and their topography in the membrane. In *Electron Transfer Chains and Oxidative Phosphorylation*, ed. E. Quagliariello, S. Papa, F. Palmieri, E. C. Slater & N. Siliprandi, pp. 55–62. Amsterdam: North-Holland Publishing Company.

HAMILTON, W. A. (1975). Energy coupling in microbial transport. In *Advances in Microbial Physiology*, ed. A. H. Rose & D. W. Tempest, vol. 12, pp. 1–53. New York & London: Academic Press.

HAROLD, F. M. (1976). Membranes and energy transduction in bacteria. In *Current Topics in Bioenergetics*, in press.

HARRIS, D. A. & SLATER, E. C. (1975). Tightly-bound nucleotides of coupling ATPases – structural and functional aspects. In *Electron Transfer Chains and Oxidative Phosphorylation*, ed. E. Quagliariello, S. Papa, F. Palmieri, E. C. Slater & N. Siliprandi, pp. 379–84. Amsterdam: North-Holland Publishing Company.

HAWKINS, J. K. (1966). Liquid state electronics. *Science Journal*, **2**, 64–9.

HILL, T. L. (1968). *Thermodynamics for Chemists and Biologists*. Reading, Mass.: Addison-Wesley Publishing Co.

HILL, T. L. (1976). Diffusion frequency factors in some simple examples of transition-state rate theory. *Proceedings of the National Academy of Sciences, USA*, **73**, 679–83.

HILL, T. L. & SIMMONS, R. M. (1976). Free energy levels and entropy production associated with biochemical kinetic diagrams. *Proceedings of the National Academy of Sciences, USA*, **73**, 95–9.

JENCKS, W. P. (1975). Binding energy, specificity, and enzymic catalysis: the Circe effect. *Advances in Enzymology*, **43**, 219–410.

KOSHLAND, D. E. (1976). The role of flexibility in the specificity, control and evolution of enzymes. *FEBS Letters*, **62**, Supplement, E47–E52.

LAGARDE, A. E. (1976). A non-equilibrium thermodynamics analysis of active transport within the framework of the chemiosmotic theory. *Biochimica et Biophysica Acta*, **426**, 198–217.

LIEBHAFSKY, H. A. & CAIRNS, E. J. (1968). *Fuel Cells and Fuel Batteries*. New York: John Wiley & Sons, Inc.

LIPMANN, F. (1946). Metabolic process patterns. In *Currents in Biochemical Research*, ed. D. E. Green, pp. 137–48. New York: Interscience Publishers, Inc.

LUMRY, R. (1974). Conformational mechanisms for free energy transduction in protein systems: old ideas and new facts. *Annals of the New York Academy of Sciences*, **227**, 46–73.

MARKIN, V. S., SOKOLOV, V. S., BOGUSLAVSKY, L. I. & JAGUZHINSKY, L. S. (1975). Nigericin-induced charge transfer across membranes. *Journal of Membrane Biology*, **25**, 23–45.

MITCHELL, P. (1962). Metabolism, transport, and morphogenesis: which drives which? *Journal of General Microbiology*, **29**, 25–37.

MITCHELL, P. (1963). Molecule, group and electron translocation through natural membranes. *Biochemical Society Symposia*, **22**, 142–68.

MITCHELL, P. (1967a). Translocations through natural membranes. *Advances in Enzymology*, **29**, 33–85.

MITCHELL, P. (1967b). Proton current flow in mitochondrial systems. *Nature, London*, **214**, 1327–8.

MITCHELL, P. (1968). *Chemiosmotic Coupling and Energy Transduction*. Bodmin: Glynn Research.

MITCHELL, P. (1970a). Reversible coupling between transport and chemical reactions. *Membranes and Ion Transport*, **1**, 192–256.

MITCHELL, P. (1970b). Membranes of cells and organelles: morphology, transport and metabolism. *Symposia of the Society for General Microbiology*, **20**, 121–66.

MITCHELL, P. (1972). Chemiosmotic coupling in energy transduction: a logical development of biochemical knowledge. *Journal of Bioenergetics*, **3**, 5–24.

MITCHELL, P. (1973a). Performance and conservation of osmotic work by proton-coupled solute porter systems. *Journal of Bioenergetics*, **4**, 63–91.

MITCHELL, P. (1973b). Cation-translocating adenosine triphosphatase models: how direct is the participation of adenosine triphosphate and its hydrolysis products in cation translocation? *FEBS Letters*, **33**, 267–74.

MITCHELL, P. (1976a). Vectorial chemistry and the molecular mechanics of chemiosmotic coupling: power transmission by proticity. *Biochemical Society Transactions*, **4**, 399–430.

MITCHELL, P. (1976b). Possible molecular mechanisms of the protonmotive function of cytochrome systems. *Journal of Theoretical Biology*, **62**, 327–67.

MITCHELL, P. & MOYLE, J. (1974). The mechanism of proton translocation in reversible proton translocating adenosine triphosphatases. *Biochemical Society Special Publication*, **4**, 91–111.

MOYLE, J. & MITCHELL, P. (1973). The proton-translocating nicotinamide-adenine dinucleotide (phosphate) transhydrogenase of rat liver mitochondria. *Biochemical Journal*, **132**, 571–85.

OSTER, G., PERELSON, A. & KATCHALSKY, A. (1971). Network thermodynamics. *Nature, London*, **234**, 393–9.

PAULING, L. (1956). The future of enzyme research. In *Enzymes: Units of Biological Structure and Function*, ed. O. H. Gaebler, pp. 177–82. New York & London: Academic Press.

PEARSON, P. H. & BRIDGER, W. A. (1975). Isolation of the α and β subunits of *Escherichia coli* succinyl coenzyme A synthetase and their recombination into active enzyme. *Journal of Biological Chemistry*, **250**, 4451–5.

PERELSON, A. S. (1975). Network thermodynamics, an overview. *Biophysical Journal*, **15**, 667–85.

PRESSMAN, B. C. (1968). Ionophorous antibiotics as models for biological transport. *Federation Proceedings*, **27**, 1283–8.

RACKER, E. (1970). The two faces of the inner mitochondrial membrane. *Essays in Biochemistry*, **6**, 1–22.

ROBERTSON, R. N. (1968). *Cambridge Monographs in Experimental Biology 15: Protons, Electrons, Phosphorylation and Active Transport*. London: Cambridge University Press.

WEBER, G. (1974). Addition of chemical and osmotic energies by ligand-protein interactions. *Annals of the New York Academy of Sciences*, **227**, 486–96.

WEBER, G. (1975). Energetics of ligand binding to proteins. *Advances in Protein Chemistry*, **29**, 1–83.

WIKSTRÖM, M. K. F. (1975). Energy-linked conformational changes in cytochrome *c* oxidase of the mitochondrial membrane. In *Electron Transfer*

Chains and Oxidative Phosphorylation, ed. E. Quagliariello, S. Papa, F. Palmieri, E. C. Slater & N. Siliprandi, pp. 97–103. Amsterdam: North-Holland Publishing Company.

WIKSTRÖM, M. K. F., HARMON, H. J., INGLEDEW, W. J. & CHANCE, B. (1976). A re-evaluation of the spectral, potentiometric and energy-linked properties of cytochrome *c* oxidase in mitochondria. *FEBS Letters*, **65**, 259–77.

WILLIAMS, R. J. P. (1974). The separation of electrons and protons during electron transfer: the distinction between membrane potentials and transmembrane gradients. *Annals of the New York Academy of Sciences*, **227**, 98–107.

WYMAN, J. (1965). The binding potential, a neglected linkage concept. *Journal of Molecular Biology*, **11**, 631–44.

WYMAN, J. (1975). The turning wheel: a study in steady states. *Proceedings of the National Academy of Sciences, USA*, **72**, 3983–7.

INDEX